Astrobiology

Astrobiology An Introduction

THIRD EDITION

Kevin W. Plaxco and Michael Gross

JOHNS HOPKINS UNIVERSITY PRESS

Baltimore

Printed in the United States of America on acid-free paper
9 8 7 6 5 4 3 2 1

Johns Hopkins University Press
2715 North Charles Street
Baltimore, Maryland 21218-4363
www.press.jhu.edu

Library of Congress Cataloging-in-Publication Data

Names: Plaxco, Kevin W., 1965– author. | Gross, Michael, 1963– author.
Title: Astrobiology : an introduction / Kevin W. Plaxco and Michael Gross.
Description: Third edition. | Baltimore : Johns Hopkins University Press, 2021. |
Includes bibliographical references and index.
Identifiers: LCCN 2020036307 | ISBN 9781421441283 (hardcover) |
ISBN 9781421441290 (paperback) | ISBN 9781421441306 (ebook)
Subjects: LCSH: Life—Origin. | Exobiology.
Classification: LCC QH325 .P488 2021 | DDC 576.8/3—dc23
LC record available at https://lccn.loc.gov/2020036307

A catalog record for this book is available from the British Library.

Special discounts are available for bulk purchases of this book. For more information, please contact Special Sales at specialsales@jh.edu.

Johns Hopkins University Press uses environmentally friendly book materials, including recycled text paper that is composed of at least 30 percent post-consumer waste, whenever possible.

Contents

Preface

Space, as they say, is big. After staring at a single patch of sky for 23 days scattered over the course of a decade, for example, the *Hubble Space Telescope* spied more than 5,500 galaxies in an area equivalent to a fraction of a square millimeter held at arm's length (fig. P.1). Extrapolated across the entire sky, this corresponds to some 100 *billion* galaxies in the observable Universe. Each of these, in turn, contains an average of 400 *billion* stars. Space is really, *really* big.

If the Universe is so big, with its 50 *sextillion* stars, does it naturally follow that there must be life out there beyond the Earth? Or are we alone in the cosmos? The answer in popular culture, replete with its UFO sightings and benevolent—or hostile—space aliens, has always been a resounding "yes!" Even scientists are susceptible to the romance: during the heady days of the late 1960s and early 1970s, there was, briefly, a burst of enthusiasm for the scientific discipline called exobiology, aimed exclusively at the study of extraterrestrial life. But because there is no known life beyond the Earth, cynics quickly tagged it as a subject with no subject matter, and exobiology fell out of fashion.

Still, the question remains: are we alone? A new perspective on this question has arrived with astrobiology, a field that, in contrast to exobiology, studies life on our planet in the context of the possibility of life elsewhere. Astrobiology focuses on broader, perhaps more fundamental, and certainly more tractable questions about the relationship between life and the physics and chemistry of our Universe. Not surprisingly, astrobiology tends to focus on life on Earth—which is, after all, the only example we have on hand—but it attempts to understand this single example in the broadest contexts of the Universe. Using

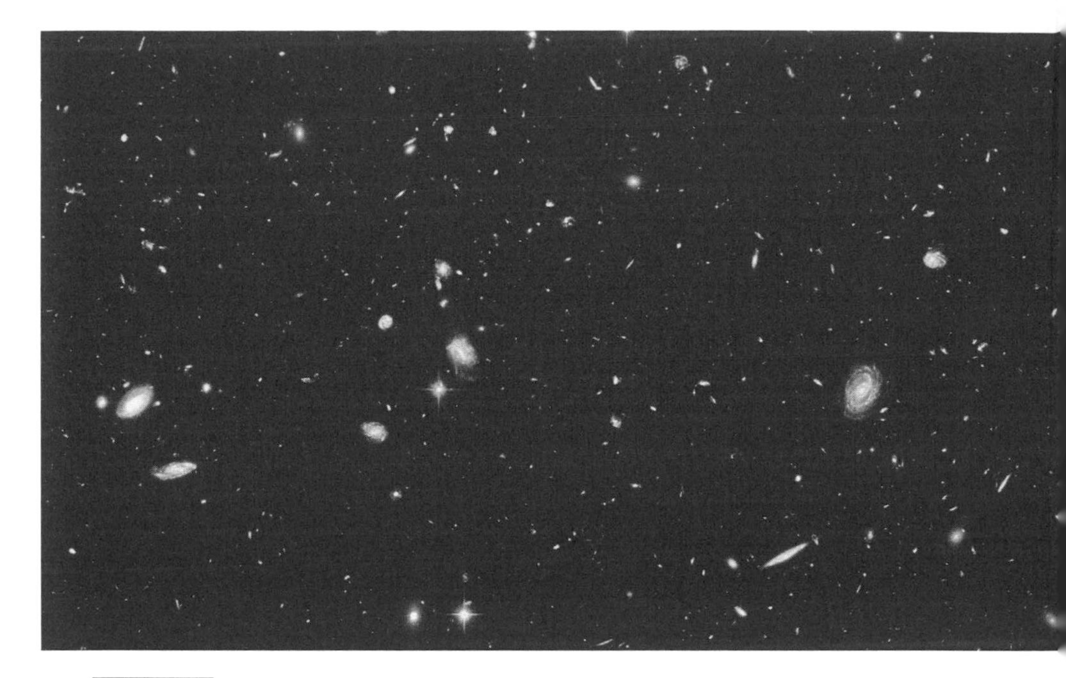

Figure P.1 Galaxies like grains of sand. Every one of the 5,500 spots and smudges in this image is an entire galaxy in its own right. This image covers approximately 1/15,000,000 of the entire sky (roughly equivalent to a 1 × 0.5 mm rectangle held at arm's length) and thus represents but a tiny fraction of the more than 100 billion galaxies in this amazing and expanding Universe. (Courtesy of NASA/STScI/ESA)

our extensive (if incomplete) knowledge of Terrestrial biology, astrobiology addresses three broad questions about life in the Universe:

1. What are the physical properties that allow our Universe and our planet to support life?
2. How did the origins and evolution of life transpire on Earth, and how might they be transpiring differently elsewhere?
3. Where else in our Universe might life have arisen, what might it be like, and how can we find it?*

Much of the worth of astrobiology lies in the fact that these are perhaps the most fundamental questions addressed by science today;

*And, in homage to Douglas Adams (1952–2001), author of *The Hitchhiker's Guide to the Galaxy*, "Can I get it to buy me a drink?"

they address the most profound issues regarding who we are, where we came from, and whether we are alone in this vast cosmos. Additional value arises from the exceptionally interdisciplinary approach that these questions demand; astrobiology encompasses a variety of scientific disciplines, ranging from cosmology, astrophysics, astronomy, geology, and chemistry to, of course, biology itself. In deference to the complexity that this lends, we endeavor in this book to outline the current status of astrobiology with only the absolutely necessary amount of scientific detail.

We begin, in chapter 1, by defining the object of our study: life. Although we can easily distinguish living from nonliving systems here on Earth, setting up a definition that strives to include all living systems imaginable in the Universe and to exclude all nonliving systems requires careful consideration. Next, in chapter 2, we need to investigate how the stage was set for life to arise in our Universe. How did the Universe come into being, and which of the crucial factors in its origins and early history distinguish it from other possible universes that may be unable to harbor any life at all? From the vastness of the Universe we zoom inward, in chapter 3, to explore the tiny blue dot that is our home planet and ask similar questions about it. Why did the third rock from the Sun become positively infested with life, while its neighbors did not? Which conditions are necessary for a planet—any planet, anywhere in the Universe—not only to become and remain habitable but to give rise to life in the first place? Zooming in closer still, chapter 4 looks at the molecular world present at the surface of the young Earth and investigates the chemical conditions and potential chemical pathways that set the stage for life to originate here.

Once the Universe, the Earth, and the molecules are all set and ready to go, the question then becomes: how did it actually happen? How did the inanimate become the animate, changing a habitable planet into an inhabited planet? The short answer is that we still don't know. However, there are partial answers to some of our questions and constraints regarding the possible answers to others, giving us a chance to spin some, necessarily rather speculative, scenarios in chapter 5. Following the chronological history of life on our planet, in chapter 6 we move on from the first spark of self-replicating, evolving life to the first cells and organisms. Again, we know very little about what really happened, but our current knowledge allows us to put constraints on what could have happened on Earth and risk an educated guess or two about what conditions might be necessary for it to happen (or have happened) anywhere else.

The veil begins to lift somewhat when we come to the last common ancestor of today's organisms: a single-celled organism named LUCA that was already quite evolved and had DNA, RNA, and hundreds of different proteins. From that point onward, molecular phylogeny (a word you'll grow to respect, if not love) and, increasingly, paleontology can help us decipher the history of life on Earth as a proxy for life in our Universe, as outlined in chapter 7.

Is there life elsewhere in the Universe? Today, researchers are pondering this question with a much more concrete and, in some ways, more optimistic outlook than they did 45 years ago when the *Viking* missions, with the first landers to operate long term on the surface of another planet, searched for life on Mars. This newfound optimism is based on the accumulating examples of organisms thriving in what we humans would consider extremely hostile conditions, such as high pressures, water well above its nominal boiling point, and extremely salty, acidic, or alkaline brines. The discovery of these Terrestrial extremophiles, which we detail in chapter 8, has had ripple effects within the field of astrobiology; the discovery of life in many seemingly uninhabitable environments on Earth has radically expanded our perceptions of what might constitute a possible habitat on other planets. The search for life elsewhere in the Universe has thus become intimately connected to the study of diverse habitats on Earth—an important element of the definition of astrobiology that sets it apart from the earlier, "extraterrestrials only" focus of exobiology.

Having explored the origins and the limits of life on Earth, we then expand our considerations beyond the home planet. In chapter 9, we ask: if life can thrive around hydrothermal vents in the deepest depths of our oceans, in the driest, coldest valleys of the Antarctic, and even deep within the Earth's crust, then which other places in the Solar System previously deemed inhospitable might actually be inhabited? Likewise, what does this tell us about potential habitats outside our Solar

System? Ultimately, speculation on where and even whether extraterrestrial life exists frustrates the scientific mind unless it can be followed up with actual investigation of the potential habitats concerned. Thus, in chapter 10, we review the history of space exploration and conclude our brief overview of astrobiology with a survey of past, present, and future searches for extraterrestrial life.

Since the second edition of this book appeared in 2011, exploration of the Solar System and discoveries in the field of extrasolar planet exploration have advanced significantly. Earthly extremophiles have turned up in yet more remote and hostile environments, and even the eternally elusive quest for the origins of life has seen some small progress. We have revised the entire text both to reflect scientific advancements and to eliminate weaknesses that attentive readers of the earlier editions have kindly brought to our attention. It is our hope that by the end of this refreshed and improved edition, both cynics and enthusiasts alike will be convinced that, unlike exobiology, astrobiology has an identifiable subject matter accessible to direct study and furthers our collective understanding of our place in the Universe.

But, before you go, a note on nomenclature: throughout this book we capitalize the words *Universe*, *Solar System*, *Sun*, and *Moon* to denote when we're speaking about *our* particular universe, solar system, sun, and moon. Similarly, we use *Terrestrial* to refer to something of or related to our planet, and *terrestrial* to refer to rocky, Earth-like planets in general. By *extraterrestrial*, though, we mean specifically not of our planet. All clear?

Acknowledgments

We would like to thank the many friends and colleagues who were kind enough to edit the earlier editions of this book and just generally keep us on our toes—with such a highly interdisciplinary field on our hands, we couldn't possibly have done it on our own. Those deserving particular recognition include Stan Awramik, Rob Geller, Jason Hollinger, George Karas, Roger Millikan, Stan Peale, John Perona, Lisa Plaxco, Zoe Plaxco, Susannah Porter, Bill Sargent, Frank Spera, and Charlie Strouse, all of whom were kind enough to put time and effort into correcting our errors of fact and language. We would also like to thank Jayanth Banavar, Steve Benner, Dave Deamer, Frank Drake, Reza Ghadiri, Leslie Orgel, Stan Peale, and John Spencer for helpful discussions regarding aspects of their work and Ken Oh and Aaron Rowe for preparing many of the figures.

Additional thanks go to Robert Antonuchi, Stan Awramik, Omer Blaes, Irene Chen, Rob Geller, Bradon Greene, Nick Hud, David Horning, Olivia Hwang, Jerry Joyce, Ben Mazin, Susannah Porter, Frank Spera, Andrea Stith, Jack Szostak, and Kip Thorne for scientific input and editorial advice regarding this edition and to Hui Li and Jade Dutcher for help with the new figures. Finally, Kevin Plaxco extends specific words of thanks to his Chem 147 students, who were, and remain, the inspiration for this work.

Astrobiology

Chapter 1

What Is Life?

Erwin Schrödinger (1887–1961), reluctant cofounder of quantum mechanics, 1933 Nobel laureate in physics, and author of a famous thought experiment involving cruelty to cats, was used to speaking his mind. So much so that, after the Nazis came to power in 1933, he resigned from his chaired professorship at Berlin University and emigrated first to Oxford, then to his native Austria, from where he was exiled again after the Anschluss. In 1939, the government of neutral Ireland invited him to take up a chair of theoretical physics at the newly founded Dublin Institute for Advanced Studies, which became his home for the next 17 years.

One of the obligations of Schrödinger's new job was to give an annual public lecture. In 1943, he held a series of three such lectures at Trinity College Dublin where he discoursed on the topic "What Is Life?" Two aspects of life, heredity and thermodynamics, took center stage in these talks. Schrödinger framed these as the basic questions of how life creates "order from order" and how it creates "order from disorder." In his analysis of genetics (order from order), he estimated the number of atoms contained in a gene ("genes" as physical things still being a relatively abstract concept). He proposed that the genetic information might be encoded in something resembling an aperiodic crystal: a combination of a regular structure with stable, information-bearing variations—an idea that, given modern knowledge of the structure of DNA, seems startlingly prescient. In the second half of his discourse, Schrödinger clarified that organisms can create ordered arrangements of molecules and cells by creating even greater disorder in the environment (e.g., by converting ordered carbohydrate molecules into less ordered water and carbon dioxide), keeping life in line with the immutable second law of thermodynamics.

Schrödinger's lectures, and the brief book that followed, were hugely influential, marking a significant turning point in how scientists look at biology. At a time when there was no such thing as "biophysics," Schrödinger was asking a question that is simultaneously simple and profoundly deep: how do the *physics* of the Universe constrain its *biology*? In the decades following Schrödinger's lectures, many aspects of this question have been resolved, such that today, in this opening chapter, we can take a stab at not only defining what life is but also listing some of its most fundamental requirements.

Definition of Life

Life scientists have long skirted the issue of defining what life is using the same approach US Supreme Court Justice Potter Stewart (1915–1985) once used in a case about public standards of decency: although, he admitted, he could not precisely and unambiguously define pornography, "I know it when I see it." If we want to embark, though, on a deep and rigorous evaluation of life's origins and their relationship to the underlying physics of the Universe, this sort of empirical approach is not going to cut it. All the more so if we are interested in alternatives to the approach life (and its origins) took here on Earth. That is, in order to understand what range of forms life could have taken elsewhere in the Universe, we will have to start with a defensible working definition of what life *is*. So, fellow travelers: just what is life?

The most striking property that distinguishes living systems from the inanimate world is their ability to copy themselves, an attribute scientifically described by the term *self-replicating*. Among *Homo sapiens*, the process is more colloquially captured in the phrase "find a partner, settle down, and have kids." The fact that living things copy themselves is so central to all of biology that some wit once pointed out that "life is just a DNA molecule's way of reproducing itself." All of biology, from the slimy mats of bacteria living in your sink drain through to warring nations, can be described as tools for or consequences of the ceaseless lust of our genes to copy themselves.

Another key limit to our discussion is to define life as a *chemical system* (as opposed to a mechanical or electronic system). Over decades, writers of science fiction and putative nonfiction that extrapolates current (nano)technological trends into the future have suggested that self-replicating, microscopically small robots may someday clean out our arteries, degrade our toxic waste, and generally make themselves useful. Irrespective of the accuracy of these predictions, it seems likely that the physical laws of the Universe allow the creation of mechanical beings able to copy themselves, thus meeting our first criterion for life. Cyberspace hosts similarly viable organisms, known as worms. Although such "malware" requires a computer, an internet connection, and typically some poorly written software in order to reproduce, one might argue that these simply constitute its ecosystem, just as we require an ecosystem for our reproduction. When working out a universal definition of life, it's not entirely clear that these "artificial" examples don't count. Perhaps all the more so given that, as technology progresses, the boundaries between biological, mechanical, and electronic systems continue to erode.

Considering the origins and distribution of life in the Universe, however, it is difficult to imagine that living computer viruses or mechanoid life could have arisen spontaneously from nonlife. Engineered things employ parts larger than molecules; after all, systems consisting of molecular parts are, by definition, *chemical.* Before the creation of the first organisms, these parts would have to be moved around by Brownian motion, the random fluctuations of molecules moving to and fro. And even at the molecular level, Brownian motion isn't all that fast: if you open a bottle of perfume in still air, how long does it take for the fragrance to diffuse across the room? At the mechanical level, it is even slower: Brownian motion slows with the square root of mass, meaning it would take far longer than the age of the Universe for a bucket of watch parts to spontaneously assemble into a functioning watch, much less into a machine capable of self-replication. Thus, while mechanically based life and computer viruses might ultimately arise through the design efforts of intelligent, chemical life forms, it seems unlikely that either could arise spontaneously, absolving us from including self-replicating robots in our working definition of living beings.

A final, but critical, element in our definition of life emerges from the observation that not all self-replicating chemical systems are alive. Wildfires are, for those of us who live in California, an all too frequently occurring example of a self-replicating chemical system that is not alive. Specifically, fire is a self-replicating network of thermally induced radical reactions* that, given more fuel and oxygen, readily produce "offspring" that can likewise reproduce (fig. 1.1). Fire, though, is not alive in any sense that is useful to our discussion. Crystals can likewise "direct" the formation of more copies of themselves. Take, for example, a supersaturated solution—that is, a solution in which a solute is, transiently, at a higher concentration than its equilibrium (i.e., long-term) solubility. If we were to smash a growing crystal into smaller pieces in this solution, each of those pieces would serve as a nucleus and grow into a new, larger crystal. Clearly, then, replication alone is insufficient to distinguish between the animate and the inanimate. What is missing? In a word, evolution.

Living beings produce offspring in their own image via the replication of their genetic material. But this replication is not absolutely perfect: random genetic mutations produce inheritable differences that may improve or impede the offspring's viability. These inheritable differences give natural selection a chance to shape the fate of future generations and the evolution of a species. Evolution, the ability to adapt under selective pressures, is a fundamental property of life and clearly distinguishes it from inanimate, if sometimes self-replicating, materials. A crystal makes perfect copies of itself. The first crystal of quartz that condensed out of a giant cloud of gas and dust that form the Solar System some 4.57 billion years ago is identical to the quartz that crystallized last week in the Corning glassware plant in upstate New York. Crystals and crystallization are changeless and thus are incapable of evolving into new and more complex forms. They are not alive.

*Radicals are carbon compounds that contain an odd number of electrons. As electrons are most stable when paired up with a partner of opposite spin, radicals are highly reactive. We'll see them again in chapter 4.

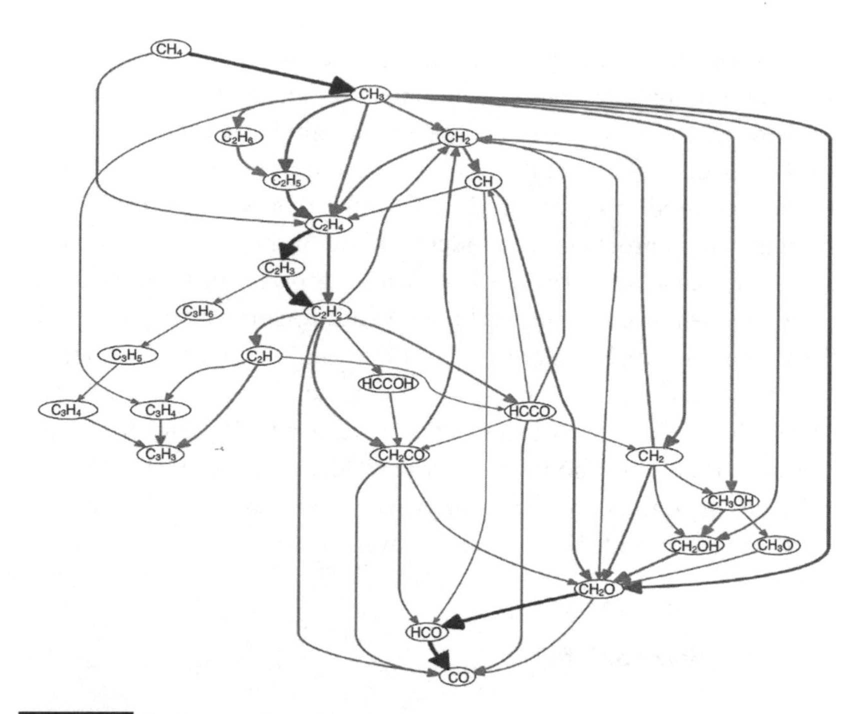

Figure 1.1 Fire is a complex, self-replicating reaction network. It is not, however, alive. Shown is a schematic of the carbon species involved in the burning of methane (CH_4) at low oxygen to produce carbon monoxide (CO). Not shown are the highly reactive hydrogen- and oxygen-containing radicals—molecular species with an odd number of electrons—that both produce these reactions and are produced by them, driving this self-replicating behavior.

As middle-aged academic types well past our reproductive years, we feel obliged to add a note here: our definition of life involves self-replicating *systems*, not necessarily self-replicating *individuals*. Mules, for example, are generally sterile, and yet anyone who has tried to lead one down to the bottom of the Grand Canyon will attest to their stubbornly animate nature. But mules are always part of a larger system (one that includes reproductively competent female horses and male donkeys) that, as a whole, is capable of reproducing and evolving. More generally, reproduction often requires the combined efforts of two organisms and not just pairs of males and females; viruses and their hosts are an example. By the same token, evolution, too, generally acts on the level of populations rather than on individuals. Neither observation, however, degrades the usefulness of our definition.

Now that we've settled on a specific definition of what life is, the next question is: what are the minimal—and thus fundamental—conditions required to support it? In answering this, we must inevitably be somewhat parochial: our understanding of the conditions under which life can arise and evolve is almost certainly going to be flavored by deeply held preconceptions based on our understanding of life on Earth. But as long as we are aware of this underlying bias, we can at least tackle each of the seemingly necessary conditions in as unbiased and logical a fashion as is (even the word itself is telling) *humanly* possible. That is, we should cast our net wide, making an effort not to mistake constraints predicated on what we observe regarding life on Earth for constraints that are truly universal.

The Chemistry of Life

A whopping 98% of the matter in our Universe is either hydrogen (H) or helium (He). Life as we have defined it, however, requires chemistry far more complex than that available to these two atoms, the chemistry of which is limited to the hydrogen molecule (H_2), "protonated hydrogen" (H_3^+), and helium hydride (HeH^+). And while the latter two are stable in extreme isolation in the laboratory and have, in 1996 and 2019, respectively, been confirmed to form in interstellar space, they cannot be prepared in bulk due to their propensity to react violently with any other type of molecule they contact. Even taking into account our potentially narrow-minded, Earth life–centric biases, it seems certain that a self-replicating chemical system cannot be built based on such meager chemical complexity. But, as we'll see, while hydrogen and helium were formed in great abundance in the first minutes of the Universe, the formation of heavier, more complex atoms was rather a more delicate matter.

What atoms are required for life? Here we must begin to speculate, but not without a lot of well-established science to back us up. There are, after all, only a finite number of elements in the periodic

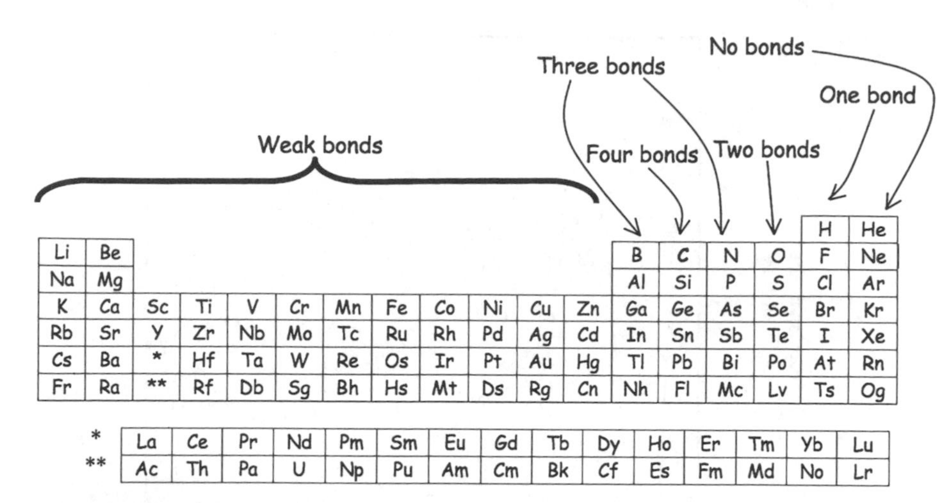

Figure 1.2 The atomic building blocks of life. Even a quick perusal of the periodic table leads us to the conclusion that relatively few elements are likely to support the complex chemistry required to form life here on Earth, or anywhere. In particular, only carbon (C) is tetravalent (forms four bonds), forms strong covalent bonds with itself and with its neighbors in the table, and is kinetically stable. These properties render the element uniquely well suited to support the complex chemistry presumably necessary for life.

table (fig. 1.2), and many of these are poorly suited to support life for a fair list of reasons. Consequently, many of the 90 or so naturally occurring elements can be ruled out. So many, in fact, that in the end there may very well be only a single element—carbon (C), the basis for *all* life here on Earth—that is able to support the complex chemistry presumably required to create a self-replicating chemical system. The easiest way to appreciate the special, perhaps even unique, qualities of carbon is to compare it with silicon (Si), the chemically similar "cousin" that sits immediately below it on the periodic table.

Some of the properties that suit carbon so well to its central role in Terrestrial life are shared or even exceeded by silicon. For example, silicon, like carbon, is tetravalent—that is, each atom forms four bonds, allowing for the formation of a rich array of complex molecular structures. And, while a silicon-silicon bond is weaker than a carbon-carbon bond, the discrepancy is only about 35%. Consistent with this, both silicon and carbon can form long molecular chains: compounds of silicon and hydrogen, called silanes, with up to 28

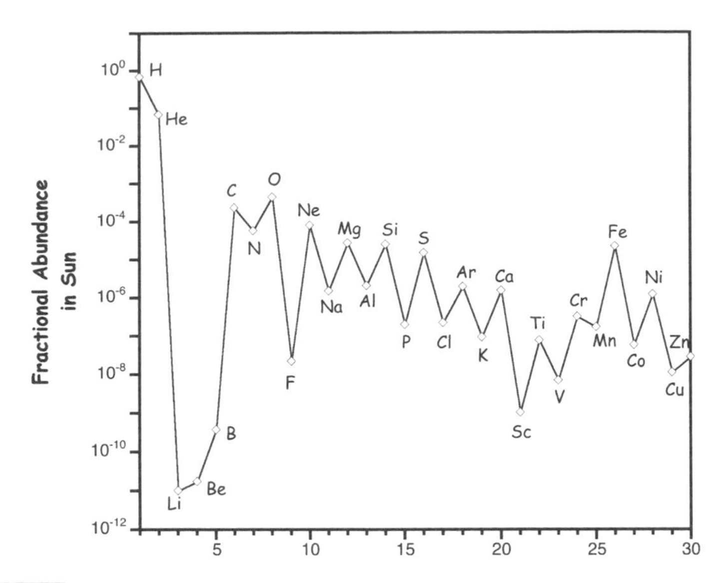

Figure 1.3 The relative abundance of the various elements available in the pre-solar nebula—and thus, ultimately, available to serve as the building blocks of the planets—can be discerned by from the relative abundances of elements in the Sun, which are shown here. Of note, the Sun is relatively enriched in elements heavier than helium relative to other stars its size. Nevertheless, this qualitative pattern is common among stars of the Sun's generation (more on this in chapter 2).

consecutive silicon-silicon bonds have been reported in the scientific literature. Likewise, while carbon is the fourth most common element in the Solar System (fig. 1.3), silicon is far more common in the Earth's crust, where it is second only to oxygen (O) in terms of its abundance (fig. 1.4). Nevertheless, silicon simply cannot support the same rich chemistry as its "upstairs" neighbor on the periodic table. The problem lies in both the *thermodynamics* (equilibrium stability) of silicon's interactions with other atoms and the *kinetics* (rates) with which those interactions form and break.

A perusal of the relative stability (thermodynamics) of various bonds formed by carbon and silicon provides the first fodder for our arguments. Because typical carbon-carbon, carbon-nitrogen, and carbon-oxygen single bonds are all equally stable (to within about 10%), nature does not have to invest much energy when swapping between, say, a carbon-carbon bond and a carbon-oxygen bond, or

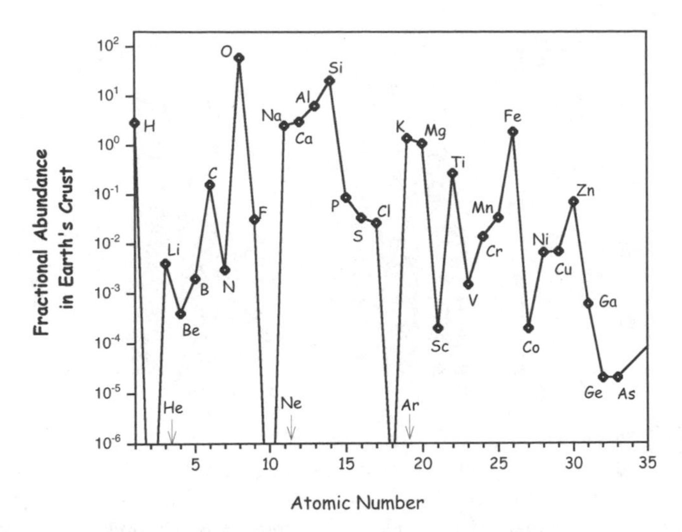

Figure 1.4 An ingredient list for life on rocky, terrestrial planets. Shown is the relative abundance of elements in the Earth's crust. Note that this pattern differs dramatically from the relative abundances in the Sun (a proxy for the Solar System at large) shown in Figure 1.3.

vice versa (fig. 1.5). The same is also true for carbon when it participates in double or triple bonds; typical carbon-carbon, carbon-oxygen, and carbon-nitrogen double bonds are all more or less equally stable (to within 30%), as are carbon-carbon and carbon-nitrogen triple bonds. Better yet, a typical carbon-carbon double bond is approximately twice as stable and a triple bond very nearly three times as stable as the corresponding single bond (fig. 1.6). That is, their stabilities *per bond* are all about the same, and thus, again, nature can shift carbon between various single-, double-, and triple-bonded configurations (e.g., converting two single bonds into one double bond) without much change in energy. This attribute is a distinct advantage when it comes to generating the sort of complex chemistry likely to support life; because of this property, it is relatively easy for biology to shuffle carbon from molecule to molecule across a bewildering array of compounds (fig. 1.7).

This situation with silicon is quite different, and far less favorable. A single bond between two silicon atoms, for example, is about 40% less stable than a typical silicon-nitrogen single bond and *less than*

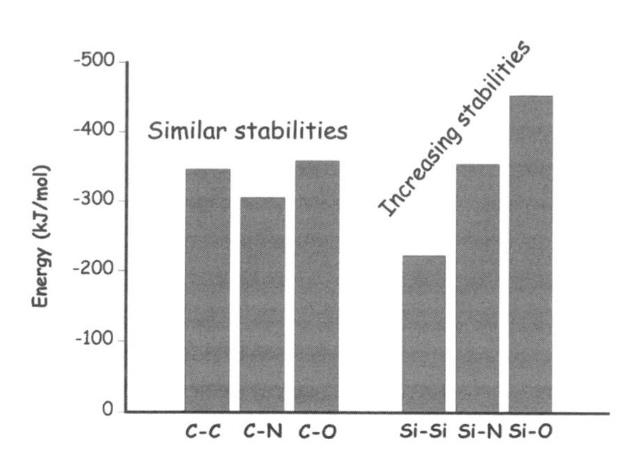

Figure 1.5 Single bonds between carbons or between carbon and nitrogen or oxygen are typically all of similar stability, and thus biology does not have to expend much energy to substitute one for another. This is far less true for silicon, which strongly prefers to be bound to oxygen atoms rather than to atoms of its own kind.

half as stable as a typical silicon-oxygen single bond (fig. 1.5). As a consequence, silicon readily oxidizes to silicon dioxide, limiting the chemistry available to this atom whenever oxygen is present. And oxygen is almost always present, it being both the third most common atom in the Universe and the single most common atom in our plan-

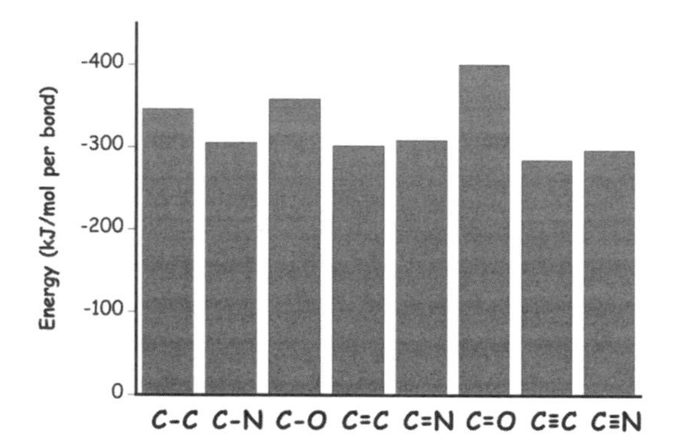

Figure 1.6 Multiple bonds (double bonds, triple bonds) between carbons or between carbon and nitrogen or oxygen are typically of similar stability *per bond*, and thus biology does not have to expend much energy to substitute, say, two carbon-carbon single bonds for one carbon-carbon double bond. Silicon, in contrast, simply does not form stable double or triple bonds.

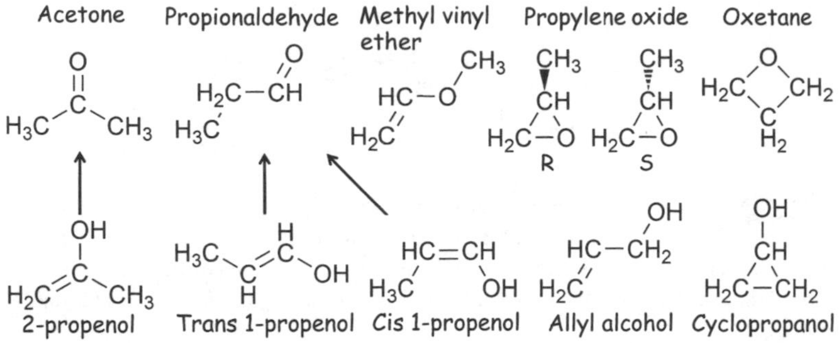

Acetone Propionaldehyde Methyl vinyl ether Propylene oxide Oxetane

2-propenol Trans 1-propenol Cis 1-propenol Allyl alcohol Cyclopropanol

Figure 1.7 The diversity of stable carbon compounds is unparalleled. Shown, for example, are the 11 isomers (i.e., different bonded arrangements) that share the C_3H_6O empirical formula of acetone. Save for the three "vinyl" alcohols (a hydroxyl group on a carbon involved in a carbon-carbon double bond) at the lower left, which quickly transfer a hydrogen to form the carbonyl (C=O) compounds above them, all these molecules are stable enough that they can be purchased commercially and shipped and stored at room temperature.

et's crust. The case for silicon grows even worse when we consider double and triple bonds. Molecules containing silicon-silicon double bonds and even a silicon-silicon triple bond have been prepared in the laboratory, but they are so unstable that they spontaneously decay unless kept at cryogenic temperatures (i.e., in liquid nitrogen at –200°C) or "protected" by big, bulky groups blocking the silicon atoms from attack by other molecules. This is because, unlike the carbon-carbon double bond, which is almost twice as stable as a carbon-carbon single-bond, a silicon-silicon double bound is *less* stable than a silicon-silicon single bond. To understand why this is so, we're going to have to spend a bit of time thinking about atomic and molecular orbitals.

The four outermost electrons of a carbon atom, the electrons that can form bonds, reside in four atomic orbitals, one designated 2s and three designated 2p. These orbitals are each capable of holding two electrons, and thus there are four "spaces" left over for binding partners to fill, which is energetically favorable.* In a double-bonded carbon molecule, such as ethylene ($CH_2=CH_2$), the geometry of the bonds

*All the available orbitals are already filled in a neon atom, hence its near complete inability to bond to anything else. The same also holds true for the outer orbitals of helium, argon, and the rest of the "noble gases" that share the column with neon over on the right-hand edge of the periodic table.

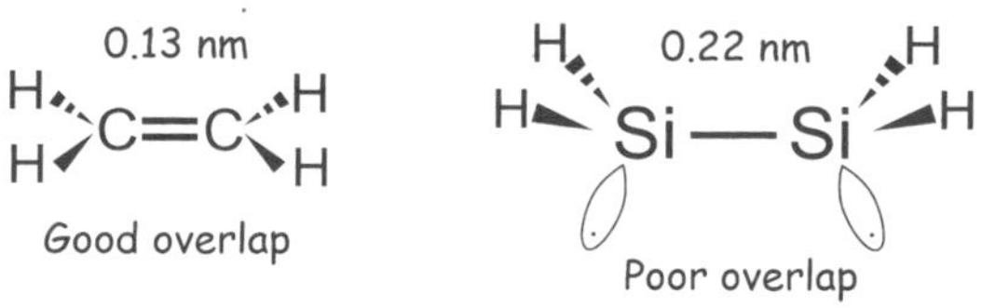

0.13 nm
Good overlap
0.22 nm
Poor overlap

Figure 1.8 A single bond between two carbon atoms is short enough that the two electrons in the remaining p orbitals still overlap, causing the creation of a stable second (double) bond. Single bonds between two silicon atoms, in contrast, are much longer, limiting overlap of the p orbitals and thus precluding the formation of a stable double bond.

suggests that the distinction between the 2s orbital and two of the 2p orbitals has disappeared: these orbitals "hybridize" to form three sp^2 *molecular* orbitals. These hybrid orbitals, arranged in a plane with identical (120°) angles between them, form sigma molecular orbitals that bond them to neighboring atoms—in ethylene, that is with the two hydrogens and the other carbon (fig. 1.8). This latter bond, the bond between the two carbons, is approximately 0.13 nm (a nanometer is a billionth of a meter)—short enough that the remaining 2p atomic orbitals on the two carbons, which are sticking out of the plane at a right angle, overlap one another sufficiently to form a second bond (called a pi bond) that is nearly as strong as the first (sigma) bond. Silicon, the next atom below carbon in the periodic table, also has four bonding electrons, but they are in 3s and 3p atomic orbitals, which are significantly farther from the nucleus than the 2s and 2p orbitals of carbon. The greater diameter of these orbitals stretches the silicon-silicon bond distance to 0.22 nm, long enough that there is no longer any significant overlap between the two remaining 3p orbitals. Because of this, the p-orbital electrons tend to stay as single, solitary electrons rather than paired electrons working together to form a bond (fig. 1.8). This leaves silicon-silicon double bonds unstable and highly reactive.

As if the poor thermodynamics of silicon bonding weren't bad enough, silicon-containing molecules are also *kinetically* unstable. For example, in an *equilibrium* (thermodynamic) sense, both methane (CH_4) and silane (SiH_4) are unstable in air as all three elements involved are much more stable in their oxides (water, carbon dioxide, and silica). Like most carbon compounds (fig. 1.7), however, methane is *kinetically* stable. For example, although methane burns vigorously in air, before the striker on your stove ignites it, the methane and other like molecules in the natural gas do not spontaneously burst into flame. Indeed, at room temperature, methane reacts so slowly that its half-life in air is approximately seven years, thus allowing methane from, say, rice paddies to build up in the atmosphere and contribute to greenhouse warming (it absorbs infrared light much more effectively than does carbon dioxide, and thus, molecule for molecule, methane is about 20 times more potent as a greenhouse gas). Silane, in contrast, spontaneously bursts into flame upon contact with air (fig. 1.9), reflecting the much more rapid reaction kinetics of silicon compounds relative to the corresponding compounds of carbon. To see why this is true, we again have to think about orbitals.

To swap one of the substituents attached to a carbon for another substituent, we have two choices. We can add the new substituent, making a five-bond (pentavalent) carbon, and only then break the bond to the substituent being replaced. Or, we can first break the bond

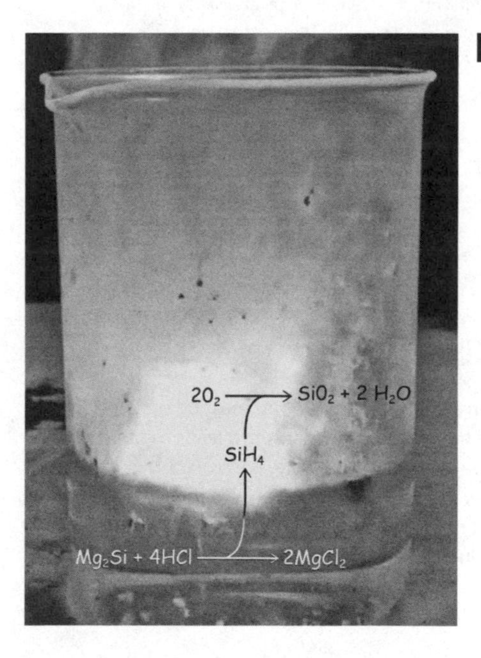

Figure 1.9 Unlike methane (CH_4), which has about a seven-year half-life in our atmosphere, silane (SiH_4) instantly bursts into flame upon contact with air, clearly illustrating the far greater kinetic reactivity of silicon compounds.

to the substituent undergoing replacement to produce a three-bond (trivalent) carbon intermediate before then adding the new substituent (fig. 1.10). And while the trivalent carbon intermediate required by the second mechanism is very high in energy, thus slowing such reactions (because collisions with the requisite energy are rare), the energy of the pentavalent intermediate of the first mechanism is higher still. The problem is that the lowest-energy orbital available to accept bonding electrons to make that fifth bond is a 3s orbital, which is quite high in energy. In contrast, the lowest-energy empty orbitals in silicon are 3d orbitals, which aren't much higher in energy than the 3s and 3p orbitals that this element normally uses to "perform its chemistry." Because of this, silicon is relatively stable when bonded to five substituents; the fifth substituent simply forms a bond with a low-lying, empty 3d orbital before the bond to the old substituent is broken (fig. 1.10). The stability of pentavalent silicon reduces the energy required for silicon to react, thus making its reactions many, many orders of magnitude faster than those of the equivalent carbon compounds.

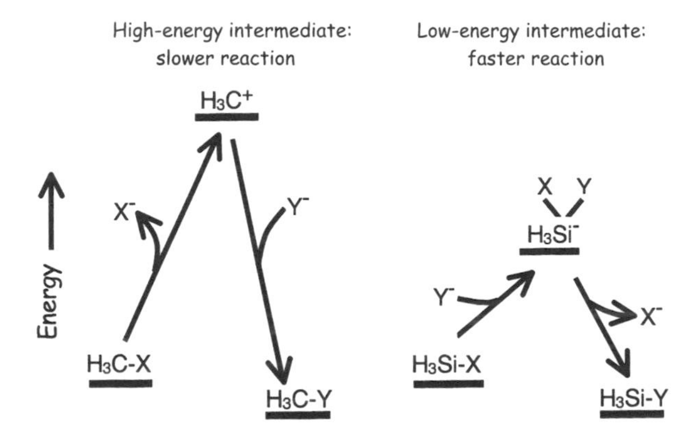

Figure 1.10 To replace a substituent on a carbon atom, a bond must be broken. If bond breakage happens at the beginning of the reaction mechanism, it leads to the formation of a highly unstable, trivalent (three-bond) intermediate. Because of this, carbon tends to react only slowly; relatively few molecular collisions are energetic enough to provide the necessary energy. Unlike carbon, however, silicon supports the formation of up to five bonds. The resultant low-energy intermediate renders silicon far more rapid in its reactions.

And there we have it. Carbon presents a fairly level playing field in which nature can shuffle around carbon-carbon, carbon-nitrogen, and carbon-oxygen single, double, and triple bonds without paying too great a cost to convert any one of these into another. Equally important, carbon's reaction kinetics are slow and thus, once parked in a given molecule, it tends to stay there. Silicon, in contrast, reacts vigorously whenever oxygen is present, rapidly moving "downhill" (energetically speaking) to form very stable silicon-oxygen bonds. Given all this, it's no wonder that on the order of 100 million unique carbon compounds have been described by chemists, which is more than all the described non-carbon-containing compounds put together. And this is not even counting all the carbon-containing biological entities, such as proteins and chromosomes, of which there are even more due to evolution's ceaseless creation of diversity.

Do you still think there's a chance that, somewhere out there, beasts based on silicon are having this same debate, but in reverse? Well, then, consider one final issue: the second bond in a carbon-oxygen double bond is more stable than the first, and thus carbon is perfectly happy to make double bonds to two oxygen atoms to form the triatomic—and gaseous—oxide, carbon dioxide (O=C=O). Silicon, as we explored above, does not form double bonds and, thus, to satisfy its desire to form four bonds, each silicon atom must link up with four, rather than two, oxygen atoms. The problem is, oxygen wants to form two bonds, a situation that can be satisfied only if each oxygen links to two silicon atoms to form a bridge between them. Because of this, silicon dioxide forms a rigid, three-dimensional network that does not dissolve without breaking bonds. The contrast is stark: carbon forms a wide range of molecules that can be solid (e.g., cellulose), liquid (e.g., alcohols), or gas (e.g., carbon dioxide) all under the same conditions of temperature and pressure, allowing the atom to be easily transported and exchanged between these states (e.g., the excretion of carbon via the exhalation of carbon dioxide). Carbon can likewise form many compounds that are highly soluble in water (e.g., carbohydrates) and many compounds that are only sparingly so (e.g., lipids), permitting both its transport and use in structural and organizational roles. In short, we eat water-soluble carbon compounds such as sugar and other carbohydrates, oxidize them, and then excrete the water-soluble gas carbon dioxide. None of these properties hold for silicates, which are highly insoluble solids such as quartz (pure silicon dioxide) and thus would be exceedingly difficult to mobilize and eliminate.

Can we categorically say that life cannot be based on silicon? Of course not. Perhaps we have a "carbon bias" and are simply not creative enough to see how silicon could support life. It is clear, however, that silicon is less well suited to support complex chemistry, and thus it seems much, much less likely that silicon-based life could form than carbon-based life. Thus, if aliens ever do visit us, the smart money says that we should welcome them with carbon-based cakes and not with silicon-based rocks.

Given the above, it's pretty clear that carbon wins over silicon regarding their suitability to serve as the basis for complex chemistry. But what of the 90 or so other naturally occurring elements? They fare even worse than silicon. At the far right of the periodic table (fig. 1.2), the noble gases, helium (He), neon (Ne), and the like, participate in effectively zero chemistry. Nor can we build a complex chemistry based on the halogens, such as fluorine (F) and chlorine (Cl), as these atoms generally bond to only a single partner, greatly limiting their chemistry. Oxygen and the similar atoms below it in the third-to-last column of the periodic table likewise make only two bonds, and nitrogen (N) and boron (B) and their respective columns make three—numbers that are also handicaps relative to carbon's tetravalency. Moving farther to the left, we have the transition metals, including iron (Fe), nickel (Ni), and gold (Au), and then the alkaline earth metals, such as calcium (Ca) and magnesium (Mg), and the alkali metals, including sodium (Na), potassium (K), and their brethren. None of these atoms, however, form strong covalent bonds—that is, bonds in which electrons are shared between two atoms (in molecular orbitals). Instead, when these elements bond they tend to transfer electrons almost entirely to themselves or over to their bonding partners, producing ionic bonds, so called because the transfer of electrons produces charged atoms or groups of atoms called ions.* These ionic bonds, however, are far less stable than covalent bonds, thus limiting the ability of metals to participate in chemistry as complex as that seen for the nonmetal elements on the right-hand side of the periodic table. In short, while carbon may not be the only element that can support life, any theory about alternative life forms that relies on elements other than carbon is significantly suspect.

The Solvent of Life

Life almost certainly requires a liquid *solvent*. More specifically, although chemical reactions can, of course, take place in the gas and solid phases, each of these has disadvantages relative to liquid-phase reactions. Gas-phase chemistry, for example, is limited to molecules that are volatile enough to stay in the gas phase. This likely rules out molecules of sufficient complexity to support self-replicating chemical systems. Likewise, even though molecular species can diffuse through solids to give rise to chemical reactions, such diffusion is extremely slow and would seem unlikely, if not unable, to support life. Consistent with these arguments, the human body contains around 70% water, highlighting the fact that this is the solvent employed by life on Earth. But is water the only liquid that can plausibly serve in this role? Once again, even a cursory exploration of the list of small molecules that one can make using the Universe's most common atoms suggests that, of all the entrants on the list of potential "biotic solvents," water may well be the only reasonable choice; many of water's properties render it particularly well suited to be a biological solvent (table 1.1). So many properties, in fact, that its ability to form the basis of biochemistry may be unique.

Some of the "ideal" properties of water are well known, and others, while less so, are no less critical for life on Earth. An example is taught

**Ion* comes from the Greek *ienai*, meaning "to go," which in turn originated from the name of a Greek tribe, the Ionians; in Ancient Greece, the Ionians migrated from Thessaly to Peloponnese and to the western coast of Asia Minor. Moving ions conduct charge, such as in salt water.

Table 1.1
Physical properties of potential biological solvents

Solvent	Formula	Liquid range at 1 atmosphere (°C)	Molar density (mol/L)	Heat capacity (J/g°C)	Heat of vaporization (kJ/g)	Dielectric constant	Density ratio: solid to liquid
Water	H_2O	0 to 100	55.5	4.2	2.3	80	0.9
Hydrogen fluoride	HF	−83 to 20	48.0	2.6	1.3	84	1.8
Ammonia	NH_3	−78 to −34	40.0	4.7	1.4	25	1.2
Methane	CH_4	−182 to −161	26.4	3.3	0.3	2	1.1
Hydrogen	H_2	−259 to −253	35.0	0.008	0.2	0.2	1.3

to almost every elementary school student: water is one of the few substances—and the only known *molecular* substance (i.e., not a monatomic element)—that expands when it freezes. Because of this, ice floats. If, instead, the ocean was filled with, for example, liquid ammonia (NH_3), which is, after all, the fourth most common molecule in the Universe, its winter pack "ice" would sink, where it would be insulated from the summer's warmth and prevented from seasonally melting. With each passing year, more and more of the ammonia ocean would be locked up in the solid until, in a time frame quite rapid by geological standards, the planet would freeze over with only a thin seasonal veneer of liquid on the surface. Could such a frozen ocean support the origins of life? Perhaps. But a permanently liquid ocean, with its ability to transport nutrients and modulate temperature, seems more likely to do the trick.

Water also has an extraordinary ability to absorb heat without much of a rise in its temperature, and even more so when it evaporates to form water vapor, which is why water and steam are used as carriers for heat in central heating systems, radiators, and hot water bottles. In more precise terms, the heat capacity of liquid water is 4.2 J/g°C, meaning it requires 4.2 J of heat energy to raise the temperature of 1 g of water by 1°C. This value is about three times that typical of rock or metal. And a whopping 2.3 kJ (i.e., 2,300 J) is required to evaporate 1 g of liquid water (at its boiling point) to form water vapor, which means that water absorbs a great deal of heat when it evaporates under warm conditions and releases (and thus transports) an equivalent amount of heat to a cool place when it condenses. Water's high heat capacity, which is significantly higher than that of most other common substances, and high heat of vaporization, which is the highest of *any* known substance, help to moderate the Earth's climate—a seemingly critical event in the origins and evolution of life that we'll cover in more detail in chapter 3. On a related note, thanks to its unique ability to form extended hydrogen-bonding networks, water remains liquid over an unusually broad temperature range spanning 100°C (at one atmosphere pressure), thus helping to ensure that, even if the climate does fluctuate radically, liquid solvent will remain available to support life.

In addition to these important physical properties, the chemical properties of water seem to render it uniquely suited as the basis for the sort of complex chemistry necessary to create living things. For example, the dielectric constant of water is around 80, which is significantly higher than that of any other cosmologically abundant liquid. This means that two oppositely charged ions in water are attracted with one-eightieth the force they would feel in a vacuum. Because of this, water can shield charged ions from one another, allowing them to be readily taken into solution where they can perform chemistry. Water also has the highest molar density of any molecular liquid; 55.5 mol of water molecules are crammed into each and every liter of the stuff. Here *mole* (abbreviated mol) refers to a unit used by chemists to describe a fixed number of molecules. In this regard, "a mole" is completely analogous to "a dozen." Because molecules are much, much, much smaller than eggs, however, the chemist's dozen is much, much, much larger than a baker's dozen: 1 mol corresponds not to 12 or 13 molecules but to 6.023×10^{23} molecules. Thus, there are 3.34×10^{25} water molecules in 1 L of liquid water; no other liquid packs anywhere near as many molecules in a given volume.

Because of its extraordinary molar density, the energy required to arrange water molecules (the solvent) around any molecules dissolved in it (the solute) is quite high (many water molecules need to be rearranged to make room for each cubic nanometer of solute), and thus water tends to force many types of dissolved molecules to organize themselves, minimizing this cost. This organizing effect is called the hydrophobic effect (from the Greek *hydro*, meaning "water," and *phobos*, meaning "fear"). This effect, which is why, for example, "oil and water don't mix," plays a critical role in the formation of biomolecular structures on Earth. Lastly, hydrogen and oxygen are the first and third most cosmologically abundant atoms, rendering water second only to hydrogen gas (H_2) as the most common molecule in the Universe.

Of course, the fact that water is well suited for life on Earth doesn't automatically rule out that life elsewhere might be based on some other solvent. Or does it? It would be hard to find an alternative as no other liquid has even a fraction of the seemingly favorable attributes of water. Hydrogen fluoride (HF) comes close. Compared with water, it has a slightly higher dielectric constant (84, to water's 80) and thus is at least as good a solvent for ionic materials, a slightly wider liquid temperature range of 103°C (at atmospheric pressure it freezes at −83°C and boils at 20°C), and a comparable molar density (48.0 versus 55.5 mol/L for water). But as fluorine is cosmologically quite rare—it is about 100,000 times less abundant in the Solar System than oxygen—it seems very unlikely that there are little purple fish happily swimming in seas of liquid hydrogen fluoride on the planet Zap 7.

Given the above, we are probably safe ruling out hydrogen fluoride as a potential solvent for life. And none of the other molecular liquids formed by the cosmologically abundant elements (methane, ammonia, etc.) come anywhere near as close to the ideal properties of water as does hydrogen fluoride (check out table 1.1 again): their liquid ranges are extremely small, their ability to solvate ionic materials is poor, and their ability to regulate climate is more limited. From such considerations emerges the near certainty that life not only has an absolute requirement for a liquid solvent, but that water is by far the most "qualified" solvent to fulfill that role. This is not to say that life cannot have arisen based on other solvents, simply that the origins of life face a much more significant hurdle in the absence of this remarkable and abundant liquid.

Energy for Life

In addition to the *matter* from which it is constituted, life also requires *energy* to drive the reactions that underlie its metabolism and replication machinery. This is obvious to the chemist as living organisms create an implausible amount of order out of disorder, such as when the randomly distributed molecules of carbon dioxide, water, and nitrate fertilizer end up in the highly nonrandom structure of a plant. According to the second law of thermodynamics, living things can achieve this increase in order only if even more entropy (roughly speaking, a measure of disorder) is created elsewhere. By using energy from an external source, life can swap energy for entropy: the living organisms get the order, while the rest of the Universe pays the price and becomes less ordered.

Thus, life requires an external disequilibrium (an "ordered" state), whose tendency to drive toward a more equilibrated ("disordered") state can be exploited by life for use in organizing its molecules into some pattern capable of reproduction. One of the more abundant sources of disequilibrium in the Universe is the much higher temperature of stars than of the Universe at large. Because of this, the copious number of high-energy photons emitted by a star can be absorbed by the surface of a (much cooler) planet. Here on Earth, plants take advantage of this disequilibrium and use it to feed the striking disequilibrium that is our biosphere. For example, as forest fires remind us, the presence of combustible wood in an atmosphere containing oxygen is a clear deviation from chemical equilibrium with respect to a mixture of water and carbon dioxide (the products of combustion). Indeed, this reaction is out of equilibrium by *more than 500 orders of magnitude*! Were it at equilibrium, there should not be a single "wood" molecule on the entire planet.* We animals, in turn, take advantage of the latter disequilibrium when we oxidize the carbohydrates in a nice al dente bowl of pasta to generate the energy we use to run our metabolic processes. We should note, however, that while these two "lifestyles" (photosynthesis, and the oxidation of photosynthesis-derived carbohydrates with photosynthesis-derived oxygen) represent the dominant disequilibria that drive life on Earth, they are not the only sources of energy known to support biology—we'll hold off on this topic, though, until chapter 8, where we delve into this issue more deeply.

*Makes you appreciate the challenge that firefighters are up against, doesn't it?

Other Constraints on Life

Chemistry, solvents, and thermodynamics aside, there are at least a handful of other constraints that must be met in order for life to arise and thrive. We'll be discussing these in detail in the following chapters, but let's take a moment here to look them over.

Life probably requires a condensed state on which—or in which—to form and evolve: it needs land or a liquid ocean. The reason for this was outlined above: molecules of sufficient complexity to form life are inevitably too dense to stay suspended in an atmosphere. This is, of course, a much bigger constraint on the *origins* of life than on its ability to thrive after it has arisen; if the surface of the Earth were to slowly become uninhabitable, it is a pretty good bet that at least some bacteria would adapt to full-time living in cloud droplets. Indeed, some Earthly bacteria may already have done so. But the limited mass transport and limited size of condensed bodies that can occur in the gas phase make this realm an exceedingly unlikely one for the *origins* of life. This effectively rules out life on gas giant planets, such as Jupiter, which lack any solid or even liquid surface—if life cannot *arise* on a planet that, like Jupiter, lacks a surface, then it likewise cannot evolve into giant, hydrogen-filled Hindenburgoids that live their whole lives in the atmosphere. And of course, life is all the more unlikely to have formed in interstellar space. This didn't stop the cosmologist Fred Hoyle (1915–2001) from writing a wonderful sci-fi novel, *The Black Cloud*, about an intelligent, interstellar cloud that pays the Solar System a visit and accidentally wipes out most of humanity. The oversight was understandable: having been born and raised in the cold vacuum of space, it simply hadn't occurred to the cloud that life was possible on the hot, crowded surface of a planet. Hoyle was a proponent of a theory that the Universe is infinitely old, and thus he was able to pass on the question of how life might arise in space by stipulating that life, too, has simply existed forever. As we'll see in chapter 2, though, Hoyle's cosmology has been proven wrong; the Universe, and thus life, had a beginning. And where might the latter beginning have begun? It looks as if the surfaces (or oceans) of rocky, solid planets are probably the only hope.

Life also presumably requires *time*. And the narrower the range of conditions under which life could arise in the first place, the more unlikely it is that sufficiently stable environments would remain within this range long enough for life to develop. The Universe is a dangerous place. The luminosity of a star changes and, with it, the temperature of any planets warmed by its light. Planets are sometimes struck by asteroids so large that the energy imparted by the impact can boil oceans and sterilize entire worlds. Atmospheres escape into space. Rotational axes tilt, plunging planets into million-year winters. Supernovae explode with the power of a billion suns, wiping out life on any planet within a few hundred light-years. Considering these risks, it is clear that not all environments in the Universe that are capable of supporting the formation of life will remain stable long enough for life to arise at all, much less gain a secure footing.

Conclusions

Given a universe containing the right ingredients, the ingredients required for life to originate seem relatively straightforward. All we need is some water, carbon on a solid planetary surface (or in its oceans), an energy source, some time, and we're off. But is it that easy? What is required to produce a water- and carbon-bearing planetary environment that provides energy sources and yet is stable over eons? And how often are these conditions met? And if we find these conditions, how likely is it that life will arise? The following chapters explore each of these critical questions in turn.

Today, more than 75 years after Schrödinger's lectures, many of the questions he asked about life have been answered in delightful and amazing detail. What we know less about is how life arose on Earth, and how—and whether—it might have arisen elsewhere, the topics to which we will now turn.

Further Reading

What is Life?

Schrödinger, Erwin. *What Is Life?* Cambridge: Cambridge University Press, 2012.

Expectations of and Constraints on Life in the Universe

Schulze-Makuch, Dirk, and Louis N. Irwin. *Life in the Universe: Expectations and Constraints.* 3rd ed. Berlin: Springer-Verlag, 2018.

The History of Origins-of-Life Research

Fry, Iris. *The Emergence of Life on Earth: A Historical and Scientific Overview.* New Brunswick, NJ: Rutgers University Press, 2000.

Chapter 2

Origins of a Habitable Universe

Humanity's first communications satellite, named *Echo 1*, was little more than a passive mirror. Launched in 1960, the satellite consisted of a metalized balloon, about 70 m in diameter, placed to reflect radio waves from a transmitter on one side of the Atlantic to a receiver on the other side. After a few years, the *Echo* program was replaced by the first "active" communications satellites, which detect and electronically amplify signals before sending them on to the recipient, rendering the sensitive antennas and receivers built for the earlier system redundant.

In 1965, Arno Penzias and Robert Wilson, radio physicists working for AT&T's Bell Labs in New Jersey, realized that a semiretired radio receiver built for the *Echo* program and located at nearby Crawford Hill might be of use for the astronomical detection of radio waves emanating from our galaxy. To use the receiver to measure these presumably very faint signals, however, they first had to characterize, and eliminate, the various sources of electronic noise that were sure to plague the instrument—which, after all, had not been designed to serve as a telescope. Not surprisingly, when they pointed the 6 m diameter, horn-shaped antenna at "empty" presumably radio-silent parts of the sky to calibrate it, they detected a faint radio frequency "hiss" that they assumed was due to instrument artifacts. However, when systematically trying to "fix" the antenna and its associated amplifiers, eliminating one by one any potential sources of electronic noise, the hiss persisted. Ultimately, the physicists began to suspect that pigeons that were roosted in the horn might be the source of the offending static. But even after chasing them away and cleaning up years' worth of droppings (the life of a physicist is not as glamorous as it may appear), the hiss remained. They were flummoxed; for more than a year, the source of the problem eluded them. Finally, they started to consider the possibility that, although the hiss remained the same no matter where in the sky the antenna was pointed, the noise might not be instrument noise but might instead reflect some authentic astrophysical phenomenon. Upon hearing this, several colleagues suggested that they call Bob Dicke (1917–1997) at Princeton University, just an hour to the south. This, it turns out, was a fortuitous idea. With his colleagues Jim Peebles and David Wilkinson (1935–2002), Dicke had just written a paper outlining an important prediction of a theory regarding the origins of the Universe. This theory, they said, predicted that the entire Universe would be filled with microwave radiation at precisely the frequency and intensity observed in the horn antenna.

Unbeknownst to Dicke and his team, a similar theory had been described as far back as 1948 by the Hungarian-born American physicist George Gamow (1904–1968) and his student Ralph Alpher (1921–2007), who theorized that the Universe formed from an initially superdense, superhot state from which today, billions of years later, it continues to expand and cool. And although this key theoretical advance was ignored for many years after its publication, the paper describing it is now well known, both for its prescient scientific prediction and its being an example of Gamow's famously puckish sense of humor. Specifically, Gamow added the name of his friend and Cornell colleague Hans Bethe (1906–2005) to the paper simply because it amused him that the authors' names, Alpher, Bethe, and Gamow, would then parallel the first three letters of the Greek alphabet.

The theory that Gamow and company and Dicke and company had independently derived was based, in part, on observations made by astronomers in the first decades of the last century. By 1917, the American astronomer Vesto Slipher (1875–1969) had shown that nearly all the many spiral-shaped nebulae (from the Latin word for "cloud") astronomers had spied in the heavens with their telescopes were "red shifted." That is, their spectral lines—atom-specific wavelengths of emitted light—were shifted to longer wavelengths than those seen in laboratories on Earth. A possible reason for such a systematic red shift was that all the nebulae were moving away from us; the Austrian physicist Christian Doppler (1803–1853) had described a century before how frequency changes with the motion of the source, an effect now known as a Doppler shift. But at the time, it was not even clear what the nebulae were, much less why they would nearly all be receding. Even the very notion of that was an anathema; since Copernicus had shown that the Earth orbits around the Sun and not, as the Catholic church had taught, the other way around, a central precept of science had been that there is nothing particularly "special" about our place in the Universe (see sidebar 2.3). And yet, here it appeared that we were the center of some monstrous offense that the rest of the cosmos was fleeing from.

The question of what spiral nebulae are was put to rest by the British-American astronomer Edwin Hubble (1889–1953). In 1924, Hubble was in charge of the largest and best telescope in the world at the time, the "100 inch" (2.54 m) Mount Wilson telescope sitting in the mountains above the then small town of Los Angeles. From that perch, he turned the telescope's unprecedented resolving power on Andromeda, the largest nebula by apparent size and thus likely one of the closest. Doing so, Hubble was able to resolve individual stars, confirming earlier speculation that Andromeda and its sister spiral "clouds" are galaxies like our own Milky Way. He was even able to identify a cepheid variable, a class of star whose variations in brightness correlate with its absolute brightness. A comparison of the cepheid's apparent brightness with its absolute brightness indicated that Andromeda is nearly a million light-years away, a distance much greater than that to the farthest stars in the Milky Way, thus offering further confirmation that Andromeda is a galaxy in its own right.* By 1929, Hubble had estimated the distances to two dozen "galactic nebulae," finding that they are galaxies in their own right. Moreover, upon comparing these distances to Slipher's red shifts, Hubble discovered something startling: the rate with which other galaxies are receding from us is proportional to how far away from us they are.

*Hubble's value for the absolute brightness of his cepheid was off; Andromeda is actually 2.54 million light-years from Earth.

The mystery of *why* they are receding, however, remained unanswered in Hubble's time.

Gamow and Alpher's theory (and Dicke and company's later, independent work) posited an answer to the riddle of the receding galaxies. Specifically, they proposed that the Universe is expanding uniformly from an initially ultradense state, nicely rationalizing both Slipher's and Hubble's observations: space, itself, was expanding, and thus from the perspective of every galaxy, it would appear that all other galaxies are receding (i.e., there is nothing special about our galaxy), and the more space there is between two galaxies (i.e., the farther apart they are), the more space there is expanding between them and thus the faster they recede from one another.

Thinking deeper, the theorists realized that if this "big bang" theory* were correct, the early Universe would not only have been unimaginably dense but also unimaginably hot and the heat of Universe's fiery origins should still be observable today, albeit "cooled down" as relic radiation at microwave (centimeter) wavelengths. Consider this: as we look farther and farther away in distance, the finite speed of light ensures that we are looking at events that happened further and further back in time. And if you look far enough away (i.e., far enough back in time), you can see the big bang itself—photons from it are arriving on Earth from all directions, even today. But farther away also means greater red shift, and so, although the light of the big bang, which originally corresponded to a very hot object indeed, is all around us, it should be red-shifted so much as to appear cold. Based on then current estimates of the age of the Universe, Gamow predicted that this relic image of the big bang should glow like a blackbody at a temperature of approximately 4 K (4°C above absolute zero). And the radio hiss observed from New Jersey? Based on Peebles and Wilkinson's observations, it corresponded to a blackbody with a temperature of about 5 K (both the theoretical prediction and

*The name was coined by the theorist Fred Hoyle, who, as mentioned in the prior chapter, was a proponent of the competing, but now disproven, "steady state" theory of the origins of the Universe, which postulated, despite the expansion first observed by Hubble, that the Universe remains statistically unchanged as new matter is constantly created from nothing, keeping the density of the Universe constant despite its propensity to expand. In coining the phrase "big bang," a term Hoyle thought derisive, he was trying to belittle the hypothesis. He seems to have failed.

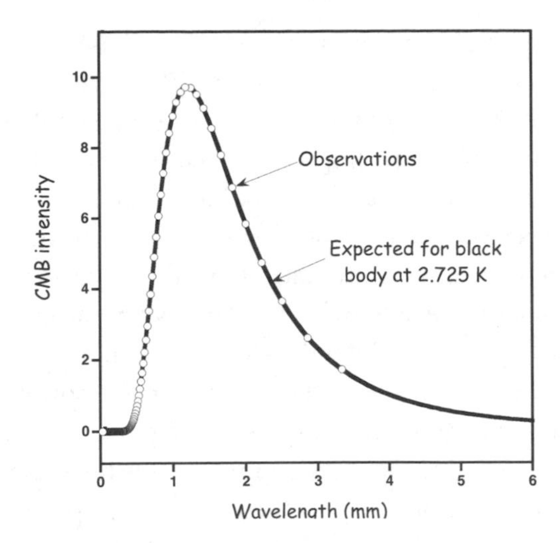

Figure 2.1 Echoes of the big bang are seen in the cosmic microwave background (CMB). The radio-wave photons that make up this spectrum are the red-shifted, cooled remnants of the hot sea of photons that filled the Universe at the time of recombination (discussed later in the chapter), some 370,000 years after its origins. As shown by the fitted line, the cosmic microwave background now exhibits the spectral characteristics of a blackbody (a perfect radiator) at a temperature of precisely 2.725 K. These data were obtained by NASA's Cosmic Background Explorer (COBE) satellite.

the observed temperature have since been refined to 2.725 ± 0.002 K; fig. 2.1). Rather than the prosaic hiss of pigeon droppings, the physicists Penzias and Wilson were hearing the red-shifted hum of the big bang itself, now termed the "cosmic microwave background."

The Big Bang

Our contemporary understanding of physics is sufficiently advanced such that cosmologists have been able to refine the big bang model into a detailed description of the origins of the Universe that, it is generally thought, is fairly accurate and detailed to as far back as within 10^{-34} seconds of the origins of time itself, and cosmologists are actively trying to push that back another billionfold, to as far back as 10^{-43} seconds (see sidebar 2.1). This is all the more impressive when one considers that, according to the current best estimates, these events happened 13.799 billion years ago (give or take 21 million years). Here we describe some of the many observations that compellingly support this hypothesis and explore the impact that the Universe's big bang birth has had on the origins and evolution of life within it.

From our perspective as astrobiologists, the "interesting bits" started a million trillion trillion times later than the 10^{-34} seconds beyond which cosmologists are now trying to probe, when the Universe was, in relative terms, an ancient millionth of a second old. At this point, everything in our Universe, all the matter—and energy—now in you, in Pluto's moon Charon, and in the most distant stars we see in the heavens, was compacted together in a dense, unimaginably hot plasma estimated to be at a temperature of 10^{13} K (at these sorts of temperatures, the Kelvin scale is equivalent to the Celsius scale). At this temperature, the mean energy per photon is higher than the energy bound up in the mass of a proton or neutron (which can be calculated from their mass by Einstein's $E = mc^2$), and, thus, when two such photons collide, they can spontaneously convert into a proton-antiproton or neutron-antineutron pair. Conversely, when proton-antiproton or neutron-antineutron pairs collide, they annihilate one another, producing—you guessed it—two high-energy photons. Before the first millionth of a second, the rate at which neutrons and protons were produced equaled the rate at which they were destroyed. After this point, however, further expansion of the Universe led to further cooling until, eventually, no new neutron-antineutron or proton-antiproton pairs were formed. (Protons and antiprotons are about 0.1% less massive than neutrons and antineutrons, which is enough to ensure that, when these pairs "froze out" of the mix of photons and nucleons, protons and antiprotons outnumbered neutrons and antineutrons by a factor of five.)

Although at this point the relic big bang photons lacked sufficient energy to create proton-antiproton and neutron-antineutron pairs, the opposite was not true. That is, existing nucleon-antinucleon pairs remained quite capable of annihilating one another to produce two γ-ray photons. But, for reasons that remain perhaps one of physics' bigger unsolved mysteries, the "particles" outnumbered the "antiparticles" at this point by one part in a few billion. That is, for every an-

Sidebar 2.1

Cosmic Inflation

Gamow's and Dicke's big bang model explains a great many things about the current state of our Universe, but by the 1970s, it was clear that there were still a few "loose threads."

One such thread was the fact that the cosmic microwave background is surprisingly isotropic: no matter which direction we look, the microwave signals we see are the same to within one part in 100,000. On the one hand, this is easy to explain: it simply means that every region of the early Universe was at the same temperature. This, however, implies in turn that the regions were all in "communication" with one another, such that any differences in temperature could equilibrate away. The problem is, the observable Universe is so large today that light has not had time to cross from one side to the other. How, then, could the Universe have been small enough that opposite ends of it could be in communication early during the big bang but now be so large that such communication is cut off? This is not a trivial problem: while the early Universe was smaller, meaning things were closer together, it was also *younger*, meaning light did not have as much time to cross it. If we naively run the Gamow-version of the big bang backward, we find that the decrease in time wins out over the decrease in distance, such that there never should have been a time when opposite sides of the Universe were in communication.

In the late 1970s and early 1980s, Alan Guth, Alexei Starobinsky, and Andrei Linde found a possible solution in the idea that some 10^{-36} seconds after the start of the big bang, space itself underwent a massive, exponential growth phase, swelling from the size of a proton to the size of a grapefruit at a rate millions of times faster than the speed of light. (Remember: it is space itself that is expanding, not the matter in it, and thus the Einsteinian "speed limit" does not hold.) During this "inflationary epoch," the part of the Universe that we can now observe would have occupied a volume so small that everything inside it could equilibrate. Inflation then expanded this tiny, thermally equilibrated, protein-sized volume of space an astounding one quadrillion-fold (one followed by 15 zeros) in an infinitesimal fraction of a second. Because of this inflationary expansion, points in the Universe that had been in communication became separated by distances great enough that they were no longer in communication, explaining the conundrum.

Given its ability to explain of the homogeneity of the Universe and a number of other, equally nagging problems associated with the original big bang cosmology, cosmic inflation is generally agreed to be a compelling, if not yet proven, addition to the big bang story. A relevant additional note about cosmic inflation is that it predicts that the Universe as a whole is much, much larger than the observable Universe. The observable Universe is everything around us within a radius of 13.8 billion light-years, that is, the distance over which light has been able to travel since the dawn of time itself. In order to achieve the homogeneity we observe today via an inflationary mechanism, the entire Universe must be at least 10^{78} times larger (in volume and, presumably, mass) than the volume we can observe. This is a *very* big number; it is only 100 times smaller than *the total number of protons in the observable Universe*. That is, for every 100 protons in our observable Universe, there may be a whole other volume of space as large and as complex as our entire observable Universe. Moreover, some models of cosmic inflation predict that inflation continues in some larger "meta-universe," giving birth to an infinity of universes beyond our own.

nihilation, a pair of photons was added to what is now the cosmic microwave background (which is estimated to contain 90% of all the photons ever emitted over the history of our Universe), and for every few billion such photons, a single proton or neutron was left to create the atoms in our Universe. Without this tiny asymmetry between matter and antimatter, there would be no matter left in the Universe, so this cosmological mystery speaks to our very existence.

The electron and its antiparticle, the positron, are 1,836 times lighter than a proton. Thus, not until the Universe was 14 seconds old and the temperature had fallen to a much more "moderate" 3 billion kelvins, did electron-positron pair formation freeze and the total number of electrons (again, for some reason, electrons outnumbered positrons by one part in a few billion) settle down to its current value.

At this point, the Universe consisted of a hot, dense sea of electrons and nucleons. Nucleons, that is, neutrons and protons, not nuclei. The strong nuclear force, which holds together neutrons and protons to form atomic nuclei is weaker than the thermal energies seen above a billion kelvins, so any conglomeration of neutrons and protons that might have been transiently formed quickly dissociated under the onslaught of the highly energetic collisions taking place in this hot, unimaginably dense state.

It wasn't until the Universe was about a minute and a half old that it cooled below a billion kelvins, the temperature at which the mean thermal energy of its matter (nucleons, electrons, and photons) was sufficiently low that neutrons and protons came together to form the first composite nucleus—deuterium, an isotope of hydrogen containing one neutron and one proton—faster than thermal collisions broke it apart. Deuterium, however, which is denoted ^{2}H to reflect its atomic mass of two, is less stable than either tritium (^{3}H, consisting of one proton and two neutrons) or helium-3 (^{3}He, consisting of two protons and a neutron), and thus deuterium readily fuses to form these nuclei (fig. 2.2). Such fusion could be with a free neutron or a free proton, but the energy released would then have nowhere to go and would most likely tear the newly formed nucleus back apart. In contrast, when a deuterium fuses with another deuterium, the resulting composite nucleus can release a free proton or a free neutron that

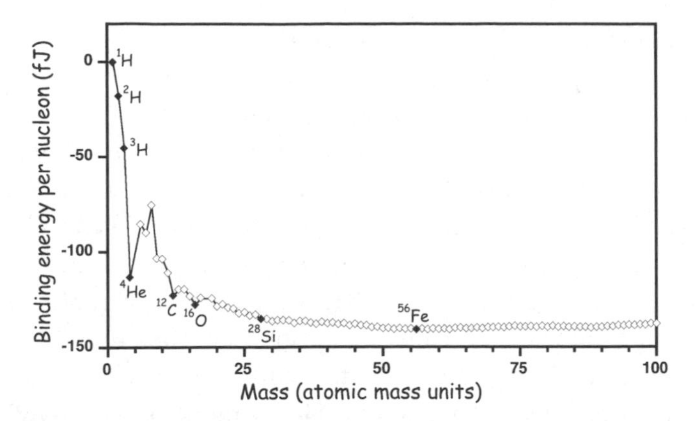

Figure 2.2 A plot of nuclear binding energies (a measure of the stability of a given nucleus) shows that helium-4 (^{4}He) is significantly more stable (further "downhill") than hydrogen (^{1}H), deuterium (^{2}H), and helium-3 (^{3}He). Because of this, the fusion of these nuclei to form ^{4}He dominated nucleosynthesis during the first few minutes of the big bang. It is also the primary energy source for stars. The dinuclear fusion of ^{4}He with itself or with any of the lighter nuclei is prohibited by the instability (relative to ^{4}He) of all the nuclei between the masses of four and 12. Fusion reactions to form heavier elements require instead the trinuclear fusion of three ^{4}He to form carbon-12 (^{12}C) in a single reaction. Once this barrier is surmounted, additional dinuclear reactions can continue until iron-56 (^{56}Fe) is synthesized. Further fusion consumes rather than produces energy.

carries the energy away, producing either tritium or ^{3}He. Tritium and ^{3}He, in turn, are less stable than helium-4 (^{4}He, two neutrons and two protons), and thus they rapidly undergo fusion with deuterium to create ^{4}He, again with the release of either a free neutron or a free proton to carry away the excess energy.*

And then? And then nothing but more of the same. Between the ages of 1.5 and 3 minutes, the Universe converted 20% of its original ("primordial") hydrogen into ^{4}He, leaving behind tiny traces of deuterium, tritium, ^{3}He, and lithium. And then, before it produced significant quantities of any heavier nuclei, the fusion stopped. A plot of nuclear stabilities (see fig. 2.2) reveals why: none of the nuclei consisting of five to 11 nucleons are as stable as ^{4}He; it is not until carbon-12 (^{12}C) that we find the first nucleon that is more stable. But the

*Hydrogen bombs are, in fact, tritium and deuterium bombs; because there are spare nucleons to carry off the energy liberated in fusion, these nuclei fuse much more efficiently than hydrogen, rendering it easier to detonate such a bomb.

formation of this nucleus requires that *three* ^{4}He simultaneously collide (or, more precisely, that a highly unstable beryllium-8 nucleus collide with a ^{4}He nucleus during the incredibly brief 10^{-16} second lifetime of the former). Thus, the formation of ^{12}C is a *third-order* reaction, a reaction whose rate scales with the concentration of reactants *cubed*. But, in addition to cooling, the Universe was also *expanding*. So much so that, by the time appreciable amounts of ^{4}He had formed, the concentration of ^{4}He was too low to support a third-order, trinuclear reaction at any appreciable rate. Thus, only 3 minutes after the start of the big bang and after a fifth of its initial complement of nucleons had been converted to ^{4}He, primordial nucleosynthesis ground to a halt, leaving mainly protons, neutrons (which convert into protons with a half-life of about 14.5 minutes*), and ^{4}He. Beyond this were only trace, one part in 20,000 amounts of deuterium, one part in 100,000 of ^{3}He and tritium (which decays into ^{3}He with a half-life of 12.35 years), and a few parts in 10 billion of the heavier nucleus lithium-7 (^{7}Li).

The ratios of the primordial nuclei provide a stringent test of the big bang model. The ratios of ^{1}H to ^{2}H, ^{1}H to ^{3}He, and ^{1}H to ^{7}Li produced during big bang nucleosynthesis are exquisitely sensitive to the precise density of matter in the expanding early Universe, with greater density leading to the conversion of more hydrogen into the heavier nuclei (fig. 2.3). If the density of matter in the early Universe had varied even a little, these ratios would differ very significantly from their current values. (The 92:8 ratio of ^{1}H to ^{4}He, in contrast, is a result of the 6:1 ratio of protons to neutrons at the beginning of nucleosynthesis, up from the 5:1 ratio at the end of nucleon synthesis due to the decay of neutrons into protons in the intervening few minutes. Unlike the other ratios, this one is thus relatively independent of the density of the early Universe.) Since we do not know a priori what the density of the original Universe was, we cannot use big bang models to predict what the ratios of these nuclei were at the end of the nucleosynthesis era. But, if we measure these three ratios in the current Universe (and

*Although the lifetime of the neutron has been studied for decades, and measurements from individual experiments now have precisions of better than a second, the two main experimental methods to this end ("bottle" and "beam") produce values that differ by nine seconds. It's a mystery.

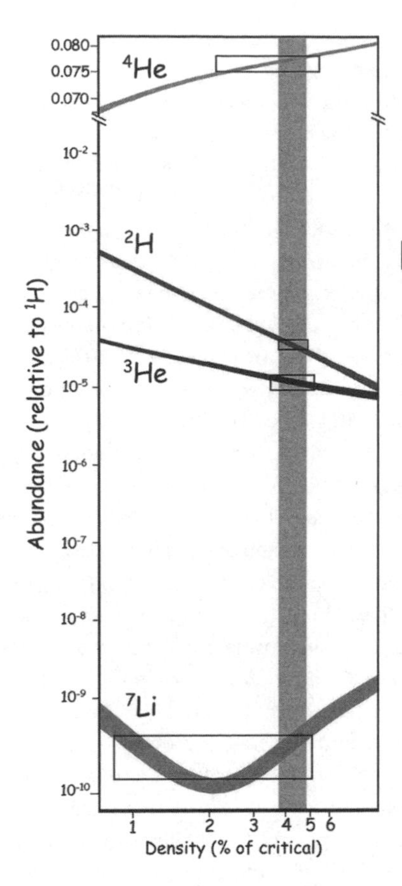

Figure 2.3 The production of deuterium (^{2}H), ^{3}He, and lithium-7 (^{7}Li) during the first few minutes after the big bang depended sensitively on the density of the Universe. The curves in this figure indicate the relative abundance of the various nuclei expected as a function of nucleon density, with the latter being presented relative to Ω, the ratio of the actual density to the "critical density" that would produce a closed universe (see sidebar 2.2). Current best estimates for these abundances, which are indicated by the boxes, are internally consistent and point to a big bang nucleon density of about 4% to 5% of the critical density (vertical grey bar). Note that the production of ^{4}He was tied to the 6:1 ratio of protons to neutrons at the beginning of nucleosynthesis and thus was only weakly dependent on density. For this reason, the ^{4}He curve is presented on a narrower, non-logarithmic scale.

correct for the fact that stars have been converting some of the hydrogen to nonprimordial helium over the intervening 13.8 billion years) and find that all point to the *same* density, this provides powerful evidence in favor of the big bang hypothesis (see sidebar 2.2). The best current measurements of these ratios do, in fact, all point, more or less, to the same density (fig. 2.3), an observation that provides strong, quantitative support for the big bang model.

Nuclear binding energy, the energy that holds protons and neutrons together to form the nucleus, is about a million times stronger than the electromagnetic forces that bind electrons to nuclei to form atoms, which is why nuclear bombs are so much more powerful than traditional explosives. Thus, even though it took only 3 minutes for the Universe to cool sufficiently for nuclear reactions to freeze, it took

Sidebar 2.2

The Density of the Universe

The big bang model makes specific, quantitative predictions regarding how various properties of our current Universe depend on the density of matter in the early Universe. In discussing this in the main text, though, we've pointedly left out the units in which the density of the Universe is measured. Astrophysicists could, of course, simply measure density in grams per cubic meter (after converting energy to mass, using $E = mc^2$). However, they prefer to normalize it to the "critical density." If the density of the Universe is above the critical density, gravity will eventually "win," causing the expansion of the Universe to slow before eventually reversing, ultimately leading to a "big crunch" billions or trillions of years from now. Conversely, if the density of the Universe is below the critical density, expansion will continue unabated for eternity. Finally, if the density of the Universe is perfectly balanced at the critical density, the expansion will slow to zero at infinite time in the future, but it will never reverse. In tangible terms, the critical density is about 10^{-30} g/m^3, or the equivalent of about six hydrogen atoms per cubic meter of space.

Actual density of the Universe divided by this critical density yields the cosmological parameter Ω. If Ω is greater than one, the Universe contains enough mass to stop the expansion; if Ω is less than one, the Universe will continue expanding (and cooling) forever. The ratios of the primordial nuclei, which as described in the text are a sensitive measure of the density of protons and neutrons (called baryonic matter) in the Universe during the era of nucleosynthesis, are consistent, with an Ω between 0.04 and 0.05; that is, the density of baryonic matter in the Universe is 4% to 5% of the critical value (see fig. 2.3). As discussed later in the chapter, anisotropies (inhomogeneities) in the cosmic microwave background also provide a completely independent measure of the density of protons and neutrons in the early Universe, this time at an age of 370,000 years. Measurements of these anisotropies collected by the *Planck* spacecraft produce an $\Omega = 0.0490 \pm 0.0002$ and an open universe. Finally, just looking up into the sky and "counting up" all the visible stars and the glowing hydrogen and helium plasma between the stars gives us an $\Omega \approx 0.015$, but as this approach is necessarily quite crude, that value is considered "within error" of the values produced from nuclear ratios and anisotropy measurements. The close convergence of these three entirely independent measures of the density of protons and neutrons in the Universe is such compelling evidence in favor of the detailed big bang hypothesis that the theory is no longer seriously in doubt.

The fact that we can measure the baryonic density of the Universe, though, doesn't mean that we understand the ultimate fate of the Universe. The first hints of a "problem" along this line came in 1933 when the Swiss astronomer Fritz Zwicky (1898–1974), then at the California Institute of Technology (Caltech), used Doppler measurements to determine the velocities of various galaxies within a tight grouping known as the Coma Cluster. Many of these galaxies, he found, were moving so rapidly that the gravity of the observable mass in the cluster couldn't possibly hold them together. Yet bound together they seemed to be. Zwicky postulated that the additional gravity was provided by some form of invisible matter—he called it "dark matter"—but his hypothesis did not gain much traction at the time.

Several decades later, Vera Rubin (1928–2016) of the Carnegie Institution in Washington, DC, used Doppler measurements to determine the speed with which stars were orbiting around the centers of some *60* nearby galaxies. In every case, she, too, found something unexpected: stars far from the center of their host galaxy, which should be orbiting slowly, were orbiting just as rapidly as stars nearer the center. Her calculations suggested that this would occur if the galaxies contained a "halo" of dark matter surrounding the visible portion that was *10 times more massive than all the stars in the galaxy combined*.

The density of dark matter can also be deduced from high-precision measurements of the cosmic microwave background, which indicate that dark mater mass outnumbers baryonic (atomic) mass by a factor of five. Despite decades of research, however, the nature of dark matter remains a complete mystery. All we know is that it is there and that it cannot be ordinary atomic nuclei (or it would affect all the parameters described above). Still, some candidates have been put forth, such as neutrinos. These subatomic particles, first postulated in 1930 by Wolfgang Pauli (1900–1958) and long thought to be massless, may actually possess a small mass and be numerous enough to make a difference. Many other, much more exotic particles have also been postulated to account for this astonishingly large extra mass.

And it gets worse. Just as researchers had accepted that the majority of the Milky Way (and of the rest of the Universe) is made up of dark, nonnuclear matter about which we know nothing, a second major deficit turned up on the balance sheets. Observations of distant supernovae have confirmed that the expansion of the Universe *is accelerating*, an effect that requires a universal mass-energy content *two times higher* than the combined amount of ordinary and dark matter. The unknown energy that makes up the difference has been dubbed "dark energy." Considering the behavior of the Universe as a whole, the balance sheet, as it looks now (based on the *Planck* spacecraft data), is as follows:

4.9% protons and neutrons (baryonic matter)
26.8% dark matter
68.3% some kind of unknown energy

Put simply, our scientific knowledge so far can detect and account for a bit less than 5% of the mass in the Universe!

Tallying up the total baryonic matter, dark matter, and dark energy content of the Universe pushes Ω to within experimental error (of about 1%) of one (inflationary cosmology predicts that it should be this value, by the way, which is another feather in the cap of that theory). This would, by the arguments above, suggest that the expansion of the Universe will slow, but never reach zero. Dark energy, however, is *repulsive*, and thus causes the expansion of the Universe to accelerate. Our future is thus one of increasingly distant galaxies filled with old, dying, and dead stars. Sorry to break this to you.

much longer to cool to the point where electrons and nuclei could join together to form stable atoms. This "recombination event" (a misnomer, given that electrons had not previously been combined with nuclei) occurred when the temperature dropped below about 3,000 K, the temperature at which the energy of background photons was low enough that they would no longer rip electrons from nuclei. Current estimates are that the Universe cooled to this temperature some 370,000 years after the big bang.

The recombination event forever altered the relationship between photons and matter in the Universe. Before recombination, the Universe consisted of plasma, a cloud of naked nuclei and electrons. Photons are scattered by such charged particles and, in being scattered, exchange kinetic energy with them, ensuring that the photons and matter in the pre-recombination Universe reflected the same temperature. The scattering would also have made the pre-recombination Universe opaque, much the same way water droplets scatter light, rendering fog opaque. Neutral hydrogen and helium, in contrast, are transparent to the visible and infrared photons that filled the Universe at the time and thus, after recombination, the scattering stopped. This also put a stop to the thermal equilibration between matter and the primordial photons, and the two parted company. The matter evolved into galaxies, stars, and, eventually, us. The primordial photons, in contrast, sped off into space. Some of those, which were emitted from recombining gas that was 13.8 billion light-years away from our current location, are just arriving today, red-shifted a thousandfold by the expansion of the Universe, lengthening their wavelength by the same factor of 1,000, from wavelengths corresponding to a blackbody at the approximate 3000 K temperature of recombination to wavelengths corresponding to the 2.725 K of the cosmic microwave background.*

Fine details of the pervasive cosmic microwave background provide further, compelling evidence in favor of the accuracy of our un-

*In the olden days, when our television sets were hooked up to antennas rather than the internet, we had the pleasure of seeing these ourselves. About 1% of the "snow" seen when an antenna-linked, analog TV is tuned to a nontransmitting channel represents relic big bang photons red-shifted to television transmission frequencies.

derstanding of the big bang. Before recombination, the photons, protons, and electrons in the Universe were in thermal equilibrium: because the photons were scattered by, and thus exchanged momentum with, the charged particles making up the plasma, anything that affected the temperature of matter affected the energy of the photons. And the early Universe was filled with something that affected the temperature of its matter: sound waves—not exactly music, but density fluctuations obeying similar physics. Under the influence of gravity, tiny fluctuations in the density of the matter tended to grow—the denser a region, the stronger its gravity and the better it pulled in matter from surrounding regions of space. And with this infalling matter came a small increase in temperature as gravitational potential energy was converted into kinetic energy. Before recombination, however, the scattering of primordial photons—which, as you'll recall, outnumber protons and electrons by a billion to one—tended to tear apart any denser regions. Under such circumstances, random density fluctuations (and with these, temperature fluctuations) grew, dissipated, and grew again in a process that produced waves akin to a violin string responding to random strikes by emitting a note and its various harmonics. The wavelengths and amplitudes of waves are dependent on the density of the medium in which they occur, and, so, if we could measure the wavelength of these primordial waves, this would provide a second, independent measure of the density of the early Universe. As it turns out, we can.

The recombination event decoupled the big bang photons from the nuclei and electrons in the Universe, and thus a "fossil record" of the conditions in the Universe at the time of recombination is imprinted on the cosmic microwave background. If the Universe at recombination were perfectly homogeneous (if the plasma had the same density everywhere), the cosmic microwave background we see today would be *isotropic*, from the Greek *ísos*, meaning "equal," and *tropos*, meaning "turning." That is, the background would correspond to the same temperature in any direction we look. If, instead, the matter in the early Universe were filled with waves, the cosmic microwave background would also exhibit fluctuations corresponding to the density fluctuations (and their commensurate temperature fluc-

Figure 2.4 A full-sky map of the cosmic microwave background shows miniscule, 10-parts-per-million variations in the 2.725 K mean temperature of the background. The angular scale of these fluctuations is a measure of the size—and thus the wavelength—of the acoustic perturbations that filled the Universe during recombination, some 370,000 years after the big bang. Recent high-precision measurement of the size of these oscillations by the *Planck* spacecraft are consistent with a nucleon density of 4.90 ± 0.02% of that necessary to produce a closed universe (see sidebar 2.2), which falls within the narrow range of densities consistent with the observed relative abundances of the light nuclei (see Figure 2.3). (Courtesy of NASA/WMAP Science Team)

tuations) that these waves set up, and the cosmic microwave background would be slightly *anisotropic*. In the early 1990s, the predicted small (10 parts in a million) anisotropies were finally detected by the *Cosmic Background Explorer* (*COBE*) satellite in a discovery that won the lead scientists on the project, George Smoot of the University of California, Berkeley, and John Mather of NASA's Goddard Space Flight Center, the 2006 Nobel Prize in Physics.

High-resolution follow-on studies using the *Wilkinson Microwave Anisotropy Probe* (*WMAP*), named after David Wilkinson, mentioned in the introduction to this chapter, and launched in 2001, and the *Planck* mission, launched in 2009, provide a means of accurately measuring the wavelengths and amplitudes of the sound waves that created the cosmic microwave background anisotropies when the Universe was just a few hundred thousand years old (fig. 2.4). These measurements, in turn, produced an estimate of the density of nucleons in the early Universe *that is within experimental error of the density derived—*

completely independently—from the known ratios of hydrogen to the heavier primordial nuclei (see fig. 2.3). That two completely independent estimates of the nucleon density of the Universe derived from the big bang model would converge on precisely the same value provides extremely strong support for the big bang theory, and the scientific community now considers this model rock solid, with only the details remaining to argue about.

Stars and Galaxies

By the time it was 370,000 years old, the Universe consisted of relic big bang photons zipping through transparent clouds of primordial, free hydrogen and helium atoms. Soon after, though, the first chemistry was born as atoms bound together to create the first molecules. As helium has a much stronger affinity for electrons than hydrogen does, it was the first to form neutral atoms. By combining with a still ionized hydrogen atom (i.e., a proton), these early helium atoms formed the first covalent bonds, resulting in the molecular ion helium hydride (HeH^+), a molecular ion so unstable it would have destroyed itself by violently reacting with any other molecule had there been any other molecules for it to react with at the time. Eventually, though, things cooled down enough to allow molecular hydrogen (H_2) to form, a far more stable molecule that then became the most common thing in the Universe.

The very limited chemistry available to hydrogen and helium is obviously not very promising fodder for life to arise; life requires both far more complex molecules than these two atoms can muster as well as seriously concentrated forms of disequilibrium (energy) to drive the replicative reactions of those more complex molecules. How did these come about? Initially they arose from the subtle, parts-per-million inhomogeneities produced by those early density waves and reflected today in anisotropies in the cosmic microwave background.

Despite their scant, 10-parts-per-million level, the inhomogeneities in the early Universe had a profound impact on its later evolution. The slightly denser regions exerted an equally slightly stronger gravitational pull than the less populated ones, and thus these regions began to accumulate even more matter. This new matter amplified the originally small density fluctuation, further accelerating the infall of matter. Within a few hundred million years after recombination, the once nearly homogeneous, post–big bang cloud of hydrogen and helium had been pulled into trillions of lumps, each a few billion times more massive than the Sun. These "protogalaxies" were the seeds of the galaxies we know today; over the next few billion years, merging protogalaxies would form the billions of galaxies that now make up the observable Universe, the sphere of space centered on the Earth that is close enough that light from objects within it has had time to reach us since the big bang (see fig. P.1 in the preface).

But we have gotten ahead of ourselves. When the Universe was young, it was filled with nothing but hydrogen and helium, contracting here and there to form the first protogalaxies. It wasn't until the Universe was several hundred million years old that the next astonishing thing happened: the first stars were born, shedding fresh light on the Universe, which had descended into darkness as the primordial photons red-shifted toward infrared and then microwave wavelengths.

Stars are simply enormous piles of hydrogen and helium compressed under the weight of their own mass to such densities, pressures, and temperatures (the latter due initially to the kinetic energy associated with all that mass falling inward toward the center) that lighter atoms fuse to form heavier atoms through reactions like those last seen when the Universe was but 3 minutes old. Due to an interesting interplay between chemistry and physics, the first-generation stars are generally thought to have been quite massive. This was first realized by James Jeans (1877–1946), who, in 1940, turned his keen, physics-oriented mind toward the clouds of gas from which stars are born, which are called "proto-stellar" nebulae (again, *nebula* being the astronomer's generic word for "thing that looks like a cloud"). Doing so, Jeans found that, in attempting to compress the gas, gravity is fighting against pressure. If the mass is too low, the gas's pressure prevents collapse. Above a certain, minimum mass, however, gravity wins. The minimum mass, now known as the Jeans mass, is proportional to the square of the initial temperature of the gas (higher temperatures = higher pressures = greater mass required to ensure collapse). But because of this, it also depends on the chemistry of the cloud. Hydrogen and helium, the only significant components of the early Universe, are transparent to infrared and visible light, and transparent things are inefficient radiators (the ideal radiator, in fact, would be perfectly black). Because of this, the earliest proto-stellar clouds tended to heat up significantly as they collapsed, causing them to rebound and dissipate instead of collapsing to form stars. In the early, hydrogen-and-helium-only Universe, Jeans estimated, a proto-stellar cloud could not collapse to form a star unless it was hundreds of times the mass of the Sun. The first generation of stars born after the big bang were truly massive.

More massive stars require more vigorous fusion to overcome the inward pull of gravity (more on this below), leading to greater energy output and higher surface temperatures. In fact, the very first stars were so prodigiously hot that they put out copious amounts of UV radiation. This UV radiation allows us to date the time of their formation: the light put out by these stars was the first thing since the recombination event that was higher in energy than the energy of an electron bound to a proton to form hydrogen. Thus, as the first stars ignited, electrons once again found themselves ripped away from their atoms. Like sunlight burning off a morning fog, the light of these early stars "re-ionized" the clouds of neutral hydrogen and helium created by recombination and turned them once again into a plasma of free electrons and nuclei. Using the *Hubble Space Telescope* to peer at the most distant observable objects—which is the equivalent to looking back 13 billion years to what was happening just 800 million years after the big bang—astronomers have observed the spectral fingerprints of neutral hydrogen, suggesting that the reionization was not yet complete at that time.

The Heavy Elements

The first-generation stars, known for rather arbitrary, historical reasons as "population III stars," were composed solely of the hydrogen and helium synthesized in the big bang. This is not the stuff of which life can be made; life requires chemistry, and, again, the only chemistry based on hydrogen and helium is the formation of molecular hydrogen (H_2). Life is based on heavier atoms, which astronomers (erroneously, from a chemist's perspective) refer to as "metals." Where did these metals, so critical to the origins of life, come from? They are produced in the life and death of stars.

The center of the Sun, to pick our own star as an example, is a toasty 16 million kelvins, a temperature at which the kinetic energy of protons is sufficient to overcome the electrostatic repulsion between two like-charged protons and allow fusion to occur. As we discussed above, the nuclei intermediate in mass between ^{1}H and ^{4}He are unstable relative to ^{4}He, and thus the deuterium, tritium, and ^{3}He that are formed as intermediates are quickly consumed, with the net result being the production of ^{4}He, two electrons (to balance the charge), and a great deal of energy.* This energy, of course, ultimately provides the disequilibrium on which most Terrestrial life is based. It also prevents the Sun from collapsing under the weight of its own massive bulk.

The outward pressure caused by fusion in the Sun's core counteracts gravity's incessant attempts to cause the star to collapse further, and a "truce" is set up in which they are perfectly balanced. And although this truce can last a long time, it is nevertheless still temporary; in another 6 billion years, for example, the Sun will consume all the hydrogen in its core (it's currently down to just 33% hydrogen), fusion will stop, and gravity will begin to win. Since the gravitational pull of a larger star is, of course, larger, the counteracting pressure induced by fusion must also be higher. Thus, the equilibrium between the forces that want to tear the star apart and those that want to collapse it generally settles at higher values, meaning higher temperature, higher density, and higher rates of fusion. For this reason, and perhaps counterintuitively, larger stars burn faster and live shorter lives than smaller stars: if the mass of a star is doubled, the rate of fusion must increase tenfold to overcome the increased gravity, thus shortening the star's life fivefold (twice the fuel burning at 10 times the rate). A tenfold increase in mass leads to a 3200-fold increase in

*That said, the 275 W/m^3 produced in the Sun's core (much less its outer layers) is far less than the 1,250 W/m^3 produced by human flesh. The Sun, however, is rather larger than the human body.

the rate of fusion and a 320-fold decrease in lifetime, such that a star 10 times the mass of the Sun will barely live to see 30 million years.

And what happens when a star runs out of hydrogen fuel in its core? As fusion slows, the outward pressure it produces decreases, and gravity begins to win this stellar tug-of-war. Eventually, the temperature and pressure in the shell of unfused hydrogen immediately surrounding the star's core rise sufficiently to cause fusion in a zone just outside the core. With the sudden production of energy so near its surface, the outer layers of the star expand dramatically; when this happens to the Sun, it will swell enough to engulf the terrestrial planets, and the Earth will have ended its 11-billion-year run. We suppose there is a philosophical point in that, as well, but what to make of it, we'll leave to the reader.

This bloated phase of a star's life history, called the red giant stage, lasts for only a few hundred million years. Meanwhile, the star's helium-rich core continues to contract and, with that, heat up. When it gets hot enough, two new reactions kick in: the fusion of three ^{4}He to produce ^{12}C and the fusion of ^{4}He with ^{12}C to produce ^{16}O. Because these larger nuclei are more highly charged than the hydrogen nucleus, they repel one another more strongly, and so their fusion requires core temperatures of at least 150 million kelvins. Such high temperatures are only reached in stars at least 80% as massive as the Sun; smaller stars never ignite helium fusion, so, after completing their red giant stage, they cool and contract into white dwarfs before eventually dying.

Helium fusion differs significantly from hydrogen fusion in that the former requires the simultaneous fusion of *three* nuclei rather than two. This occurs because, as we've noted, none of the nuclei intermediate in mass between ^{4}He and ^{12}C are stable relative to ^{4}He (see fig. 2.2), and thus the formation of nuclei intermediate between them does not produce energy. Remember that ^{12}C was not formed during the expanding big bang because trinuclear collisions were too rare at the low densities found by the time the fusion of hydrogen to helium had progressed significantly. In fact, even in the highly compressed core of the post–red giant Sun, the density should still be too low for the efficient formation of carbon.

The deeper problem with trinuclear reactions is that the vast majority of nuclear collisions are nonproductive. For example, the first fusion step in hydrogen burning—the fusion of two protons to form deuterium (and an electron)—takes place only a few times for every *trillion* collisions. This startlingly poor efficiency occurs because the fusion reaction liberates energy, and this excess energy tears the newly formed nucleus apart, reversing the reaction. The low efficiency isn't much of a problem, though, for dinuclear reactions; dinuclear collisions occur frequently enough that, even if very few of them are productive, fusion hums along at a reasonable pace. This is not true for trinuclear reactions, for which low efficiency would be prohibitive. Considering this problem in 1939, Hans Bethe, whom, six years later, Gamow would humorously add to his big bang paper author list, concluded that "there is no way in which nuclei heavier than helium can be produced permanently in the interior of stars under present conditions. We can therefore drop the discussion of the building up of elements entirely and confine ourselves to the energy production, which is, in fact, the only observable process in stars."* Bethe was probably right in focusing on energy production in stars as this is what won him the 1967 Nobel Prize. But he was wrong about their inability to create carbon.

Pondering the carbon issue, the astrophysicist Fred Hoyle (author of the "life in space" novel mentioned in the previous chapter) later realized that, in order for carbon to exist, the carbon nucleus must contain a *resonance*. That is, the carbon nucleus must be able to form an "excited state" exactly equal in energy to (i.e., is "resonant" with) the energy liberated by the fusion of three ^{4}He nuclei. Nuclear excited states such as this are analogous to the electronic excited states of molecules; that is, states in which electrons have been "excited" into higher-energy, previously unoccupied atomic or molecular orbitals. The energy required for an electron to jump from its "normal," low-energy orbital into one of these higher-energy, excited orbitals is the reason that many molecules absorb light and are thus colored. In the case of this particular nuclear excited state, the energy it absorbs prevents the energy liberated by fusion from rupturing the nascent nucleus, increasing the reaction's effi-

*Hans Bethe, "Energy Production in Stars," *Physical Review* 55 (1939): 434–56.

ciency to one in 10^4, a 10-millionfold improvement over the efficiency estimated for the reaction in the absence of such a resonance.

Hoyle predicted this resonance in 1953, and it was confirmed just a few months later by Ward Whaling (1923–2020) and Willie Fowler (1911–1995) at the California Institute of Technology (Caltech), the latter of whom received the 1983 Nobel Prize for his work on nucleosynthesis. To test Hoyle's prediction, the Caltech researchers bombarded ^{14}N with a deuteron (^{2}H nucleus) to produce ^{12}C and an α-particle (^{4}He nucleus). Per Hoyle's prediction, the resulting α-particles emerged with two energies: some had the energy expected for the reaction, but others had less energy, with the difference corresponding to the energy left behind to form the excited carbon nucleus at exactly the energy level predicted by Hoyle.

Although he did not pitch it in these terms at the time, Hoyle's argument, it was later realized, was an early invocation of what is now known as the anthropic principle (fig. 2.5). That is, Hoyle did not necessarily know the precise details of the structure of the carbon nucleus, but because such a resonance had to occur in order for carbon to be formed, and because carbon had to be common for us to be here to ponder this issue, *there must be such a resonance.*

More broadly, the anthropic principle simply notes that living observers *will always find themselves in universes capable of supporting life*, an observation that profoundly influences what our existence tells us about the broader relationship between life and the Universe. Specifically, physicists should not be surprised to find the physical parameters of our Universe, including, for example, the parameters that go into defining the carbon nuclear resonance, seemingly perfectly "fine-tuned" to allow for living physicists to arise, no matter how "coincidental" such tuning may appear. More generally, if the Universe were not suitable to support life, living observers like us would not have arisen to study the Universe and raise the point in the first place. We touch on this issue (which we discuss in detail in sidebar 2.3) time and again throughout the rest of this book, so you may as well get comfortable with it.

Leaving aside, for the moment, the discovery of carbon's nuclear resonance, let's return to the story at hand. The helium in the Sun's

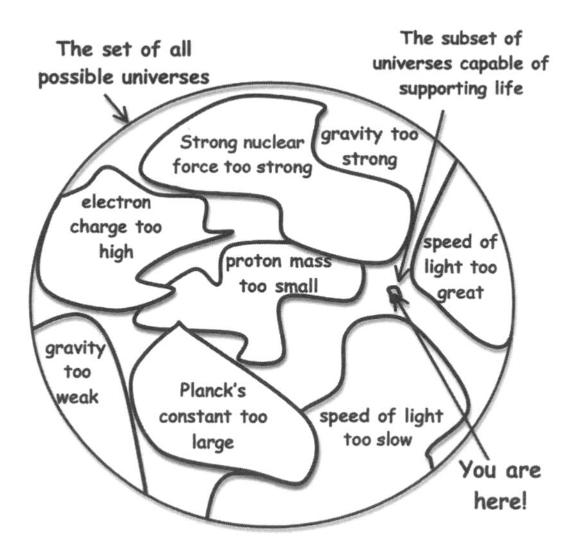

Figure 2.5 How many ways can a universe be constructed, each with different values of the many fundamental constants that describe it? And of the myriad possible universes, how many are capable of supporting life? The most likely answer to the latter question is "a vanishingly small fraction." Nevertheless, it is *not* simply a lucky coincidence that *our* Universe just happens to be one of these rare, "special" universes that are habitable.

core will run out only a couple of billion years after helium fusion begins, rather significantly less than the 11-billion-year span of hydrogen fusion. The shorter duration of helium fusion occurs for two reasons. The first is that helium fusion produces less energy per nucleon than hydrogen fusion (see fig. 2.2), and thus, in order to balance the Sun's gravitational contraction, the helium core must fuse more vigorously than the earlier hydrogen core. The second is that, because helium fusion requires higher temperatures than hydrogen fusion, it is limited to a smaller volume nearer the Sun's center, so there is less helium that can fuse. Thus, after only 2 billion years, helium fusion will stop and the Sun will consist of a carbon-rich core surrounded by a helium-rich shell, surrounded in turn by a thick outer shell of primordial hydrogen and helium. When this happens, the Sun is doomed; bad news for any surviving Earthlings who may have escaped to Saturn's moon Titan when the Earth was destroyed during the onset of the red giant stage a couple of billion years earlier.

Sidebar 2.3

The Anthropic Principle

A fundamental precept of science since the Renaissance has been the *Copernican principle*. In 1543, Nicolaus Copernicus (1473–1543) published his *De revolutionibus orbium coelestium libri VI*, which argued that the Earth is not at the center of the Universe, an assertion that revolutionized science. (Indeed, our modern, political usage of the word *revolution* stems from the book's title!) When Copernicus's heliocentric model of the Solar System eventually overturned the earlier view of an Earth-centered cosmos, all theories based on *exceptionalism*, an assumption that things are different on the Earth than elsewhere in the Universe, became suspect. No longer was man a privileged observer. Even the seemingly benign assumption made by Isaac Newton (1642–1727) that some observers could be said to be at rest in an absolute sense and thus to define a universal reference frame, fell, when, in 1905, Albert Einstein (1879–1955) published the theory of special relativity.

In astrobiology, however, the Copernican principle itself must be held suspect for the very fact of our existence has profound consequences, consequences that require we at least partially abandon this long-cherished principle. This idea, called the *anthropic principle*, comes in several flavors.

The simplest version is termed the "trivial anthropic principle." Starting from the observation that complex, carbon-based life exists on Earth allows us simply to deduce that the Universe is more than a billion years old, because it takes at least a billion years for stars to form, go supernova, and seed the Universe with carbon. Anthropic arguments of this magnitude, although they may at first glance appear trivial (hence the derogatory name of this version of the principle), have contributed significantly to the history of scientific thought. A classic example comes from the late nineteenth century, when, based on the known temperature of the Earth and the rate at which a sphere of Earth's volume and specific heat would cool, Lord Kelvin (1824–1907) estimated that our planet is only about 10 million years old. Even at that time, however, biologists were in a position to strongly argue that this was insufficient time to allow for the evolution of life's extraordinary diversity. Thus, invoking the trivial anthropic principle (albeit a phrase not yet coined), biologists were certain that something was missing from Kelvin's arguments. What was missing, though, would not be known until the discovery of radioactivity at the start of the twentieth century, a discovery that Kelvin lived to see. We now know that the radioactive decay of uranium, thorium, and potassium in the Earth's core has kept it hot for billions of years after the heat of accretion would have been lost.

Our very existence constrains not only the physical parameters possible on our planet but also the range of parameters that describe the Universe as a whole, an idea called the "weak anthropic principle." In this chapter, we have already discussed an example of the weak anthropic principle: the fact that carbon-based organisms exist implies there must be a nuclear resonance in ^{12}C that allows the efficient production of this nucleus by trinuclear reactions. In 1953, before we were able to calculate much about the

(*continued*)

Sidebar 2.3 *(continued)*

detailed structure of the nucleus, the British cosmologist Fred Hoyle and his coworkers used the existence of carbon to surmise the existence of the ^{12}C nuclear resonance. In a sense, this is also a trivial conclusion: carbon exists, so this resonance must, in turn, also exist. But on reflection, one is struck by the enormity of this coincidence. Were the charge on the proton or the strength of the strong nuclear force to differ by even a fraction of a percent, this resonance would not exist, and neither would life. Does this imply that some higher being must have designed nuclear physics specifically so that life can exist? Not necessarily.

Imagine, for example, some process that produces a large number of universes, each with wildly differing physical parameters and laws (see fig. 2.5). The weak anthropic principle merely states the obvious: *living observers will always find themselves in a universe whose physical properties are consistent with the existence of life*. We should not be surprised that the physical properties of our Universe are consistent with life—*no matter how coincidental this may seem*—because the Universe has to be consistent with the existence of life in order for us to be here to observe the fact that life exists. An important point is that this "observational selection effect" severely limits our knowledge of how probable, or improbable, the formation of life was. We exist, so the probability of life arising in our Universe was not zero. But the probability could be infinitesimally close to zero. Because *we have to exist in order to be having this discussion*, we do not know how "lucky" our origins were, only that they did happen. Using the Solar System as an analogy, Martin Rees, an astrophysicist at Cambridge University, describes the importance of this observation as follows: "If Earth were the only planet in the universe, you'd be astonished that we just happened to be exactly the right distance from the Sun to be habitable" (quoted in *TIME*, November 29, 2004). But that absurd improbability becomes much less absurd when taken in context. The Universe almost certainly contains hundreds of trillions of planets. With so many to choose from, it's much less improbable that at least one would be habitable. And the fact that we find ourselves on a habitable planet says nothing at all about how common they are beyond the fact that at least one exists. This point has such important implications for astrobiology that it bears repeating a third

That is, although the Sun will begin to contract again after the helium burning stage, it doesn't have enough mass to achieve the pressures and temperatures required to ignite the fusion of carbon and oxygen. The Sun will thus slowly collapse under its own weight, forming a white dwarf that cools over billions of years into a black, cold ember.

Of course, for the purpose discussed here—namely, how to enrich the Universe with heavier elements—metals locked in some cold, dark dwarf star are of little use. During the red giant phase, however, some of the heavier elements do leak out. The rate of helium fusion is time, using the words of Francis Crick (1916–2004), codiscoverer of the structure of DNA, from his *Life Itself: Its Origin and Nature*: "We cannot decide whether the origin of life on Earth was an extremely unlikely event or almost a certainty, or any possibility in between these two extremes."

Although the weak anthropic principle argues that we can say nothing more precise than the probability of life arising is not zero, some authors argue that the generation of life is a physical imperative. This "strong anthropic principle" states that the physical properties of (all) universes must be consistent with the formation of life; indeed, some interpretations of quantum mechanics argue that existence itself is impossible without an observer to observe it ("If a tree falls in the forest without anyone around to hear it . . ."). Unfortunately, it is hard to test such a hypothesis. The entire basis of the weak anthropic principle is the argument that life-free universes cannot be observed, and it is precisely the observation of life-free universes that would be required to disprove the strong anthropic principle.

And how seriously should we take the idea of multiple universes existing simultaneously in some enormously larger meta-universe? It is perhaps not as far-fetched as you might think; astrophysicists have been moving toward that conclusion for reasons entirely divorced from the anthropic principle. For example, as noted at the end of sidebar 2.1, some variants of inflationary cosmology predict the formation of an infinite number of universes: even after the inflation died down in our region of space, theorists believe, it should have continued (indeed, should still be continuing) in other regions. And thus, while our part of the meta-universe pinched off, slowed down, and evolved into the cosmos we see around us, the rest is still spawning an infinite number of new universes—universes that might be ruled by physical laws differing wildly from those we live under. Some super-astronomically small fraction of those universes will be habitable, and some still smaller fraction will be inhabited. Inhabited, no doubt, by creatures marveling about the improbability of all the physical laws of their universe being so perfectly tuned to ensure their existence.

Crick, Francis. *Life Itself: Its Origin and Nature*. New York: Simon and Schuster, 1981.

Lemonick, Michael D., and J. Madeleine Nash. "Cosmic Conundrum." TIME. November 29, 2004.

extremely sensitive to temperature: it is proportional to temperature to the *fortieth power*, so even a 2% change in temperature will *double* the rate of fusion. As a result, any small increase in temperature accelerates helium burning, causing the temperature—and thus the fusion rate—to rise further. This increase in temperature, however, also causes the outer shell of the star to expand, which reduces the pressure in the core, ultimately slowing fusion. This oscillation drives the outer regions of the star to pulse in and out until, eventually, the motion becomes so vigorous that the star tosses its outermost atmosphere into

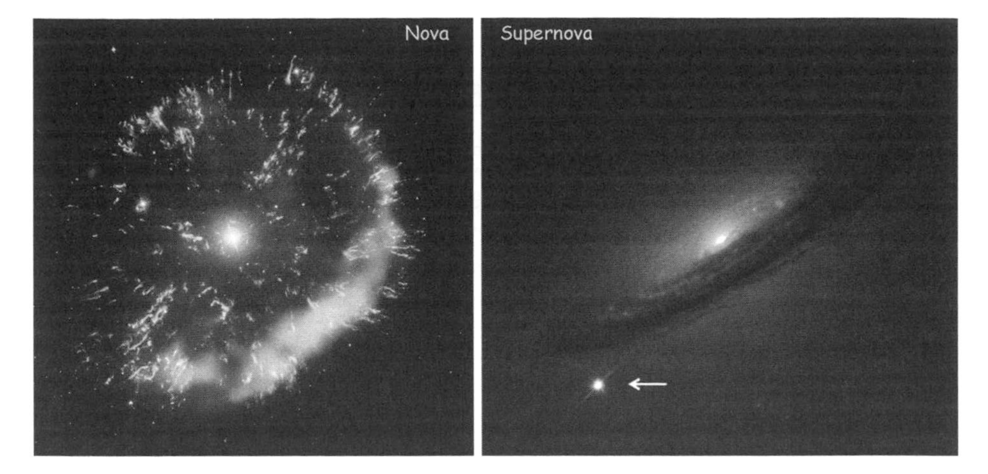

Figure 2.6 Stars seed the Universe with heavier elements, some as novae, others as supernovae. Novae occur when stars at least 80% the mass of our Sun evolve into red giants, which shed their outer atmospheres into space. Shown on the left is GK Persei, the remnants of a nova 1,500 light-years away that was, for a few weeks in 1901, the fourth brightest star in our sky. In contrast, a star of at least eight times the mass of the Sun ultimately collapses and rebounds in an enormously more powerful "supernova" explosion, which can shine as brightly as all the other stars in its host galaxy together. Shown by the arrow on the right is supernova 1994D, which took place in the galaxy NGC 4526, 55 million light-years from Earth. (Left image courtesy of NASA/CXC/RIKEND/STSci/NRAO/D, Takei et al.; right image courtesy of NASA/STScI)

space. The resulting explosion, which seeds the Universe with the heavier elements that the star has produced, is so bright that it often shows up in the sky as a new star, and thus its name: *nova*, from the Latin word for "new" (fig. 2.6, left).

We are all made of stardust. But stars such as the Sun stop their fusion reactions at helium burning and thus are not the source of many key biological elements. Stars much larger than the Sun, however, go on to create heavier atoms. In stars at least eight times heavier than the Sun, for example, the rise in pressure and temperature is sufficient to ignite the fusion of carbon with its own kind and with oxygen-16 (produced by the fusion of carbon and helium) to form magnesium-24, magnesium-23 (releasing a free neutron), sodium-23 (liberating a free proton), and silicon-28. Given that these reactions require still higher temperatures and pressures (the larger charges associated with carbon and oxygen nuclei require higher temperatures to overcome the associated larger electrostatic repulsions), they occur in a still smaller volume of the central core and continue for an even shorter time than the helium fusion era. In fact, the fusion phase of carbon and oxygen lasts only about a thousand years! And when the core becomes depleted of elements? You guessed it. More contraction, higher core temperatures, and fusion that forms heavier nuclei, with each fusion reaction phase lasting for shorter and shorter periods. This leaves behind concentric layers of hydrogen, helium, carbon, and so on, like a weird, spherical layer cake.

But this cycle does not continue ad infinitum. Eventually, silicon-28 fuses to form iron-56 (^{56}Fe), but if you look at the chart of nuclear binding energies (see fig. 2.2), you'll see that, at this isotope, we've hit rock bottom. Any further fusion *consumes* rather than liberates energy. In just *hours*, the silicon fusion reaction burns to completion, leaving behind a small, iron-rich core in which no further fusion is possible. Catastrophic collapse, postponed so long, can be averted no longer.

When energy production in the core stops, the core collapses under its own gravity, which is so strong that electrons are forced into protons to produce neutrons: what was once a billion-kelvin ball of plasma a few million kilometers across rapidly collapses to form a neutron star just 10 km in diameter. The outer layers of the star, no longer supported by the core, fall into this mass of neutrons *within seconds*. The resulting rebound causes a massive shockwave that ricochets through the outer, lighter, *still fusible* layers, heating and compressing them and producing a massive pulse of fusion. This, in turn, produces an extraordinary density of free neutrons, which, because they are neutral and need not overcome the electrostatic repulsion of the nucleus, avidly combine with any nuclei with which they collide to produce massive amounts of heavier, more neutron-rich nuclei. Many neutron-rich nuclei, however, are unstable and rapidly decay, typically by emitting electrons. This converts the excess neutrons into protons, raising the atomic number (while retaining the atomic mass) of nuclei and producing many of the stable nuclei heavier than iron. With the rebound of the infalling material off the core, and the intense burst of fusion as the lighter, outer layers collapse inward, this new material spews into space at speeds approaching a few percent of the speed of light. Within seconds, a star that was many times as massive as the Sun is torn asunder in a titanic "supernova" explosion that, for a short time, is comparable in brightness to all the rest of its galaxy (fig. 2.6, right). The core of the star, in contrast, remains behind as either a neutron star or a black hole. That is, if the mass of the remaining core is less than 2.4 times that of the Sun, the force of its gravity crushes its electrons into its protons, producing a solid mass of neutrons a few tens of kilometers across, but if its mass is more than 2.4 times that of the Sun, its gravity can overcome even neutron-neutron repulsion, and it collapses further to form a black hole.

While supernovae seed galaxies with elements up to and a bit beyond iron's atomic mass (what with all those neutrons whizzing about, some additional, energetically "uphill" nuclei are formed during the final, cataclysmic explosion), it had long proven difficult to account for the abundances of the still heavier elements. Their source became clear, however, in August 2017 when the Laser Interferometer Gravitational-Wave Observatory (fig. 2.7, left) observed a gravitational wave signature indicative of the merger of two neutron stars that had hitherto been in close orbit around one another.* Subsequent telescopic studies at visible, x-ray, and γ-ray wavelengths identified the formation of 16,000 Earth-masses of neutron-rich, high-mass nuclei in the matter splashed out by this "kilonova" (so named because it is about 1,000 times brighter than the nova described above), which occurred in the elliptical galaxy NGC 4993, some 140 million light-years from Earth (fig. 2.7, right). We now believe that such mergers, which happen about once every 10,000 years in the Milky Way, are responsible for the creation of about half of all matter heavier than iron.

Some 100 million high-mass stars have died during the more than 12-billion-year lifetime of the Milky Way to date, going nova, supernova and, rarely, kilonova in the process. That is fewer than one-tenth

*Gravitational waves, which are sinusoidal oscillations of space itself, are detected by monitoring the distance between two test masses 4 km apart. As the gravitational wave passes, the space between the masses rhythmically expands and contracts by an amount equivalent to *one-thousandth of the diameter of a proton.*

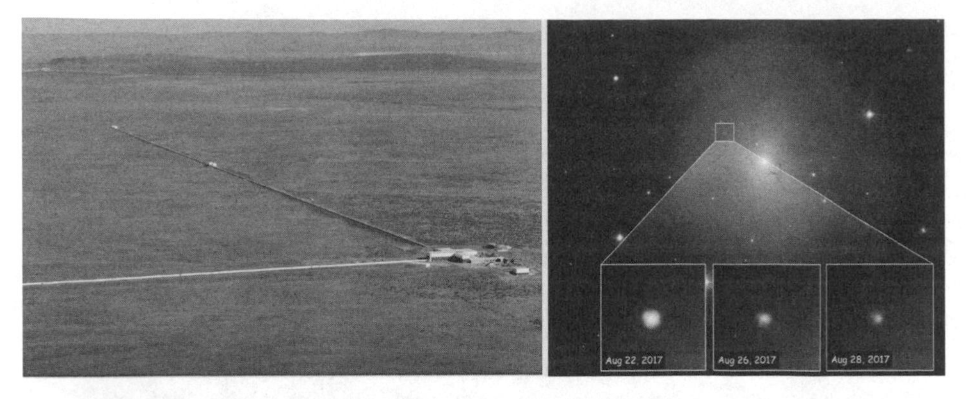

Figure 2.7 On August 17, 2017, the Laser Interferometer Gravitational-Wave Observatory (LIGO), shown on the left, detected the gravitational waves produced by two co-orbiting neutron stars, each of mass similar to that of our Sun, as they spiraled toward one another at near light speed before coalescing to form a black hole. LIGO achieved this by monitoring the distance between test masses at the ends of the observatory's 4 km long arms (shown); as the gravitational wave rhythmically warped space itself, the distances between the masses changed by an amount equivalent to a thousandth of *the diameter of a proton*. Immediate follow-up studies with optical (shown on the right are visible light images), x-ray, and γ-ray telescopes captured the radioactive decay of the neutron-rich, high atomic weight nuclei produced as the merger sprayed neutron-rich material into space. (Left image courtesy of Caltech/MIT/LIGO Laboratory; right image courtesy of NASA and ESA)

of 1% of all the stars ever born in our galaxy, but it's still been enough to convert about 2% of the original hydrogen and helium into heavier elements, leaving later-generation stars like ours relatively enriched in such atoms. Indeed, studies of the elemental and isotopic composition of the Sun indicate that it is a third-generation star (termed a "population I star"—again, for historical reasons). "Population II stars," the second generation of stars formed after the big bang, are also extremely common. In fact, spectroscopic measurements, which allow us to assay the metal content of stars by the colors of the light they emit, indicate that the majority of stars in the outer half of our galaxy are from this second generation, with membership in this population indicated by the relative paucity of "metallic" atoms heavier than helium. Careful inspection of the spectra of these stars indicates, however, that they do contain some metal, so they are not the first generation of stars. But unlike our third-generation, relatively metal-rich Sun, these stars were born very shortly after formation of the galaxy. And what of the Sun's grandparents, the population III stars formed from the primordial gas of the big bang? As noted above, the first stars to form were probably quite large because the paucity of metals in the early Universe prevented the cooling required for smaller gas clouds to collapse to form stars. Those massive, first-generation, population III stars would have gone supernova within a few million years, neatly accounting for the fact that, while we see many metal-poor, population II stars in the Milky Way, intensive searches have identified only two candidate stars that are very nearly metal free. The metal abundance in the two stars is less than 1/300,000 that of the Sun, and so these may be among the most ancient of stars. The metals so rapidly produced—and released in supernova explosions—by the population III stars would have contributed, in turn, to the metal composition of the second-generation, population II stars.

Stellar Requirements for the Origins and Evolution of Life

So far it seems easy: make a universe, fill it with hydrogen and helium, let those elements contract and ignite, and let the resulting stars go nova, supernova, and, rarely, kilonova for a couple of generations to seed space with heavier elements, eventually producing metal-rich, third-generation stars and setting the stage for life, or at least potentially habitable solar systems, to arise. But it may not be so simple. Evidence collected over the past few decades suggests that stars capable of supporting inhabitable planets may be rather rare.

The Sun is not all that typical a star. For example, even a cursory survey of the map of nearby stars reveals that ours is different. For one, the Sun is solitary. Because they are born in densely packed stellar nurseries (more on this in chapter 3), about 85% of all stars occur in multiple-star systems in which two or more stars are in orbit around their common center of mass. Given the complex gravitational tugs associated with being in orbit around—or even near—two or more stars, stable planetary orbits are more difficult to achieve in such systems.

The Sun's size also bodes well for its ability to support life while further rendering it somewhat special. The massiveness of the Sun is often underappreciated because it lies more or less in the middle of the size range between the most and least massive stars. But because they form more rarely and because they have much shorter life spans, larger stars are exponentially less common than smaller stars, and thus the Sun is in the top 10% of all stars in terms of its mass.* And the size of the Sun is a critical element in its ability to support the origins and evolution of life. Were the Sun smaller, like the red dwarf stars that outnumber it more than tenfold, it would be so cool and dim that the volume of its "habitable zone"—the region in which liquid water can form—would be positively puny.** For example, the habitable zone of Barnard's star, a typical red dwarf, six light-years away and thus the second closest stellar system to our own, extends out to only one-twentieth the distance of Mercury's orbit around the Sun. This is a problem. Specifically, the height of tides (like the nearly twice-daily rise and fall of the Earth's oceans) drops off inversely with the cube of orbital radius, an effect that has important consequences here. The friction generated as this tidal "bulge" is dragged around a planet (or moon) has the net effect of bleeding rotational energy away as heat, thus slowing the rotation (the length of the day on Earth, for example, has increased by about two hours in the past 600 million years). Ultimately, the length of the "day" would increase so much that it would match the length of the "year," causing the bulge to stop moving around the planet and the planet to spend the rest of eternity with one face locked forever toward its star and the other in perpetual cold and darkness. (By analogy, tides raised by the Earth in the Lunar crust long ago forced the Moon into such a resonance, which is why we only ever see one side of it.) For a planet in the habitable zone of a

*That is, in galaxies like ours. In elliptical galaxies, which are quite common, dwarf stars outnumber Sun-like stars by an even larger margin.

**Admittedly, this argument ignores potential habitats outside the traditional habitable zone, such as within the geothermally heated crust of a moon or planet. This is a topic we'll explore in detail in chapter 9.

red dwarf, the tides would be so large that this "tidal locking" would occur within a few hundred million years. Not, perhaps, the most likely environment to support life, because—apart from any other reason—volatiles such as water would rapidly migrate from the warm, dayside of the planet to the bitterly cold nightside, where water would freeze and remain forever unavailable. Calculations suggest that the habitable zone does not move out beyond the zone of tidal locking until a star is at least 40% the mass of the Sun (fig. 2.8).

Conversely, it is also important that the Sun is not too large, lest it burn out before life had time to arise. As noted above, the Sun is small enough that it will remain in its hydrogen-burning phase for some 11 billion years, which is plenty of time for life to arise and evolve. A star twice the mass of the Sun, in contrast, would last for less than a billion years. All in all, only about 5% of all stars lie within the narrow range between being massive enough to push the habitable zone beyond the reach of rapid tidal locking, yet small enough to remain stable for hundreds of millions of years.

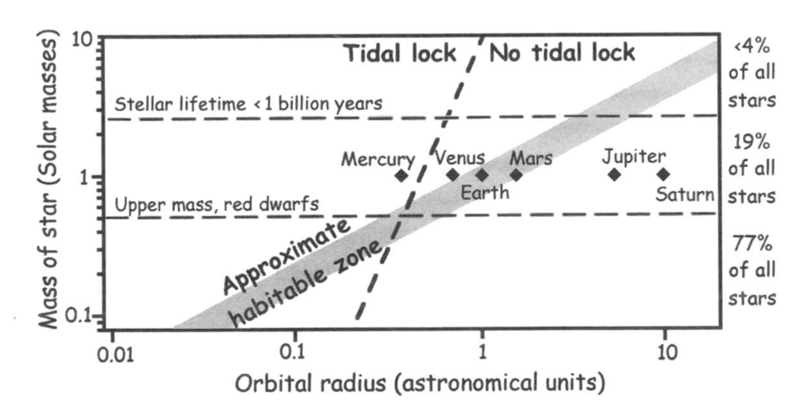

Figure 2.8 Both the habitable zone (gray band in the figure) and the tidal locking zone (in which one face of a planet is locked forever toward its star) shrink as stellar mass decreases. Because red dwarf stars are dim, their habitable zones lie within their tidal locking zones, rendering it far less likely that such stars—which are by far the most abundant stars in the Universe—can support life. For scale, our Solar System is shown. Mercury is just within the tidal locking region but, due to its highly eccentric (noncircular) orbit, is locked into a 2:3 resonance in which the planet revolves three times for each two orbits it makes around the Sun.

Galactic Requirements for the Origins and Evolution of Life

To make matters worse for potential extraterrestrials, the Sun's special nature may not be limited to its size and lack of stellar companions. It may also extend to its placement in our galaxy. Research on the topic suggests that the origins of life are in line with so many other aspects of existence in which the three secrets to success are "location, location, location."

The Sun orbits the center of the Milky Way at a distance of 26,700 light-years (plus or minus 1,300 light-years), taking 230 million years to complete each, nearly perfectly circular trip. This distance is near optimal in terms of the Sun's ability to support a life-bearing planet. Nearer to the center of the Milky Way, the density of stars becomes greater, rapidly climbing to densities so high that supernova explosions are frequent enough and close enough to sterilize planets on a rapid timescale compared with life's evolution. Nearer still to the center, the x-rays produced by the massive (four million Solar masses) black hole that resides there would destroy the complex molecules associated with life (fig. 2.9). Much farther out than the Sun's orbit, all the stars are metal-poor, population II stars. Without the heavier elements, planet formation is likely inhibited, and what planets exist are probably poor substrates for life. In combination, these conditions produce what is known as the galactic habitable zone. And whereas the galactic habitable zone does make up a fair fraction of the volume of our galaxy, it is a sparsely populated fraction. Our best estimate is that, at any given time, only approximately 10% of the stars in the Milky Way reside within its habitable zone. The existence of life on our planet, from simple microorganisms to human beings, is a result of the unique conditions that exist in this zone.

Of course, merely being in the galactic habitable zone at a given time is only part of the equation. Just as critically, in order to bear life, a star must *remain* in the habitable zone for a sufficient length of time. And most stars do not. The eccentricity of the Sun's orbit around the galactic center is extremely small; that is, the Sun's orbit traces out an almost perfect circle. Because of this, the Sun—and with it the Earth—remains at a near constant distance from the galactic center, which pro-

Figure 2.9 In 2019, astronomers of the Event Horizon Telescope project captured the first image of a black hole silhouetted against the glowing ring of plasma spiraling into it. The feat required connecting radio telescopes at eight sites around the globe (including the South Pole) to create a single virtual telescope with the resolving power of a single telescope nearly as large as the Earth itself. The black hole, which weighs in at 6.5 billion times the mass of the Sun, lies at the center of the galaxy M87, some 55 million light-years from Earth. In contrast, the black hole at the center of our own galaxy, called Sagittarius A*, weighs in at only 4 million solar masses. (Courtesy of ETH collaboration)

vides a safe haven from the potentially disruptive effects we've just described. Such low-eccentricity orbits are relatively rare, however, and the large majority of the Sun-like stars currently in our neighborhood spend a significant fraction of each galactic orbit far too close to the galactic center for comfort. Taken together with the relatively small size of the galactic habitable zone, the eccentricity of the majority of Sun-type stars is sufficiently great that fewer than 5% of the stars in the Milky Way remain permanently in its life-supporting zone. Thus, when considered together with the stellar size limits and the frequency of multiple-star systems, only a tiny fraction of the stars in the Milky Way seem to be well placed to support life (see sidebar 2.4).

Sidebar 2.4

Weighing the Probabilities

For life to arise and thrive around a star, that star almost certainly must meet several important criteria that we discuss in this chapter. These include being of sufficient size to produce a large habitable zone outside the tidal locking zone yet small enough to burn for billions of years without going supernova (1 in 20) and being a metal-rich, population I star (1 in 3) occupying a relatively low-eccentricity orbit permanently within the galactic habitable zone (about 1 in 20). The probability of meeting all three criteria simultaneously is given by the product of the individual probabilities:

$$P_{habitable} = P_{size} \times P_{metal} \times P_{GHZ}$$

Our best estimates for these parameters, described in the text, produce a $P_{habitable}$ of 0.0008; that is, about one star in 1,200 seems to be suitable. Of course, there are an estimated 100 to 400 billion stars in our galaxy and approximately 100 billion galaxies in the observable Universe, so these odds may not be that bad.

Conclusions

Our understanding of the origins of the Universe is reaching a fair degree of maturity. The big bang model is now a highly quantitative and compellingly confirmed model of how everything within and around us came into being. With this understanding of our origins, though, comes an appreciation of the vast number of things that had to be right for a universe to be habitable. If the Universe were too dense, it wouldn't have survived long enough to make us. If it were too sparse, galaxies and their associated stars and heavy elements would not have formed. Were there no resonance in the ^{12}C nucleus, there would be no heavy atoms, no chemistry, no life. And on, and on. Of course, these fundamental properties of the Universe are not the only things that had to be just right to create conditions for life. The size of the Sun, its location in our galaxy (and thus its metal content), and its low-eccentricity galactic orbit all seem to be critical aspects of the recipe that makes our planet habitable.

And what of Gamow, Alpher, Penzias, Wilson, Dicke, Peebles, and Wilkinson? Gamow died in 1968, just three years after Penzias and Wilson's striking confirmation of his theory. A decade later, the latter two shared the 1978 Nobel Prize in Physics for their discovery—although, perhaps unfairly, they did not share it with either Alpher, Gamow's then still-living student, or the Princeton group who were the first to explain the implications of that annoyingly persistent radiofrequency hiss.*

Further Reading

The Big Bang

Weinberg, Steven. *The First Three Minutes: A Modern View of the Origin of the Universe.* New York: Basic Books, 2020.

Stellar Nucleosynthesis

Bethe, Hans. "Energy Production in Stars." *Physical Review* 55 (1939): 434–56.

The Anthropic Principle

Barrow, John D., and Frank J Tippler. *The Anthropic Cosmological Principle.* Oxford, UK: Oxford University Press, 1988.

The Galactic Habitable Zone

Lineweaver, Charles H., Yeshe Fenner, and Brad K. Gibson. "The Galactic Habitable Zone and the Age Distribution of Complex Life in the Milky Way." *Science* 303, no. 5654 (2004): 59–62.

*Jim Peebles, though, did win the 2019 Nobel Prize in Physics for his many contributions to theoretical cosmology.

Chapter 3
Origins of a Habitable Planet

On a cloudy March night in 1993, Carolyn Shoemaker was peering at some photographs of the heavens that she, her husband Eugene Shoemaker (1928–1997) of the US Geological Survey, and their amateur astronomer friend David Levy had taken a couple of nights earlier from atop Mount Palomar in Southern California. The two photos, taken an hour apart, were part of a multiyear survey of comets and asteroids. The motivation behind the survey was Eugene Shoemaker's: he was the world's preeminent authority on *impacts*—that is, on the effects of meteorite and cometary strikes on the solid surfaces of the Solar System, a subject he had been studying since his days as a graduate student in the late 1950s. The survey was his attempt to quantify the number of asteroids and comets that might be expected to cross paths with the planets. But what Eugene Shoemaker really wanted was to see the effects of an impact with his own eyes. He had long dreamed of capping his career as one of the founders of astrogeology by exploring a fresh impact crater, perhaps in one of his favorite stomping grounds such as the remote deserts of Australia's outback. No significant, crater-forming impact, however, had been witnessed during the entire span of recorded human history.*

The photos Carolyn Shoemaker was viewing that night revealed a

*This said, on February 15, 2013, a 60 ton meteorite exploded in the air some 30 km above the city of Chelyabinsk, Russia. And while more than 1,000 people were injured—primarily because they, understandably, ran to their windows to see what the bright flash was only to have the blast wave shatter the glass in their faces—no one was actually hit by the stones that subsequently rained down. The only known direct hit of a meteorite on a human occurred on November 30, 1954, when a 4 kg space rock crashed through a roof, bounced off a radio console, and struck and bruised the thigh of Elizabeth Ann Hodges (1923–1972) of Sylacauga, Alabama, who had been napping on her couch at the time.

strange, fuzzy streak in the sky. "I don't know what I've got," she said, "but it looks like a squashed comet." The problem was that the fuzzy image on the film looked kind of like a comet, but it did not show the typical bright central core of a comet accompanied by a diffuse tail. Instead, the bright core of the object was an elongated blob, from which not one but several diffuse tails streamed off into space.

In the hour between the two photographs, the object had moved, a sure sign that it was within the Solar System (the stars are so far away that they appear as if fixed in space). Oddly, though, it seemed to be moving in the same direction across the sky—and with the same speed—as Jupiter, which was nearby in the sky and imaged on the same piece of film. Could the strange streak simply be a stray reflection of Jupiter's bright shine? The elongated streak did not quite line up with the overexposed spot of Jupiter, suggesting that, unprecedented as it was, the streak was real. It wasn't until the drive home that Eugene Shoemaker struck on a workable theory. What if the comet didn't just appear to move along the same line of sight as Jupiter but was actually physically near Jupiter in the three-dimensional vastness of the Solar System? If so, the enormous gravity of this planet, by far the most massive in the Solar System, could have raised such large tides in the weak cometary material as to tear it apart. What they had seen, he surmised, was not a single comet but a train of comet pieces produced by passage a bit too close to the giant planet.

As word of the discovery spread, observations of the newly named Shoemaker-Levy 9 comet poured in. (The name stems from the fact that this was the ninth periodic comet, a comet that orbits the Sun in less than 200 years, that the team had codiscovered.) With each new observation, the orbit of this celestial oddity became more precisely defined. The first observations confirmed Eugene Shoemaker's hypothesis; namely, that the comet had passed within 120,000 km of Jupiter on July 8, 1992, some 20 months before its discovery. By early April 1993, the data were plentiful enough to nail down an approximate orbit and, to everyone's surprise, it turned out that the comet wasn't in orbit around the Sun at all. It seems that this ancient resident of the outermost reaches of the Solar System had been captured into orbit around Jupiter way back in 1929, and the comet was now a moon of the giant planet. More startling still, by the end of May 1993, the data were voluminous enough to define the comet's orbit precisely, and it was found that Shoemaker-Levy 9 was not going to survive its next trip. After circling Jupiter for 65 years, the comet's next close pass, slated for July 25, 1994, was going to let Eugene Shoemaker see his dream come true—albeit not an impact on Earth, but an impact of the comet, which *his team* had discovered, onto the largest planet in the Solar System.

Researchers far and wide studied this "impact of the century" using Earth-based telescopes, the *Hubble Space Telescope*, and the *Galileo* probe, then still en route to its six-year orbital tour around Jupiter. In all, 21 impacts were observed as pieces of the comet train slammed into the deep atmosphere of Jupiter (fig. 3.1). Huge dark welts many times the size of the Earth appeared and lasted in the atmosphere of the gas giant planet for weeks. Spectroscopic studies suggest that large fractions of the estimated 1-billion-ton mass that ploughed into

Figure 3.1 The impact of the comet Shoemaker-Levy 9 with Jupiter left enormous, Earth-dwarfing scars in the planet's atmosphere (lower left). The impacts also delivered millions of tons of carbon, oxygen, and nitrogen compounds from the far reaches of space to the gas giant, highlighting the role that Jupiter has played in controlling the dynamics of the Solar System. (Courtesy of NASA/STScI)

the planet must have consisted of substances such as water (estimated at approximately 20 million tons), ammonia, and methane, collectively known as "volatiles" in planetary science due to their relatively low boiling points. And while Jupiter wasn't exactly in need of the extra bulk, the impact of Shoemaker-Levy 9 highlighted Jupiter's huge influence on the movements of stray bodies in the Solar System. With a mass 0.1% that of the Sun and more than two and a half times that of all the other planets combined,* Jupiter's gravitation is a force to be reckoned with. And as we'll see, Jupiter's influence on the motions of objects in orbit around the Sun is so significant that it played a critical role in the evolution of Earth as a habitable planet.

The Proto-Sun

As described in chapter 2, our Sun is a middle-sized yellow star. From observations of newly born stars in relatively nearby cosmic nurseries and from detailed studies of the composition of the Sun's planets, asteroids, and comets, astronomers have pieced together the story of the birth and evolution of our Solar System.

As the pre-solar nebula contracted, conservation of angular momentum forced the contracting dust and gas to spiral ever more rapidly around the cloud's center of gravity. Eventually, while the large bulk of the cloud that was to form the Solar System collapsed to form the proto-Sun, a modest proportion of the nebula fell into orbit around the rapidly forming star. As the cloud slowly collapsed under its internal gravitational pull, its gravitational potential energy was converted into kinetic energy, and the inner core of the nebula became quite hot. Eventually, the center of the nebula reached 10 million kelvins, the temperature at which the fusion of hydrogen into helium ignites, and a new star was born. The ignition of fusion occurred before the pre-solar nebula had entirely collapsed, and thus the early Sun, like all very youthful stars (which are called T-Tauri stars after a cluster of well-

*According to the Romans, a bearded old man by the name of Iuppiter was the most powerful being in the Universe. He was the god of the clear sky, thunderstorms, and rain; like Big Brother, he saw everything and was therefore also in charge of law and order.

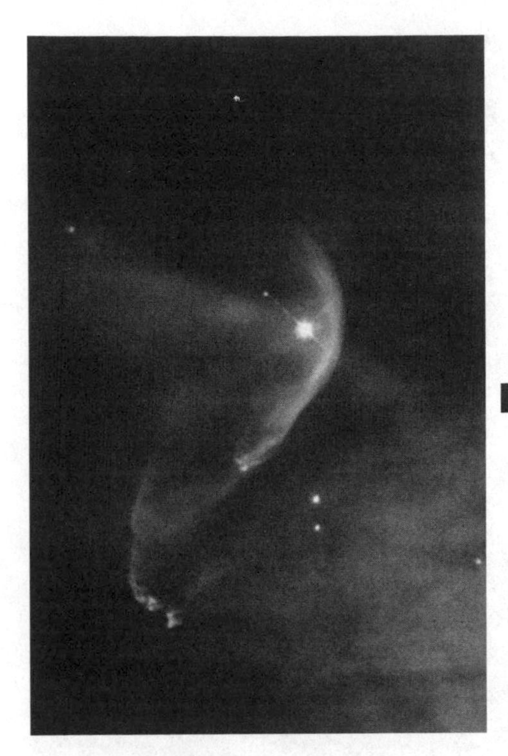

Figure 3.2 During their T-Tauri stage, young stars like this one in the Orion nebula (the nebula appears as the middle "star" in Orion's sword) begin to blow away the cocoon of gas and dust from which they were born. Here the T-Tauri solar wind is seen colliding with the net flow of the nebula, producing the visible shock wave to the right of the star. (Courtesy of NASA/STScI)

studied examples in the constellation Taurus), was wrapped in a dense cloud of nebular materials (fig. 3.2). In those early days, the Sun was rather more active than it is today; observations of dozens of relatively nearby, T-Tauri stars in the Orion nebula demonstrate that they are much hotter, rotate dozens of times more rapidly, and expel a stream of charged particles (the stellar wind) that is thousands of times more energetic than the solar wind pumped out by our now middle-aged star.

The dregs of material that remained in orbit around the new Sun moved at first with a bewildering array of orbital inclinations and eccentricities. Nongravitational forces (primarily friction with the remaining gas), though, fairly quickly established a degree of order in the movements and confined most of the material to a thin disk in the Sun's equatorial plane, an arrangement that still holds most of the matter in the Solar System today. How does this work? Just imagine one stray little rock orbiting the Sun on a path that is tilted relative to

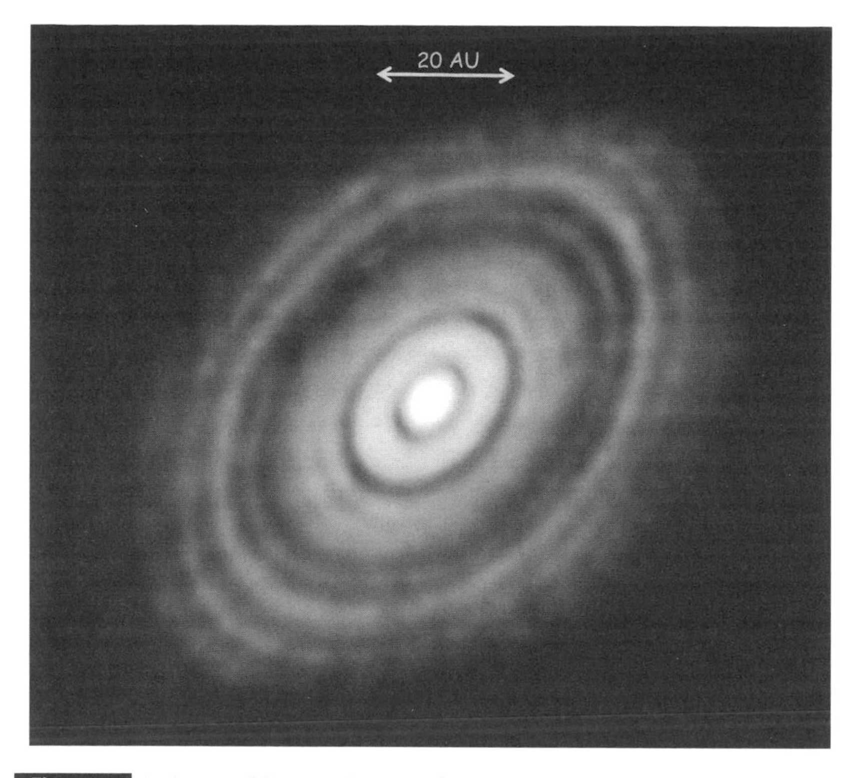

Figure 3.3 An image of the star HL Tauri and its environs collected at millimeter wavelengths sensitive to the thermal emission of dust illustrates its protoplanetary disk, which is estimated to contain about a tenth the mass of the central star. The cleared lanes in the image suggest that planetary formation is already well under way around this young, Sun-sized star, which is estimated to be just 100,000 years old. For comparison, Uranus's orbital diameter is 19 AU. (Courtesy of Atacama Large Millimeter/sub-millimeter Array/ESO/NAOJ/NRAO)

the thick main disk of gas and dust. During each orbit, it would pass through the disk once on its ascent above the equatorial plane and again half an orbit later on its descent. During each crossing, the stray rock would lose some of its out-of-plane velocity to friction with the gas and dust. Similar drag quickly ordered the early solar nebula, herding the remaining gas and dust into a thin disk nearly in the equatorial plane of the Sun. Although such a disk does not emit visible light, the young star warms it, and thus it does emit at infrared and millimeter wavelengths. By observing at these wavelengths, astronomers have been able to image such "protoplanetary disks" around a number of nearby young stars (fig. 3.3).

The Formation of the Planets

Protoplanetary disks are far from homogeneous. Heated from within by friction (in effect, the gravitational potential energy liberated as the gas and dust spiral inward) and from the light and solar wind emitted by their young central star, the inner portions of these disks reach several thousand kelvins. Farther out, though, the temperature of the disk falls dramatically: dust and gas in the outer reaches of the cloud lose less potential energy and, because the inner cloud is fairly opaque, receive less energy from the star. Thus, protoplanetary disks exhibit strong temperature and pressure gradients. The current theory of the formation of our Solar System, known as the equilibrium condensation model, attempts to explain the size, location, and composition of the Sun's planets (fig. 3.4) as a function of these temperature and pressure gradients—and the chemical heterogeneity they produced—in the Sun's protoplanetary disk. And, as we'll see, it does a remarkably good job, albeit with two glaring, and astrobiologically relevant, exceptions.

At the low pressures found in protoplanetary disks, liquids are not stable. Thus, the only relevant phase change is the sublimation of solids directly into gases and, as a complement, the deposition of solids from the vapor. Astrophysicists refer to materials with high sublimation points, such as the iron and nickel that constitute the core of our planet, as "refractory." Near the inner reaches of a disk, only the metals (used in the stricter, chemical sense now, i.e., shiny, electrically conductive elements) and most refractory oxides, such as alumina and a few aluminum-rich silicates, condense to form solid particles (fig. 3.5). Slightly farther out, less refractory, aluminum-free silicates, such as silica, also condense. At much larger distances from the central protostar, molecules we consider volatile, such as water, ammonia, and methane (the cosmochemically most abundant molecular forms of oxygen, nitrogen, and carbon, respectively) condense to form ices as the temperature continues to drop with increasing distance.

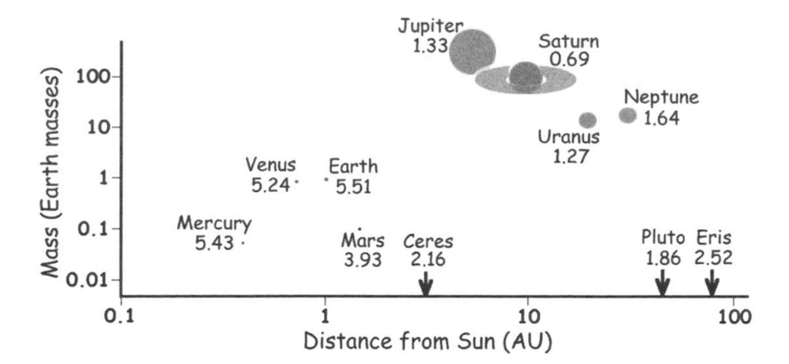

Figure 3.4 The equilibrium condensation model of the formation of the Solar System attempts to explain the observed distributions of planetary masses and compositions as a function of distance from the hot, dense center of the protoplanetary disk. Noted are planetary densities in grams per cubic centimeter (or, equivalently, tons per cubic meter), which serve as a proxy for composition. The masses of the dwarf planets Ceres, Pluto and Eris, as well as their myriad of smaller brethren, are so low that they fall well below the bottom of the graph.

These condensed materials are the fodder from which the planets in our Solar System were made. Conglomerations of them rapidly built up, first as dust, but then, as dust motes coalesced, centimeter-sized particles and meter-scale boulders. Due to gas drag, the smaller (centimeter- to meter-sized) particles of solid material that condensed from the nebula slowly spiraled inward. As they were not in purely Keplerian orbits, that is, their motions were not solely defined by gravity, but also by friction,* they suffered frequent, slow collisions, causing them to agglomerate into still larger objects. By the time this *accretion* had generated kilometer-sized bodies, an additional force kicked in and accelerated the process: the mutual gravitational attraction of kilometer-sized bodies is sufficiently large that it begins to dominate gas drag and greatly accelerate the rate at which protoplanets grow.

As described above, the composition of the planetesimals present in the early Solar System varied as a function of distance from the Sun, with metal-rich planetesimals dominating near the center, silicate-rich

*These orbits were named after Johannes Kepler (1571–1630), who, around 1605, discovered the laws governing orbital mechanics.

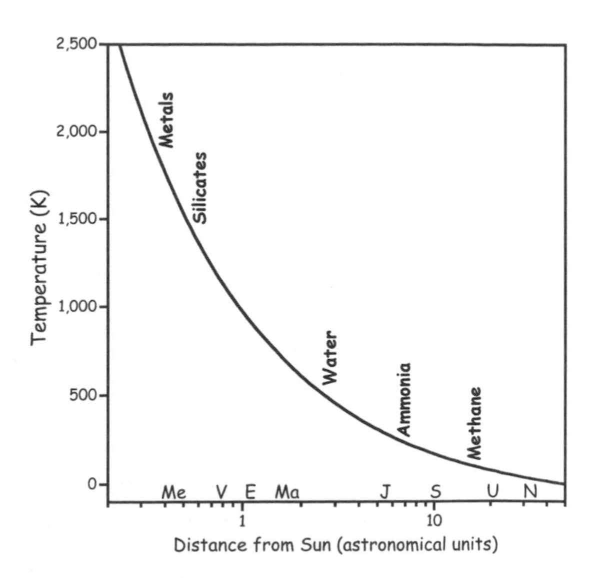

Figure 3.5 The temperature of the protoplanetary disk dropped rapidly with distance from its center. In the hot, inner reaches, only metals and the most refractory oxides could condense, producing small, dense, metal-rich Mercury (Me). Slightly farther out, silicates condensed to produce the larger metal-and-rock-rich Venus, Earth, and Mars (V, E, Ma). Jupiter (J) formed just beyond the "snow line," the point at which water—a cosmologically abundant molecule—condensed. Because of this, the proto-Jupiter rapidly grew large enough to also hold both hydrogen and helium, allowing it to swell to its current enormous size. Farthest out, around the orbits of Uranus and Neptune (U, N), the condensation of other ices occurred, including ammonia and methane. But here disk densities and particle speeds were low, preventing these planets from becoming gas giants like Jupiter.

ones at the middle distances, and volatile-rich ones farther out in the cold. This equilibrium condensation model, so named because the composition of each neighborhood was determined by the equilibrium chemistry that could occur at the temperature found there, roughly predicts the composition of each of the planets (table 3.1). And even though the predictions are only rough—because there was some scattering of particles away from where they originated—the model does a reasonable job of explaining why the small, solid, terrestrial planets*—Mercury, Venus, Earth, and Mars—are in the inner Solar System; the

*Remember, we use *terrestrial* to denote smallish, rocky planets in general, and *Terrestrial* to refer specifically to something of the Earth.

Table 3.1

Physical parameters of some prominent Solar System objects

Object	Distance from Sun (AU)	Composition	Mass (Earth = 1)	Density (g/cm³)
Mercury	0.39	Metals + alumina	0.06	5.43
Venus	0.72	Metals + alumina + silicates	0.82	5.24
Earth	1.00	Metals + alumina + silicates	1.00	5.51
Moon	1.00	Metals + alumina + silicates	0.012	3.34
Mars	1.52	Metals + alumina + silicates	0.11	3.93
Vesta	2.36	Metals + alumina + silicates	0.00004	3.46
Ceres	2.77	Metals + alumina + silicates + water	0.00014	2.16
Jupiter	5.21	Metals + alumina + silicates + water + hydrogen + helium	317.8	1.33
Saturn	9.58	Metals + alumina + silicates + water + hydrogen + helium	95.1	0.69
Uranus	19.28	Metals + alumina + silicates + water + ammonia	14.6	1.27
Neptune	30.14	Metals + alumina + silicates + water + ammonia	17.2	1.64
Pluto	39.88	Metals + alumina + silicates + water + ammonia + solid methane	0.003	1.86
Eris	68	Metals + alumina + silicates + water + ammonia + solid methane	0.003	2.52

Note: AU = astronomical unit.

"gas giants," Jupiter and Saturn, come next; the "ice giants," Uranus and Neptune, follow; and the smaller, icy outer worlds of the Kuiper belt, the scattered disk, and the Oort Cloud extend farther out still.

Near the Sun, the terrestrial planets are formed of various ratios of refractory metals and silicates. As Mercury is the closest to the Sun, it should consist mostly of refractory metals, with a smaller component of silicates. Consistent with this, Mercury's mean density of 5.43 g/cm^3 (table 3.1) suggests that its bulk composition is about half silicates (mean density a bit under 3 g/cm^3; table 3.2) and about half nickel and iron (mean density about 8 g/cm^3). The metals, which are much denser, sank to form the core, and the silicates, floating like slag on iron in a blast furnace, formed the mantle and crust. The first direct evidence in favor of this model came during the 1973 flyby of the *Mariner 10* spacecraft, humanity's first expedition to the inner reaches of the Solar System. *Mariner* discovered that Mercury possesses a mag-

Table 3.2
Properties of common materials in the protoplanetary disk

Material	Density (g/cm^3)
Primordial gases (at 25°C)	
Hydrogen (1 atm)	9×10^{-5}
Hydrogen (10^6 atm)	0.3
Helium (1 atm)	1.2×10^{-4}
Helium (10^6 atm)	0.5
Common ices (at their melting points)	
Nitrogen	0.81
Methane	0.49
Ammonia	0.82
Carbon dioxide	1.50
Water	0.92
Common silicates (at 25°C)	
Quartz	2.65
Granite	2.69
Basalt	3.01
Common metals (at 25°C)	
Iron	7.87
Nickel	8.91

netic field, suggesting that the planet's core is conducting, as would be expected for liquid iron. Mercury's iron core was later confirmed by the *MESSENGER* spacecraft, which orbited Mercury from 2011 to 2015. Studies of how the planet altered *MESSENGER*'s orbit found that a core of iron-like density extends through 85% of the planet's radius (i.e., 60% of its volume).

Stepping out from Mercury, we find the other terrestrial planets Venus, Earth, and Mars. The densities of Venus and Earth are slightly less and slightly greater than that of Mercury, respectively (table 3.1). Venus and Earth, however, are some 14 and 17 times more massive than Mercury, and the increased gravity associated with these greater masses leads to greater compression of the material in their cores. Were it not for this compression, it is estimated that the bulk densities of Venus and Earth would be about 4.2 g/cm^3, which is equivalent to about a 3:1 mixture of rock and metal. Correspondingly, the Earth's dense metal core extends through only a bit more than half of its radius, rendering it, relative to the size of the planet, significantly smaller than Mercury's. This trend of increasing rock-to-metal ratios continues with Mars, which has an observed density of 3.93 g/cm^3 and an estimated uncompressed density of just 3.3 g/cm^3, indicating that it is composed of an even larger fraction of rock (the exact size of its core awaits confirmation from the ongoing *InSight* Mars lander mission). In short, these planets formed at distances far enough from the early Sun that silicates readily condensed, and thus all three consist of a thick rocky mantle surrounding a relatively small metallic core.

Further corroboration of the equilibrium condensation model is seen in the ratios of potassium to uranium found on the rocky planets. Both elements are radioactive, rendering them easy to detect. Both also tend to remain in the crust, where they can be observed, rather than partitioning into the iron-nickel core. But uranium-containing minerals are far more refractory than those that contain potassium, and thus the ratio of these two elements reveals the temperature at which a planet's constituents condensed. The Russian *Venera 8* and *Vega 1* and *2* landers, which each spent a few tens of minutes sampling the surface before being overwhelmed by the heat, found the potassium-to-uranium ratio on our nearest neighbor to be 7,000:1. This ratio rises to about 12,000:1 for the Earth, 18,000:1 for Mars, and to 70,000:1 by the time we reach the outer asteroid belt (fig. 3.6). As the theory goes, as we move farther from the Sun and the temperature of the protoplanetary disk drops, more of the volatile potassium compounds solidify to be swept up into planets. This said, there is an oddity: at 2,500:1, the Moon's ratio is less than a fifth that of the Earth's, suggesting the stuff from which it was made is far more refractory than the material that formed our planet. More on this in a moment.

Beyond Mars lies the asteroid belt and beyond that, the outer Solar System. This area of space is dominated by Jupiter, more than 300 times as massive as the Earth, and the other gas giant planets each more than a dozen times as massive as the Earth (itself the largest of the rocky, terrestrial planets of the inner Solar System). A key step toward acceptance of the equilibrium condensation model was its ability to explain this enormous inequity. Even though we think of Jupiter as

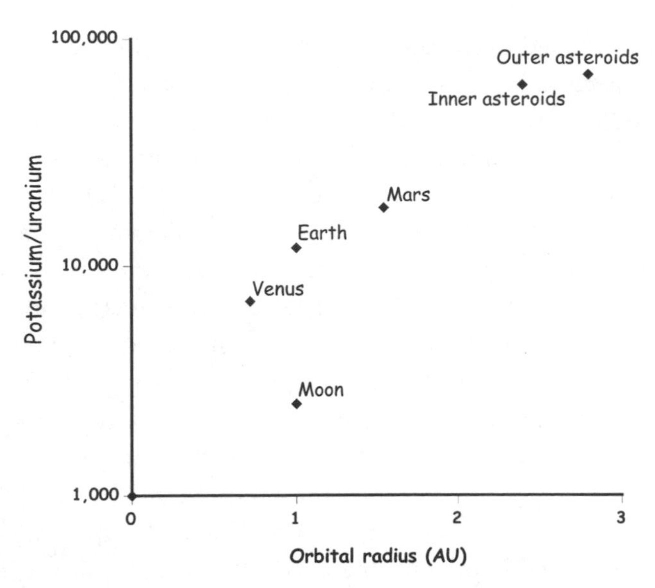

Figure 3.6 Potassium and uranium are radioactive and segregate into the rocky crusts of planets rather than sinking into the core, traits that render the two elements relatively easy to detect when robotic spacecraft come calling. Because boiling points of potassium minerals are far lower than those of uranium-containing minerals, the ratio of the two elements reflects the protoplanetary disk temperature at which various Solar System components condensed.

a gas giant, it isn't made entirely of gas, and the key to understanding its size and location lies in one of its other components: water.

As oxygen is one of the most cosmically abundant of the elements heavier than helium, water (a combination of oxygen with the Universe's most abundant element) was a major component of the solar nebula. Water, however, is volatile. In the low pressure of the planetary disk, water vapor could not form liquid water at all, but it would condense to form ice when the temperature of the disk hit approximately 150 K at the so-called "snow line." Studies of meteorites and the asteroids they arise from indicate that asteroids beyond about 2.7 AU (astronomical units; 1 AU is equal to the mean Earth-to-Sun distance of about 150 million kilometers) contain a significant fraction of water. For example, Ceres, the largest asteroid (now classified as the only "dwarf planet" in the asteroid belt), which orbits just outside the snow line at 2.8 AU, has a mean density of just 2.16 g/cm^3 and is alone estimated to contain as much water as all the Earth's oceans despite weighing in at only about 1/7,000 the mass of our planet. Indeed, the *Dawn* spacecraft, which dropped into orbit around Ceres in March 2015, even found signs of relatively recent (last few tens of millions of years) water-driven "volcanism" (fig. 3.7). In contrast, the 3.46 g/cm^3 density of the second most massive asteroid, Vesta, which orbits the sun at a distance of 2.4 AU, suggests that it is comprised predominantly of basalt surrounding a small iron-nickel core.

Because water must have been abundant in the protoplanetary disk—the density of which decreased with increasing distance from the Sun—Jupiter, which resides immediately beyond the asteroid belt, was in a prime location. At this distance from the early Sun, water ice condensed in larger amounts than anywhere else and rapidly accreted with metals and silicates to form a "proto-Jupiter" with a mass 10 to 15 times that of the present-day Earth. (A major goal of the *Juno* spacecraft, which entered a polar orbit around Jupiter in July 2016, is to better understand the nature of the Jovian core.) At this mass, gravity became strong enough to pull in and hold onto the *gases* hydrogen and helium. The terrestrial planets never reached this step; they could only accrete by collisions of solid objects as their gravity was too weak to hold onto the much more abundant hydrogen and helium that were by far the dominant components of the nebula. In contrast, young Jupiter acted like a giant cosmic vacuum cleaner, rapidly sweeping up all the material in or near its orbit until it mostly consisted of hydrogen and helium. As a result, Jupiter grew rapidly, ultimately becoming, by a factor of more than three, the most massive planet in the Solar System despite having a density of just 1.33 g/cm^3. Indeed, it is thought that Jupiter grew so rapidly that, even before it had finished growing, its gravity disrupted the accretion process nearby and prevented the formation of a planet where we now see the asteroid belt. It may also have starved the growing proto-Mars, leaving the Red Planet significantly smaller than it would have been without its giant neighbor.

The story for Saturn starts similarly, with the rapid formation a rock and water core that began to attract hydrogen and helium. But as one moves outward from the center of the protoplanetary disk, not

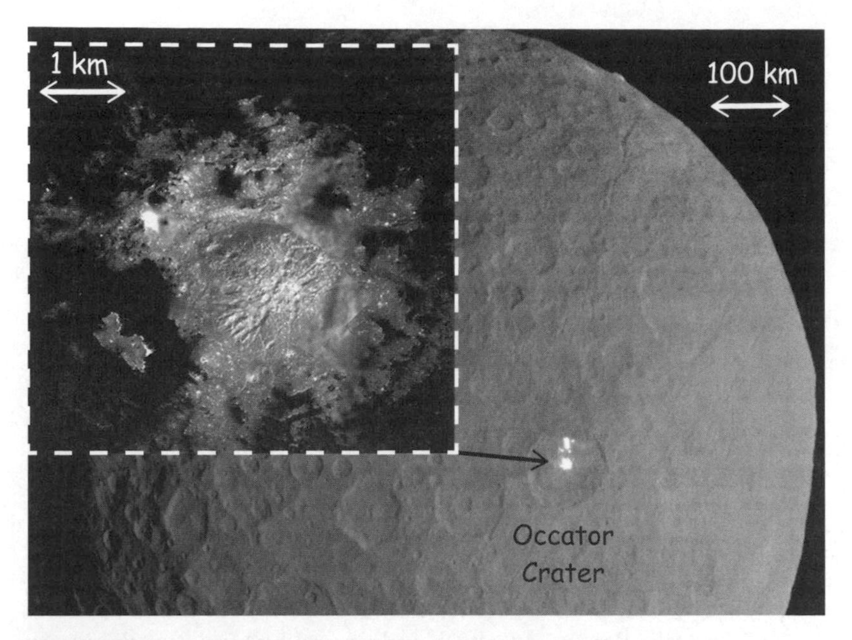

Figure 3.7 In 2015, the *Dawn* spacecraft discovered signs of relatively recent hydrothermal activity on the dwarf planet Ceres. Shown in the inset, for example, is a kilometers-wide mountain of "evaporates," which are dried salts (sodium carbonate appears to be a main component) left behind when the Occator Crater formed a few tens of millions of years ago, cracking the crust and allowing brines to erupt onto the surface. (Courtesy of NASA/JPL-Caltech)

only does the density fall, but also the orbital speed. These two effects slowed accretion, leaving Saturn unable to accrete as much hydrogen and helium as Jupiter before these gases were driven off by the intense solar winds of the T-Tauri–stage Sun. Saturn is thus less than a third the mass of Jupiter and, because its weaker gravity does not compress its gases as much, only reached a density of 0.69 g/cm^3.

The story for Uranus and Neptune is a bit different. Although born in still less-dense portions of the protoplanetary disk, these planets had the advantage of being so far down the temperature gradient that ammonia and methane (which, as the hydrides of nitrogen and carbon, respectively, are also cosmologically abundant) also condensed, allowing the planets to build up rock and ice cores several times the mass of the Earth. However, neither planet grew bulky enough to pull in and hold significant amounts of molecular hydrogen or helium before these gases were blown out of the Solar System, so their bulk densities come in at 1.27 and 1.64 g/cm^3, respectively. Given this, astronomers now categorize these near twins as “ice giants” to distinguish them from the true gas giants Jupiter and Saturn.

Still farther out from the Sun, low densities and centuries-long orbital periods greatly slowed accretion, ensuring that only small, icy bodies could form. Some of these bodies form a disk analogous to the asteroid belt, named the Kuiper belt after the Dutch-American astronomer Gerard Kuiper (1905–1973), stretching from 30 AU (approximately the orbit of Neptune) to 50 AU from the Sun. Still others form a much more irregular disk, the “scattered disk” that extends to greater than 100 AU and is comprised of icy bodies tossed into the outer Solar System by Jupiter and the other giant planets (more on this in a moment). The largest known members of the Kuiper belt and the scattered disk are Pluto and Eris, respectively, the latter of which was discovered in 2005 by Mike Brown of Caltech and named by him after the Greek goddess of strife and discord and which led to the former losing its planetary status. These two “dwarf planets” weigh in at only 0.22% and 0.27% of the Earth’s mass, or about one-fifth the mass of the Earth’s Moon. At 1.86 and 2.52 g/cm^3, the bulk densities of these frozen orbs suggest that Pluto and Eris are comprised of slightly more than and slightly less than half ice, respectively, with the rest being rock.

How rapidly did these condensation and accretion processes occur? Chondritic meteorites have given us an indication of the timescale. These rocky space travelers are the most abundant type of meteorite; they consist of the oldest, most primitive and unaltered material in the Solar System and thus offer a window into the first condensation events. The oldest chondrite precisely dated using radioisotope dating (see sidebar 3.1) is 4.567 billion years old, providing a date for the first condensation events. Most other chondrites date to within 20 million years of this age, suggesting that the condensation process was very rapid indeed relative to the age of the Solar System. Computer simulations of the condensation of the presolar nebula that take into account the physics and chemistry of the problem suggest that, once condensation into meter-sized bodies was

Sidebar 3.1

Radioisotope Clocks

Many nuclei are unstable and decay through a range of processes collectively known as radioactivity. Because the rate of decay of any one type of nucleus (an isotope) is well known and is completely independent of environmental conditions, the decay process can be used as a clock with which to date geological events.

In principle, the concept of radioisotope dating is straightforward. We simply compare the amount of a given isotope present today in a sample with the original concentration of the isotope. Using the known rate of decay (as given by the isotope's half-life), it is a simple matter to calculate how long the decay has been taking place and thus the age of the sample. But radioisotope dating is straightforward only in principle. The rub is figuring out how much of the relevant isotope was there in the first place.

For some isotopes and some decay routes, the task at hand isn't so hard. For example, the isotope potassium-40 (^{40}K) decays with a half-life of 1.26 billion years, either by the emission of an electron, which converts a neutron into a proton to form calcium-40 (^{40}Ca), or by capture of an electron by a proton to form a neutron, to produce argon-40 (^{40}Ar). Unfortunately, ^{40}Ca is the most abundant isotope of calcium and is extremely common in minerals, and thus it is not possible to distinguish the ^{40}Ca produced from the decay of ^{40}K from the ^{40}Ca initially present when the rock was formed. The ^{40}Ar produced by the second decay process, in contrast, is well suited for radioisotope dating: argon being an extremely inert gas, if any was originally present it would escape when the rock was last molten. The ^{40}Ar in a rock today, then, must have come from the decay of ^{40}K since the rock crystallized. Thus, using the known half-life of ^{40}K and knowledge of the amounts of ^{40}K and ^{40}Ar in a rock, we can estimate the time that has passed since the rock solidified to its present form.

Other isotopes also make suitable targets for radioisotope dating, but the determination of their original concentrations requires more effort. For example, uranium-238 (^{238}U) transmutes through several steps into lead-206 (^{206}Pb) with a half-life of 4.47 billion years. Some minerals, such as zircon, preferentially exclude lead ions from their crystal structure while including the better-fitting uranium ions. Thus, when a zircon crystallizes, its initial lead concentration is zero. Over time, any ^{238}U in the zircon decays into ^{206}Pb. Knowledge of the current ^{238}U content plus the current ^{206}Pb content therefore indicates the original ^{238}U content, which in turn provides a means of dating the crystallization of the material.

Another commonly employed isotope for dating is rubidium-87 (^{87}Rb), which decays into strontium-87 (^{87}Sr) with a 48.8-billion-year half-life. Here the availability of an isotope of strontium that is not the product of radioactive decay, strontium-86, provides a means of estimating the original strontium concentration. With this, and with knowledge of the current amounts of ^{87}Rb and ^{87}Sr, we can calculate the original and current ^{87}Rb concentrations and thus the time that has elapsed since the rock crystallized.

All three of these radioisotope clocks—and others—have been used to date the formation of more than 70 different meteorites and so provide a lower limit on the age of the Solar System. And while the observed dates vary over the range 4.53 to 4.58 billion years ago, the best estimate is 4.56 billion years. Given that many of these meteorites are very primitive (and thus unlikely to have had their radioisotope clocks "reset" since they condensed from the pre-solar nebula), this "lower limit" is probably within 10 million years of the actual date on which the pre-solar nebula started to condense to form the Solar System.

complete, the accretion of these planetesimals into the planets we see today required only another few million to few tens of millions of years. Given the magnitude of the construction process (after all, we are talking about the creation of an entire planetary system here), this is an astonishingly short time. But as described above, studies of T-Tauri stars indicate that, within a few million years of the start of fusion, the fierce solar wind of the early Sun would blow most of the remaining dust and gas out into interstellar space, limiting any further accretion to that involving larger planetesimals. Specifically, surveys with infrared telescopes (infrared, again, being particularly sensitive to warm dust) indicate that, while more than 80% of stars exhibit evidence of a protoplanetary disk when they're a million years old, by the time their 10 millionth birthday comes around, fewer than 1% do. In short, if it were to happen at all, the accretion of the Solar System had to happen rapidly.

The loss of protoplanetary disk material during the T-Tauri stage is quite dramatic. The total amount of, say, iron in all the Solar System's planets is equivalent to about 3% of the mass of all the iron in the Sun, implying that, if the protoplanetary disk started out with the same composition as the Sun (a good bet), then the disk's mass was at least 3% that of the mass of the Sun. Moreover, if, in addition to losing volatiles, the disk lost iron-containing dust (also a good bet), it would have been even more massive. Consistent with this, the protoplanetary disks circling young, Sun-like stars typically weigh in at about one-tenth the mass of the Sun. In contrast, the total mass of all the planets, dwarf planets, and asteroids in our Solar System today amounts to 0.14% of the mass of the Sun, meaning that 95% to >98% of the protoplanetary disk was blasted into interstellar space during the Sun's T-Tauri stage.

The Mysterious Moon

The accretion model makes for a planetary system that changes gradually and logically from the inner planets toward the outer planets. However, one particular Solar System body falls outside this logic in a number of aspects: the Moon. As we'll see, the Moon's gravitational effects here on Earth have exerted a significant influence on life on the planet, so we should have a closer look at why the Moon is there.

Until the 1990s, there was no clear consensus regarding the origins of the Earth's natural satellite. Indeed, before the 1970s, three theories had been put forth as to the Moon's origins. The first, the "wife hypothesis," suggested that the Moon originated elsewhere in the Solar System and only later was captured into orbit around the Earth. The "sister hypothesis" was based on the premise that the Moon originated in the same neighborhood as the Earth but failed to fuse with it. The third, not surprisingly called the "daughter hypothesis," postulated that the Moon, perhaps due to the extremely rapid rotation of the early Earth, was torn from the Earth. One imaginative version of the latter theory postulated that the large hole left behind after the split became the Pacific Ocean—this was well before the acceptance of plate tectonics, a paradigm that implies the Pacific is far younger than the Moon. By the late 1960s, the scientific community had largely discounted the daughter hypothesis based on the improbably high rotation rates (equivalent to about a 2-hour-long day) required before centripetal forces would overwhelm the gravitational forces that held the proto-Earth together. But there was no clear advantage to either of the two remaining hypotheses. Everybody expected the question would be settled by the early 1970s once the *Apollo* missions brought back Lunar samples with which to test the two competing models. It was not.

The *Apollo* missions and the minerals they brought back only uncovered new contradictions. The isotopic distributions found in the Lunar samples (e.g., the ratios of oxygen-16 to oxygen-17 and oxygen-18 in various minerals) are similar to those found on Earth. This ruled out the wife hypothesis because objects accreted in different parts of the solar nebula exhibit different isotopic patterns. Likewise, however, the sister hypothesis is ruled out by the observation that the *elemental* makeup of the Moon is vastly different from that of the Earth. The Moon's density, for example, is only 3.34 g/cm^3, meaning that, unlike the Earth, it lacks a substantial iron-nickel core and consists almost entirely of silicate rocks. Likewise, when compared with Terrestrial rocks, the minerals collected from the Lunar surface are strangely depleted in the lower boiling point metals, such as sodium and potassium. The ratio of potassium to uranium in the Lunar crust, for example, is only 2,500:1, indicating that the Moon is greatly depleted in the former, rather volatile element relative to the level found in the Earth's crust (see fig. 3.6).

After a couple of decades of hand-wringing about these conflicting bits of evidence, the planetary science community has now achieved some degree of consensus regarding the origins of the Moon. According to the currently in-favor theory, the Moon formed by a "daughter-like" mechanism. However, the force that disrupted the planet was not rapid rotation but, instead, the impact of a Mars-sized object late in the accretion process that very nearly ripped the Earth apart. The force of the impact would have melted both the proto-Earth and the impactor and splashed an enormous amount of liquid rock into space. Given that the Earth had already differentiated into a dense, iron-nickel core and a rocky mantle, the splashed material would have been almost entirely silicates (the impactor was presumably differentiated too, but its metal core would have sunk and merged with the Earth's). With the heat of the impact, any volatiles and even sodium- and potassium-containing minerals would have evaporated, leaving the Moon as it is today: a piece of Earth rock (explaining the Earth-like isotopic abundances) depleted of both volatile elements and metals (accounting for its differing bulk composition). And when did this cosmic collision occur? The radioactive decay of hafnium-182 into its daughter isotope tungsten-182 gives us a clue. Tungsten is a "siderophile," an element that dissolves readily in iron-nickel alloys, so any tungsten present in the proto-Earth would sink into the core. Hafnium, in contrast, remains in the crust. Measurement of the tungsten-182 currently in Lunar rocks thus provides an indirect measure of the amount of hafnium-182 that was present in the crust when the Moon formed. Comparing this with the amount of tungsten-182 in the oldest meteorites (the parent-body sources of which did not differentiate to form a core, thus leaving their primordial tungsten levels unchanged), combined with knowledge of the 9-million-year half-life of hafnium-182 (see sidebar 3.1), suggests that the Moon formed some 30 million years after the beginning of accretion. Thus, the Moon—which, as we'll see, is thought to play a vital role in the evolution of life on Earth—was formed in a fluke accident near the tail end of the accretion period.

Cleaning Up the Mess and Delivering the Goods

Next to the odd composition of the Moon, the abundance of water and other volatiles on Earth is the other striking deviation from what the equilibrium condensation model of planet formation would predict. According to this theory, the temperature of the gas and dust that condensed to form the proto-Earth was far too high to allow water to condense, and thus the Earth should not contain significant amounts of this or any of the other volatile hydrogen, nitrogen, and carbon compounds critical for life. And yet the bulk composition of our planet includes, for example, about 350 parts per million water, about 300 parts per million carbon, and about 50 parts per million nitrogen (equivalent to 0.035%, 0.03%, and 0.005%, respectively). This, obviously, is not a lot, but it is enough to support all life on our planet and cause some problems for the equilibrium condensation model. Where did these volatiles come from? Most "planetologists" believe that they came from out beyond the snow line.

The presence of Jupiter, Saturn, Uranus, and Neptune had far-reaching consequences for life on Earth. In particular, the intense gravity of these massive planets affects orbital dynamics throughout the outer Solar System, which ultimately led to the "cleanup" of most of the planetesimals remaining after the end of accretion. The idea is that, through orbital resonances or close encounters, the outer giant planets would perturb the orbits of any planetesimals in their neighborhoods. Whenever they'd toss a planetesimal in toward the Sun, this would cause the orbit of the planet doing the tossing to migrate slightly outward. And whenever they'd toss anything out away from the Sun, this would cause the orbit of the tossing planet to migrate a bit inward. Jupiter, though, threw a bit of a wrench into the works: because it is so much larger than all the rest of the planets, it alone has enough gravity to toss planetesimals entirely out of the Solar System. The fact that Jupiter tossed many planetesimals entirely "out of the ballpark" created a net asymmetry that, over the first few hundred million years of the Solar System, caused the orbits of Neptune and, to a lesser extent, Saturn and Uranus to migrate outward and the orbit of Jupiter to migrate inward. As these planets migrated, almost all the planetesimals between Jupiter and Neptune were either tossed inward, where they would eventually collide with the rocky, inner planets, or were lifted into the scattered disk, into the more distant Oort Cloud, or even entirely out of the Solar System.* This "cleanup" produced two effects that were so vital to the formation of life on Earth that, in a nutshell, without Jupiter we would not exist.

The first effect of all this planetesimal tossing was to deliver volatiles to the otherwise "dry" proto-Earth. Such delivery is still occurring today. A dramatic example of the probable extraterrestrial origins of Earth's volatile organic inventory occurred on January 18, 2000, when an approximately 60 ton carbonaceous chondrite meteorite was seen to explode over Tagish Lake in the far north of Canada's British Columbia province. As the fall occurred in the middle of winter, it was easy to snowshoe out onto the lake and collect pristine samples of fresh meteoritic material. Fortunately for those of us interested in this sort of thing, one of the few inhabitants of this remote area was an outdoorsman by the name of Jim Brook, who understood the importance of such pristine materials. Venturing out on the ice a week after the fall, he carefully collected uncontaminated samples in clean plastic bags and stored them in his freezer.**

The couple of hundred meteoritic samples that Brook and later expeditions recovered are some of the freshest extraterrestrial material we have on hand to study. Spectroscopic studies of the Tagish meteorite indicate that it is a good match for the asteroid 773-Irmintraud, which orbits in the outer reaches of the asteroid belt, well beyond the snow line. Sometime in the past 100 million years, a meteor impact broke off a chunk of this asteroid, and the gravitational perturbations of Jupiter sent the chunk earthward. Of note, the Tagish meteorite is composed of 5% total carbon and about 3% organic material, mostly as aromatic hydrocarbons. The water content of Tagish is somewhat harder to determine because the meteorite fell on snow. But typical carbonaceous chondrites are 5% to 20% water and, with comets, could

*The asteroid 'Oumuamua, which flew past the Sun in 2017, was the first known example of such an interstellar traveler, presumably flung out of some far distant Solar System by that system's Jupiter equivalent. Its name is the Hawaiian word for "scout" or "messenger."
**Now you know what to do the next time you see a meteorite land.

have been major suppliers of the Earth's oceans. Even today, giant outer planets in our Solar System continue to push volatiles inward toward the rocky inner planets.

In addition to moving volatiles inward, the era of gas giant migration and its resultant planetesimal tossing had a second profound impact on the origins of life on Earth: it put a hard stop to the process of accretion. The history of the end phase of the cleanup is readily visible on any clear, moonlit night: the impact craters left behind by these outer Solar System planetesimals have been preserved on the Moon's face due to the absence of any remodeling of the Lunar surface by erosion or plate tectonics. With the Lunar rocks they brought back, the *Apollo* astronauts obtained more quantitative evidence regarding this cleanup stage. The 382 kg of Lunar samples allowed geologists to isotopically date various surfaces on the Moon, then calculate the rate of crater formation during various epochs of the Moon's history. In doing so, they found that the cleanup phase ended with what is now called the "late heavy bombardment," some 3.8 billion years ago (fig. 3.8).

One of the last major craters formed in the late heavy bombardment was the Imbrium Basin, which, at 1,160 km in diameter, is easily

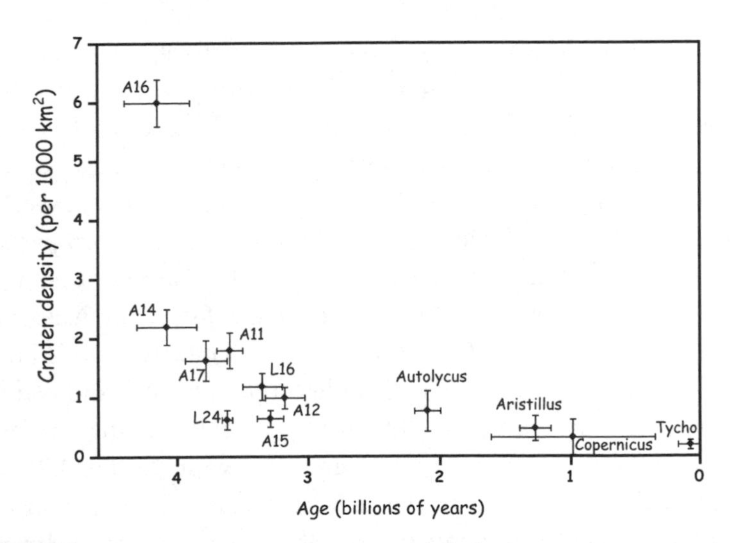

Figure 3.8 The numbers of larger craters per 1,000 km^2 on various age-dated surfaces of our Moon illustrate the drop-off of the late heavy bombardment at 3.8 billion years ago as Jupiter and the other giant planets completed their cleanup of the Solar System.

visible from Earth as the right eye of the Man in the Moon. The size of this crater amply demonstrates that these impacts delivered not only volatiles but also enormous kinetic energy. The planetesimal that formed the Imbrium Basin is estimated to have been about 400 km in diameter. An impactor of this size striking the Earth would provide enough kinetic energy not only to boil all the water in the Earth's oceans but also to vaporize hundreds of meters of the crust over the entire globe. Because impacts even significantly smaller than the one that formed Imbrium pack sufficient energy to sterilize the entire planet, it is safe to assume that we owe our existence today to the era, billions of years ago, when the outer planets largely swept the Solar System free of such impactors. Indeed, studies of the effect suggest that Saturn, Uranus, and Neptune are too small to have cleaned up the Solar System in any reasonable time frame, and so, were it not for Jupiter's vast bulk, the Earth would have been subjected to sterilizing impacts for billions, rather than hundreds of millions, of years after its formation.

The Volatile Inventories of the Other Inner Planets

The other inner, rocky bodies, of course, were also subjected to the late heavy bombardment. But even if you have a rocky planet to start with and get the volatiles delivered to your doorstep, things can still go spectacularly wrong. On innermost Mercury, for example, intense heat and relatively weak gravity allowed most of the volatiles to escape into space. Most, but not all. Mercury's spin axis is almost perfectly perpendicular to its orbital plane. Because of this, there are no seasons on Mercury, and the bottoms of craters at the poles, forever in shadow, are chilled to a temperature of a few kelvins. These craters act as cold traps and thus, as shown by both radar studies from the Earth and neutron spectrometer measurements of the hydrogen content of the surface by the *MESSENGER* Mercury orbiter, are filled with ice (fig. 3.9). The Moon yields a similar picture; the mean temperature of the Moon at its distance from the Sun is much lower than that of Mercury, but the Moon's gravity is still lower, and thus it is able to hold only the tiniest whiff of atmosphere. Indeed, the Moon's atmosphere is so thin that,

Figure 3.9 Due to its low gravity and high surface temperatures, Mercury, the innermost planet, is unable to hold an atmosphere, leaving its surface ancient and cratered. The tilt of its rotational axis, however, is so small that the bottoms of some of its polar craters (shown here is Mercury's north pole) are in perpetual shadow, where the temperature is near absolute zero (left). Any volatiles delivered to Mercury by comets and meteorites eventually make their way to these cold traps. The ice is easily identified in radar images, where it is highly reflective (right). (Courtesy of NASA/JHU-APL and National Astronomy and Ionosphere Center/Arecibo Observatory)

between rocket exhaust gasses and the astronauts venting their cabin air, each *Apollo* landing temporarily *doubled* its mass. The excess gasses, though, were lost to space over the course of days, returning the Lunar atmosphere to its normal state. Like Mercury, though, the Moon does harbor some deep, ice-filled craters at its poles.

But what about Mars and Venus? Mars we'll discuss in detail in chapter 9 as a potential abode for life. But Venus is nowhere near as hospitable. This is ironic because, in many respects, Earth and Venus are near twins. Venus is our nearest planetary neighbor, orbiting only 28% closer to the Sun than Earth. And, being only 5% smaller in diameter and 18% less massive than the Earth, Venus is also quite similar in size, suggesting that the two should have been on the receiving end of similar amounts of volatiles delivered from the outer Solar System. Consistent with this, their estimated carbon inventories are nearly identical, and their estimated nitrogen inventories are the same to within a factor of three (fig. 3.10). But here the similarities end. Whereas

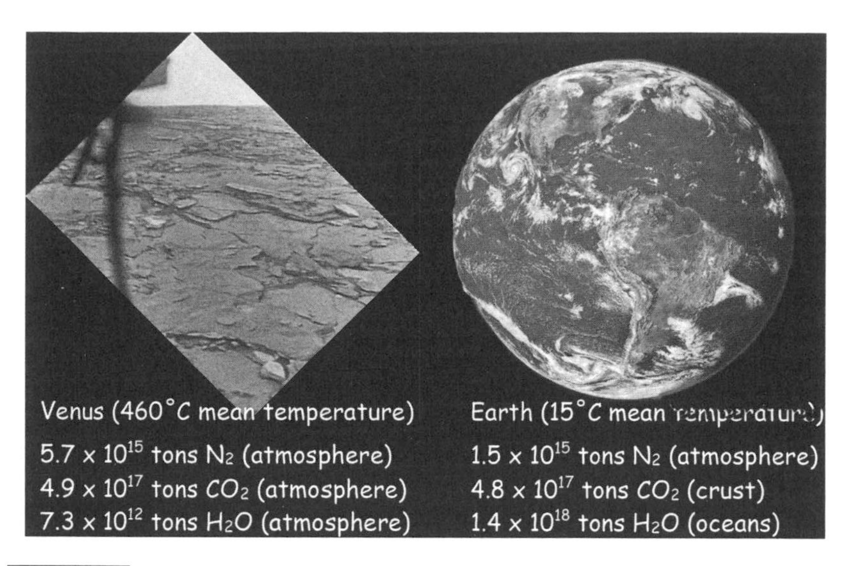

Figure 3.10 Venus is the Earth's near twin in terms of size and is our closest neighbor in space, orbiting just 28% closer to the Sun than the Earth does. The nitrogen and carbon inventories of the two are likewise similar. Venus, however, is extremely dry, and almost all of its carbon is in its atmosphere, as carbon dioxide, which pushes its mean surface temperature to 460°C. The image on the left, one of only a half dozen we have of the Venusian surface, was taken by *Venera 13*, which landed in March 1982 and survived on the surface for just under an hour. (Left figure courtesy of Russian Academy of Sciences/ Ted Stryk; right figure courtesy of NASA)

more than 99% of the Earth's carbon is stored in the crust (primarily as carbonate rocks, such as limestone), almost all of Venus's carbon is in its 96.5% carbon dioxide atmosphere, which blankets the planet with a pressure some 93 times greater than that found on Earth. Carbon dioxide, of course, is a greenhouse gas. So much so that the average temperature on the surface of Venus is more than 460°C,* hot enough to melt lead, tin, and zinc. From whence does this dramatic difference arise? The answer lies in Venus's water, or lack thereof: Venus is 200,000 times drier than the Earth.

Planets tend to lose light atoms, such as hydrogen and helium, in a process called Jeans escape, named after the same Jeans whose "Jeans mass" defines the minimum sizes of stars. Jeans escape occurs when

*In the shade. And there's no shade.

the velocity of the most rapidly moving atoms exceeds the planet's escape velocity and they are lost to space. At a given temperature, molecular (or atomic) velocity is inversely proportional to the square root of mass, and thus the light hydrogen isotope ^1H, which moves on average 41% more rapidly than the heavy isotope ^2H (deuterium), is more likely to escape, leaving the remaining water on the planet enriched in the heavier isotope. In 1979, the *Pioneer Venus* mission sent four entry probes into the atmosphere of Venus and measured, for the first time, the planet's hydrogen to deuterium ratio. At 62.5:1, the ratio *Pioneer* found is *100 times* greater than the ratio seen on Earth. Venus, in fact, is so deuterium rich that it is believed to have lost enough hydrogen to fill *several oceans'* worth of water. It is this lost water that is the ultimate cause of the vast climatic differences between the Earth and its twin.

Water plays a vital role in the regulation of carbon dioxide levels on Earth and so is critical in regulating the Earth's temperature. When atmospheric carbon dioxide levels rise (due to volcanism, for example), the greenhouse effect increases temperature, which in turn increases rainfall and the fluvial (i.e., by water) and chemical (reactions speed up at higher temperatures) weathering of rocks. This weathering releases calcium and magnesium ions from the rock, which flow into the ocean where they react with carbon dioxide (as carbonate ion) to form calcium carbonate (limestone) and calcium-magnesium carbonate (dolomite) in a process that, on Earth, removes carbon dioxide from the atmosphere on a timescale that is geologically rapid (fig. 3.11). This sequestration of carbon dioxide into rocks reduces atmospheric carbon dioxide levels, thus lowering the temperature and reducing weathering. Enough so that, were it to continue unabated, it would remove all the carbon dioxide from our atmosphere, reducing temperatures enough to drop the Earth's mean temperature below freezing.

What halts the cooling is plate tectonics. That is, the slow, inexorable creation of new oceanic crust at "spreading centers," such as the mid-Atlantic ridge, a 16,000 km volcanic cleft in the middle of the Atlantic from which new crust spreads out at a rate of about 2.5 cm a year. The island nation of Iceland sits astride the ridge, accounting for both its exuberant volcanism and its many tourist-friendly hot springs.

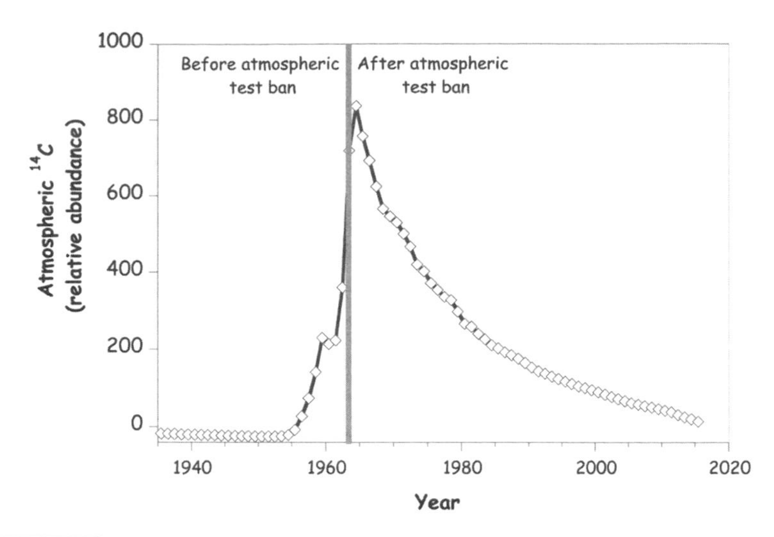

Figure 3.11 Carbon cycles fairly rapidly between the Earth's atmosphere, oceans, and crust. The rate of such turnover was provided by the aboveground testing of nuclear bombs in the years 1945 to 1963, which produced a large increase in the (trace) amounts of radioactive carbon-14 in atmospheric carbon dioxide. After the 1963 partial test ban treaty, which outlawed aboveground testing, atmospheric levels of this radioisotope relaxed back toward baseline over the span of a few decades. Some of this turnover is driven by sequestration in plants via photosynthesis, but current best estimates are that a bit more than half of this flux is driven by geological sequestration in the soils and ocean, where the carbon ultimately forms carbonate rocks.

Given that the total surface area of the planet is fixed, these spreading centers must be matched by "subduction zones," in which oceanic crust is driven down into the mantle. The Mariana Trench, the deepest spot in the Earth's oceans, is an example. It's where the Pacific plate is diving under the Mariana plate. As the Pacific plate descends and its volatile-rich sedimentary rocks heat up, they melt, rise to the surface, and produce volcanoes, the tops of which form the Mariana Islands. This volcanism releases the carbon dioxide stored in the sediments, starting the process anew. This "carbonate cycle" or "carbonate-silicate cycle" forms a negative feedback loop that maintains a constant temperature on the Earth's surface. Without liquid water, the cycle fails and carbon dioxide builds up in the atmosphere, leading to runaway greenhouse warming, as apparently happened on Venus.

The carbon dioxide cycle plays an absolutely critical role in, if not the origins of life, at least the maintenance of Earth's habitability. The issue in question is "the faint early Sun paradox," first pointed out in the early 1970s by Carl Sagan (1934–1996). Stars are relatively simple and predictable systems, and our understanding of their physics is quite mature. It leads us to the well-established extrapolation that early in its life the Sun must have been around 20% dimmer than it is today. (As fusion causes helium to build up in the Sun's core, the zone of fusion moves outward, increasing the surface temperature and therefore the brightness.) Thus a planet situated at a habitable distance from the early Sun (i.e., a distance at which liquid water could form) would, as the Sun grew brighter, heat up enough to lose its water by upper-atmosphere photolysis (a process that breaks water into its constituent hydrogen and oxygen) and turn into a Venus. This problem is exacerbated by the fact that water vapor is a potent greenhouse gas, thus accelerating the heating as the oceans begin to evaporate in earnest. This slowly increasing brightness is a universal feature of stars, and thus astrobiologists need to distinguish between habitable zones and "continuously" habitable zones. The former is the type of zone in which liquid water can form at a given period in a star's life, and the latter is the much, much narrower region in which water stays liquid over billions of years. The continuously habitable zone, although still fairly narrow, is broadened significantly by the carbon dioxide cycle; it is estimated that, for an Earth-type planet in orbit around a Sun-like star, the continuously habitable zone ranges from 0.95 to 1.15 AU. If the Earth had formed outside this tight band, life would not have flourished here for the billions of years it has. This said, *continuously* is a relative term; the Sun's steadily increasing brightness will push the inner edge of the supposedly continuously habitable zone past us in just 2 billion years. Don't get too comfortable!

The importance of the carbon dioxide cycle suggests that plate tectonics—the geological process in which crustal rocks are recycled into the mantle—plays a critical role in maintaining a habitable planetary environment. The precipitation of carbonate rocks is very efficient and over geological time would lead to the removal of effectively all atmospheric carbon dioxide. The resultant reduction in the greenhouse effect would be greatly exacerbated by the fact that snow and ice are white and thus reflective. This, in turn, would lead to further cooling, plunging the planet into a global ice age (termed the "snowball Earth"). The existing evidence suggests that this may have happened several times in the previous billion or so years of our planet's history. When the Earth's surface was covered by ice, however, the atmosphere was isolated from the oceans and atmospheric carbon dioxide levels were free to rise. All that was needed was plate tectonics, with which the carbon dioxide in carbonate rocks is recycled by volcanoes back into the atmosphere. In this way, ice-covered oceans lead to increasing atmospheric carbon dioxide levels, warmer temperatures, and melting of the snowball Earth, and the global carbonate cycle starts anew.

Liquid water, then, is required to prevent runaway greenhouse warming (*à la* Venus), and plate tectonics is required to prevent runaway sequestration of carbon dioxide leading to a snowball planet. It is already clear that liquid water is rare in the Solar System. Plate tectonics may also be rare; Mercury, Venus, and Mars do not exhibit any compelling evidence of the effect. We are not sure why the Earth exhibits plate tectonics while its near twin, Venus, does not, but, ironically, it has been hypothesized that water may be the missing ingredient on our nearest neighbor. Without water as a lubricant, the theory suggests, Venus's crustal plates are too rigid to sink into the mantle the way that Earth's crustal plates do.

This, of course, simply pushes the question one step farther: why did Venus lose its water when Earth so clearly did not? The answer to this question lies in the precise locations of the two planets. Although, because of greenhouse warming, the mean temperature on the Earth's surface is 15°C, given its albedo (reflectivity) and its distance from the Sun, its mean temperature would otherwise be well below freezing. Because of this, by 10 km above sea level, which is above most of the greenhouse gas, atmospheric temperatures reach −50°C.* Any water vapor that diffuses up toward the upper atmosphere thus even-

*At still higher altitudes, the Earth's atmosphere warms again due to the absorption of UV light from the Sun.

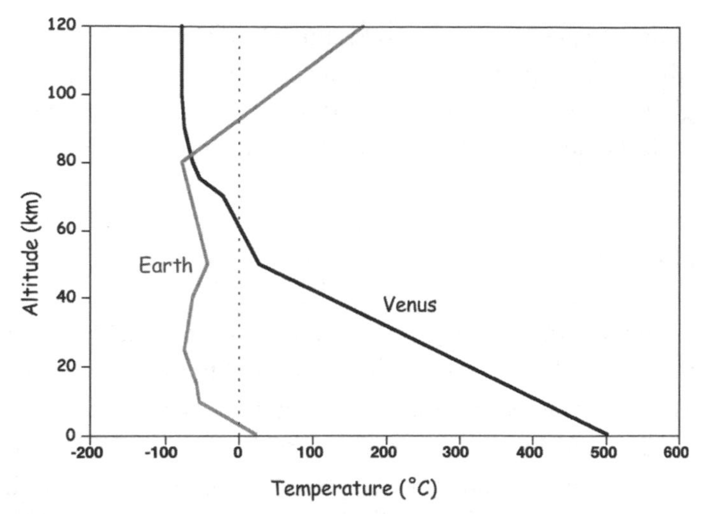

Figure 3.12 The mean temperature of an object with the Earth's reflectivity at the Earth's distance from the Sun is below the freezing point of water. For this reason, the middle layers of the Earth's atmosphere are cold enough that water condenses and precipitates at relatively low altitudes, leaving the upper atmosphere quite dry. (At extreme altitudes, the absorption of solar UV warms the atmosphere to high temperatures.) In the warmer Venusian atmosphere, in contrast, water easily diffuses to heights in excess of 60 km, where solar UV splits it into oxygen and hydrogen, with the latter being lost to space.

tually condenses and falls back as rain or snow (fig. 3.12). The mean temperature at Venus's orbit, in contrast, is above the freezing point of water, so water can diffuse far higher in the Venusian atmosphere, ultimately being subjected to the full intensity of the Sun's UV light, which tears the molecule into its constituent atoms. The oxygen, being extremely reactive, drifts back down and oxidizes Venus's surface rocks. (Radar images of the surface of Venus indicate that the planet is covered with relatively young lava flows, and thus there are plenty of fresh, reduced mantle rocks to oxidize.) In contrast, because it is very light, a small but significant fraction of the hydrogen is moving faster than the Venusian escape velocity and is lost into space. Over the course of the past 4.5 billion years, this mechanism has removed all but a tiny trace of Venus's original inventory of water. Indeed, as it turns out, Earth might also have lost a great deal of water by this mechanism.* Specifically, at 6,000:1, the ^1H to ^2H ratio of Earth is one-seventh that of Jupiter, which, because Jupiter is so massive, is not thought to have lost any hydrogen via Jeans escape, suggesting that its ratio reflects the primordial ratio of the protoplanetary disk.

Conclusions

Our planet formed in the inner reaches of the solar nebula, and thus like all terrestrial planets it is composed predominantly of refractory metals and silicates. These, however, are not the materials of which life is made. Those we owe to Jupiter, whose mighty bulk tossed icy, volatile-rich material from the outer Solar System inward to the nascent Earth. This cosmic cleanup also saved us from later catastrophic, planet-sterilizing impacts, and thus Jupiter played a significant role not only in life's origins but also in providing the billions of years of stability required for complex life to evolve. But even with these favorable conditions helped along by Jupiter, the development of an Earth-like planet can easily go off toward a state that does not support a diverse biosphere, as the example of Venus shows (see also sidebar 3.2).

The impact of Shoemaker-Levy 9 was an example of the final vestiges of this cleanup. It also marked the pinnacle of the long and successful career of Eugene Shoemaker, who, sadly, was killed a few years later in an auto accident while exploring impact craters in Australia's outback. In a fitting tribute to this well-liked and extremely influential planetary scientist, a small sample of his ashes was placed on the *Lunar Prospector* orbiter that was launched not long after his death and that later confirmed suspicions that the deeply shadowed craters on the Lunar poles contain hydrogen, no doubt as water ice. On July 31, 1999, *Lunar*

*Although the slightly higher gravity and lower upper-atmosphere temperatures of the Earth slow the escape of gases, it definitely happens here too. Because it does not get "trapped" in higher molecular weight molecules, helium, is a simpler case. The Earth's crust outgases about 3,000 tons of helium per year (produced by the radioactive decay of uranium and thorium), yet the total amount of helium in our atmosphere is only 6 billion tons. This suggests a mean residence time of just 2 million years (6,000,000 tons ÷ 3,000 tons/year) despite the fact that helium is four times heavier than atomic hydrogen and thus moves twice as slowly.

Sidebar 3.2

Weighing the Probabilities

So what are the chances of everything coming together just right to form a habitable planet? The question of the frequency with which suitable stars host planetary systems is covered in chapter 9. But this is a reasonable place to ask: of all the potential solar systems out there, what fraction includes planets capable of supporting the formation and evolution of life? We simply do not know.

One of the issues is how frequently rocky, terrestrial planets form. Again, we discuss this in detail in chapter 9, but for now let's just point out that the first successful planet-finding techniques were limited to the detection of gas giant planets. Only since 2012 have we begun to get a handle on how often Earth-sized planets orbit in the habitable zones of suitable stars. Moreover, we need to know something even more specific: what fraction of all planetary systems include both Earths *and* Jupiters? The problem is that, as noted in this chapter, without Jupiter to clean out the newly formed Solar System, the late heavy bombardment would have continued effectively unabated, with dinosaur-killing-sized impacts happening once a century or so, and larger, planet-sterilizing impacts happening a few times each billion years. How frequently Sun-like stars host both Earth-like and Jupiter-like planets remains an open question.

And if we're interested in the evolution of higher life, a second element comes into play: the Moon. Due to our planet's rotation, centripetal force deforms the Earth from a perfect sphere—Earth has a distinct bulge at the equator. The Sun's gravity pulls on this bulge and, over the course of millions of years, should be able to very significantly shift the Earth's rotational axis. If this force ever succeeded in pushing our planet's axial tilt much below its current value of 23.5°, the seasons would run amok. The Moon, however, weighing in at a quite reasonable 1.25% of the mass of the Earth and residing a mere 384,000 km away, exerts a similarly large force on the Earth's bulge. Because the Moon's pushing and shoving happens over a different time frame from the Sun's pushing and shoving, the Moon prevents the Sun's perturbations from building up and keeps our axial inclination at a nice, mild level. Were it not for this effect, the Earth's climate would fluctuate wildly over the course of a few tens of millions of years. And while this would not necessarily have prevented the formation of life, it could well have made the origins of life that much less likely.

Evidence for these putative fluctuations can be seen on our neighboring planet Mars, where thickly layered terrains near the poles are thought to represent massive climatic fluctuations arising from the wandering of the planet's axial tilt. Although its current axial tilt is an Earth-like 25.2°, Mars lacks large moons (its moons, Phobos and Deimos, are but a few tens of kilometers across), and therefore the Sun's persistent, small torques build up over time, causing the Martian poles to make massive excursions. Jack Wisdom of the Massachusetts Institute of Technology and Jacques Laskar of the Institut de Mécanique Céleste et de Calcul des Ephémérides and the Paris Observatory have independently shown that, under the influence of these torques, the axial tilt of Mars

(*continued*)

Sidebar 3.2 *(continued)*

migrates randomly over tens of millions of years from 10° to a devastating (for climate and the evolution of advanced life) 50°.

So, without our massive Moon, our climate would be unstable and life would be that much less likely to have formed and thrived here. And how improbable was the formation of our Moon? Again, we can only speculate, but the answer may well be that it was very improbable. The Moon is thought to have formed through an off-center collision between the accreting proto-Earth and a Mars-sized object. The frequency of such collisions is probably low; after all, none of the other three rocky planets have large moons. On the other hand, Charon, one of the three moons of icy Pluto, is a whopping 15% as massive as the dwarf planet (or planet—your choice!) it orbits, suggesting that the probability of such impacts, while low, may not be astronomically low.

Touma, Jihad, and Jack Wisdom. "The Chaotic Obliquity of Mars." *Science* 259, no. 5099 (1993): 1294–97.

Prospector was intentionally crashed into a crater at the Moon's north pole in an attempt to toss up a plume of dust and steam. (Although telescopic observations of this event from the Earth failed to spot the water, it was eventually confirmed during a similarly intentional crash by the *LACROSS* mission in 2009.) With this impact, Eugene Shoemaker, one of the most powerful and eloquent advocates of the benefits of astrogeological research, became the first human interred on another world.

Further Reading

The Origins and Evolution of the Earth

Lunine, Jonathan. *Earth.* Cambridge: Cambridge University Press, 1999.

The Improbability of Habitable Worlds

Ward, Peter, and Donald Brownlee. *Rare Earth.* New York: Copernicus, 2000.

Eugene Shoemaker Biography

Levy, David. *Shoemaker.* Princeton, NJ: Princeton University Press, 2000.

The Origins and Evolution of Solar Systems

Lewis, John S. *The Chemistry and Physics of the Solar System.* New York: Academic Press, 1995.

Chapter 4

Primordial Soup

In 1951, Melvin Calvin (1911–1997), a professor at UC Berkeley, who, some years later, would win the Chemistry Nobel Prize for his studies of photosynthesis (more in chapter 7), published a paper in the journal *Science* that opened with the following: "The question of the conditions under which living matter originated on the surface of the earth is still a subject limited largely to speculation. The speculation has a greater chance of approaching the truth when it includes and is based upon the ever-wider variety of established scientific fact. One of the purposes of the observations reported herein is to add another fact that might have some bearing on this interesting question."* The "another fact" his paper added? It was that upon bombarding aqueous solutions of carbon dioxide with highly ionizing radiation (helium nuclei accelerated to near the speed of light in the Berkeley cyclotron) in the presence of an iron catalyst, up to 22% of the carbon dioxide was converted into formic acid (HCO_2H) by the reactive hydrogen species the radiation produced.

About the same time Calvin was conducting his work, Stanley Miller (1930–2007), then a graduate student at the University of Chicago, was in the audience when Harold Urey (1893–1981), who had won the 1934 Nobel Prize in Chemistry for his pioneering work on separating atomic isotopes, presented his thoughts regarding the chemical processes associated with the formation of the Solar System and how this chemistry might have driven and constrained the origins of life on Earth. Approaching Urey in 1952, Miller proposed experiments aimed at simulating the chemistry of the early Earth. At first Urey

*W. M. Garrison et al., "Reduction of Carbon Dioxide in Aqueous Solutions by Ionizing Radiation," *Science* 114, no. 2964 (1951): 416–18.

discouraged him, believing that a process that took geological eons might be a bit hard to re-create during the five or six years one typically spends earning a PhD. But Miller persisted and, in the end, triumphed.

Urey's, Calvin's, and Miller's ideas were built on a hypothesis put forward in the early 1920s by the Soviet scientist Aleksandr Oparin (1894–1980) and, slightly later, independently proposed by the Scottish scientist J. B. S. Haldane (1892–1964).* Both had theorized that life arose through the slow "evolution" of chemical systems of increasing complexity. Haldane, specifically, postulated that, acting on an atmosphere of carbon dioxide, ammonia, and water vapor, UV light from the Sun could create key components of life, such as amino acids and sugars, which would have "accumulated till the primitive oceans reached the consistency of hot, dilute soup."** In their own speculations on the process, Miller and Urey concluded that the early atmospheres of terrestrial planets—like the current atmospheres of Jupiter and Saturn—were made up of methane, ammonia, and hydrogen. Clearly, too, there would have been oceans with water still warm from the accretion of the planet, and thus the skies would have been filled with water vapor, clouds, and rain.

Miller built for his studies an apparatus consisting of two connected glass reservoirs (fig. 4.1). In the smaller, lower reservoir he placed water as a mimic of the primordial ocean. He filled the larger, upper sphere with methane, ammonia, and hydrogen—Urey's hypothesized primordial atmosphere. Miller placed two electrodes in this "atmosphere," through which he passed a high-voltage discharge mimicking lightning, in place of the UV light that Urey and Haldane had originally suggested was the likely energy source and the ionizing radiation that Calvin had employed. The connections between the two containers were such that, when heat was applied to the "ocean," water vapor would rise into the lightning-filled atmosphere. A second tube connected the atmosphere and the ocean by way of a water-cooled con-

*Haldane is also known for his succinct and illuminating description of the four stages of acceptance that any new scientific theory goes through: (1) this is worthless nonsense, (2) this is an interesting, but perverse, point of view, (3) this is true, but quite unimportant, (4) I always said so.

**J. B. S. Haldane, "The Origin of Life," *Rationalist Annual* 148 (1929): 3–10.

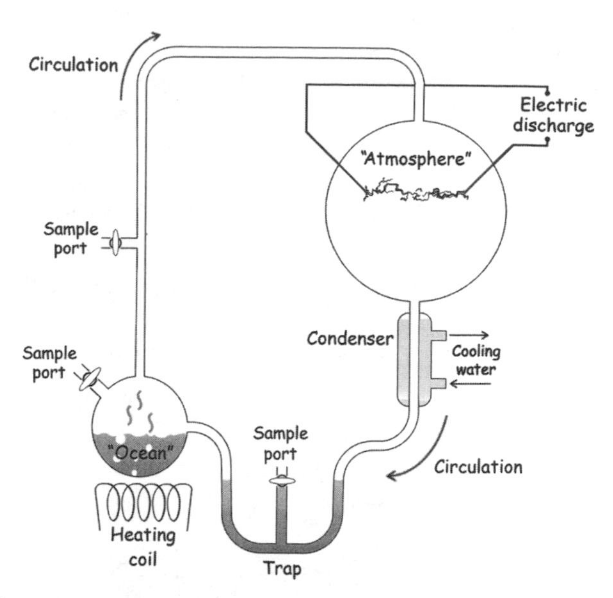

Figure 4.1 In 1953, Stanley Miller set up an experimental apparatus to simulate the conditions on the primordial Earth. After the experiment had been running just a few days, the "ocean" in the lower sphere became discolored and the "atmospheric" chamber (the upper sphere) became coated with a brown tar. Within a week, almost all the carbon originally introduced as methane had become fixed into larger, more complex molecules, including many of the amino acids on which present-day Earth life is built.

denser, with the condensed water mimicking rain and setting up a simple hydrological cycle.

Legend has it that Miller sat by his experiment day and night for a week while the simulated lightning crackled and the faux ocean boiled, condensed, and fell back as "rain." He didn't have to wait long to see something happen: after just a day or two, the once-clear liquid became yellowish and then brown, and the discharge chamber became coated in an increasingly thick tar. After five days, Miller removed samples from the "ocean" for analysis. In a striking confirmation of his hypothesis, he found the water in his reaction vessel was now a rich solution of higher molecular weight carbon compounds accounting for 10% to 15% of the total carbon he had originally put in as methane. Moreover, this mixture contained several percent amino acids, including many that were known to be the building blocks of our proteins (fig. 4.2; see also sidebar 4.1). Under what seemed to be plausible prebiotic condi-

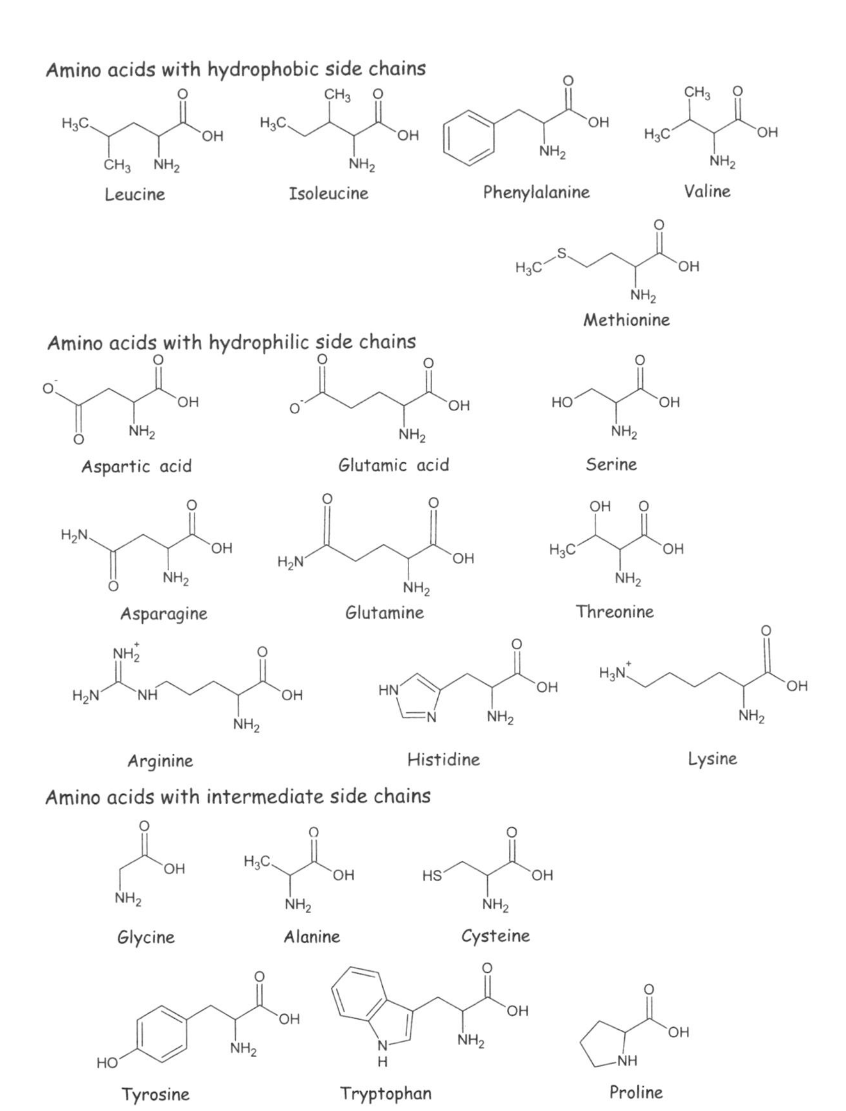

Figure 4.2 Amino acids are small organic molecules containing a carboxylic acid group ($-CO_2H$) and an amino group ($-NH_2$). The α-amino acids, shown here, are a subclass of amino acids in which the amino group is one carbon removed from the carboxylic acid group (i.e., placed on the carbon "alpha" to the carboxylate carbon). The α-amino acids play fundamental roles in Terrestrial biochemistry, both as important intermediates in metabolism and as the monomers out of which proteins are built. The 20 α-amino acids shown here, which are classified by their solubility in water (their "hydrophilicity"), are the amino acids from which our proteins are synthesized. (Note that the acidic amino acids, glutamic and aspartic acids, are often referred to as their negatively charged ions, glutamate and aspartate.) The unlabeled vertices in the structural formulas in this and later figures represent carbon atoms bound to enough hydrogen atoms to ensure that each participates in four bonds.

Sidebar 4.1

An Earth Biology Primer

An intellectual hurdle that the field of astrobiology faces is its extremely limited data set: we have only a single example of biology to study—as we discuss in chapter 6, every living thing on Earth arose from a single, fairly sophisticated, common ancestor. Given this, our views on the range of the possible must be at least somewhat suspect. Still, astrobiology being based on one example is a clear improvement over the earlier concept of exobiology, which had no examples at all: Earth life provides a tangible source of insights that we will constantly return to during our broader discussion of life in the Universe. Given this, we're going to need everyone to be up to speed with the basics of our own chemistry.

Contemporary life on Earth is based on deoxyribonucleic acid (DNA), ribonucleic acid (RNA), and proteins. In cellular life, DNA serves as the genetic material and contains the information necessary to synthesize the RNA and proteins; however, some viruses, including the coronavirus that causes COVID-19, eschew DNA and use RNA to encode their genetic information. Proteins, and to a far lesser extent RNA, are the catalysts that perform the myriad of chemical and mechanical functions required for us to thrive and reproduce. They are, in effect, the tools that our genes have invented to ensure that they—the genes—reach the next generations. (Like it or not, every aspect of our biology is focused on ensuring that our genes are handed down to our descendants.)

DNA and RNA are long, information-containing molecules in which many nucleotide building blocks (monomers) are linked together by strong, covalent bonds to form linear chains called polymers. Each nucleotide in DNA or RNA consists of a sugar (ribose in RNA, and deoxyribose, a modified ribose missing one oxygen atom, in DNA) linked to the adjacent monomers in the chain via a phosphate group. Hanging like a pendant from the sugar is one of four nucleobases: adenine (A), guanine (G), cytosine (C), or thymine (T) in DNA and the same in RNA with the exception that thymine is exchanged for the closely similar uracil (U). It is the sequence of these nucleobases, lined up along the polymer chain, that encodes the information necessary for DNA to make RNA and for RNA to make proteins. The bases occur in complementary pairings: adenine binds weakly but specifically to thymine (or uracil), and guanine to cytosine. This allows two complementary polymer strands to come together to form a double helix. It also provides a means of copying one strand of DNA into another during replication or into a strand of RNA when it is needed to synthesize a protein.

Proteins are also polymers, but their building blocks are amino acids rather than nucleotides. An amino acid (more specifically, an α-amino acid) is an organic molecule containing an amino group ($-NH_2$) on the carbon (called the α carbon) adjacent to a carboxylic acid group ($-CO_2H$). Polymers of amino acids are generated by fusing the amino group of one amino acid to the carboxylic acid group of another to form a peptide bond with the following structure:

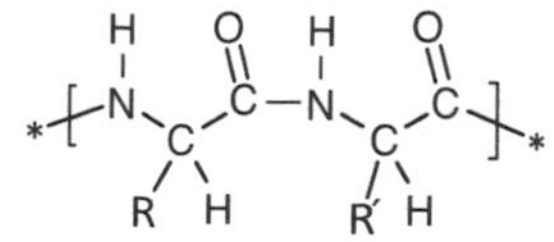

```
       H    O    H    O
       |    ‖    |    ‖
  *-[-N    -C-N    -C-]-*
        \ /      \ /
         C        C
        / \      / \
       R   H    R'  H
```

(*continued*)

Sidebar 4.1 *(continued)*

As the two functional groups involved in the polymerization are directly adjacent in the amino acid (both are on the α carbon), only the nitrogen and the two carbons closest to it form the backbone of the protein polymer. The rest of the amino acid molecule, called its side chain, is designated by the chemist's shorthand "R," which can represent any combination of atoms. Twenty different "proteinogenic" amino acids are commonly found in proteins on Earth (see fig. 4.2), with their 20 different side chains ranging in complexity from a single hydrogen atom to much larger chains and rings and spanning a wide variety of chemistries. Under the influence of selective pressures, evolution has come up with billions of different specific sequences of amino acids (around 25,000 different sequences are employed in us humans), each of which folds into a unique, three-dimensional shape that is defined by interactions between the side chains. Because proteins are, in effect, tools—they catalyze almost all the chemical and mechanical actions in a cell—these three-dimensional structures are critical to their activity. As with all tools, form defines function.

tions, Miller had shown that even the simple single-carbon, single-nitrogen, and single-oxygen inputs of methane, ammonia, and water would spontaneously form many of the most fundamental chemical components of life on Earth.

Miller's "primordial soup" study was hailed on the front page of the *New York Times* as a breakthrough that would quickly lead to a deep understanding of our origins. His result, in some regards the second part of a one-two punch following elucidation of the structure of DNA published just a few short months earlier, produced tremendous optimism that the "origins" question would soon be answered. Alas, though, while the Miller-Urey experiment provides fascinating and tantalizing insights into possible prebiotic chemistry on the early Earth and elsewhere, it seems that some of this optimism was misplaced. A good fraction of a century later, many fundamental questions regarding prebiotic chemistry—and even more about our origins—remain unresolved.

The Volatile Inventory

Miller-Urey chemistry requires an atmosphere and an ocean, both of which are composed of materials far more volatile than the silicates and metals that make up the bulk of terrestrial planets. Thus, the origin of the "volatile inventory," which we discussed in chapter 3, plays a fundamental role in defining the relevant prebiotic chemistry of a planet. And, as we'll see, the *chemical form* of the volatiles is equally critical to our story.

Applying a cursory knowledge of chemistry to a table of the cosmological abundances of the elements (see fig. 1.2) suggests some likely candidate volatiles. Hydrogen is, of course, far and away the most common element in the Universe, but carbon, oxygen, and nitrogen are also relatively plentiful. Looking over likely (i.e., more chemically stable) "permutations" of these atoms, we can come up with a list of the chemically reasonable atmospheres of young terrestrial planets (table 4.1).

Look at table 4.1 closely and you'll see that the likely volatile compounds of hydrogen, carbon, nitrogen, and oxygen are grouped into related columns. The relationship by which the compounds are organized is the degree to which they are *oxidized*—that is, the degree to which the core elements in these compounds have given up control over the electrons in a shared chemical bond, typically with an element such as oxygen (thus the term *oxidized*) that attracts the shared electrons more strongly. Conversely, if an atom forms a bond with hydrogen, it pulls in a larger share of the bonding electrons and is thus *reduced*, which is the opposite of oxidized.

On the left side of table 4.1 are the fully reduced compounds, such as ammonia (NH_3), in which the central atom (nitrogen in this case) has electron density to spare and can therefore reduce other compounds, albeit at the price of becoming oxidized itself. In an environ-

Table 4.1
Volatile "sets" potentially available for young terrestrial planets

	Reduced	Oxidized	Oxic
Carbon (C)	CH_4	CO, CO_2	CO_2
Nitrogen (N)	NH_3	N_2	N_2
Oxygen (O)	H_2O	H_2O, CO, CO_2	O_2
Hydrogen (H)	H_2, CH_4, NH_3, H_2O	H_2O	H_2O

ment dominated by reducing substances, nitrogen would be most likely to occur as ammonia, and carbon as methane (CH_4). In the center column of the table are more neutral (meaning less reduced or, equivalently, more oxidized) materials, such as molecular nitrogen (N_2), which are generally unwilling to give up or take on electrons and thus are rather chemically inert. Finally, at the far right we have the equilibrium mixture we would observe in the presence of free oxygen, which is very oxidized indeed: oxygen is the second most *electronegative* atom in the periodic table, which means that, in a bond with any other atom except fluorine (the only element that is more electronegative), oxygen will attract the shared electrons more strongly than does its binding partner. An atmosphere that is so strongly oxidized as to contain free oxygen is called an *oxic* atmosphere, a term that describes the present atmosphere of our planet. But the free oxygen in our atmosphere is due to photosynthesis, so the primordial atmosphere clearly lay to the left of the oxic column on the spectrum of possibilities shown in the table. The question we face in speculating on the origins of life is how far toward the left-hand, highly reduced end of this spectrum does the atmosphere of a newborn terrestrial planet typically lie?

Miller-Urey Chemistry and the Early Earth

Although the Earth provides the best storehouse of information on the evolution of terrestrial planets, we have no direct record of the conditions prevalent during our planet's earliest era; given Earth's extraordinarily active geology, no rocks (and only a few small crystal grains) are known that unambiguously date from her first half billion years. Thus, we are left to speculate on the conditions present immediately after the accretion of a terrestrial planet.

Urey was one of the first to consider this. He reasoned that, while the crust and mantle of the terrestrial planets consist of fairly oxidized silicates, their *bulk* composition is *reducing*. That is, they are primarily composed of atoms and molecules that tend to give up electrons in chemical reactions, typically by forming bonds to very electronegative elements such as oxygen. The majority of the Earth's mass, for example, is contained in its core, which is metallic iron and nickel. Free metals such as these are reduced. (Indeed, the word *reduced* originates from compounds that could *reduce* ore into free metal; we now know this occurs because the reducing compound provides electrons to the ores, which are typically oxides, and thus oxidized, to produce the electron-rich free metal.) If the process by which the Earth differentiated into a dense metallic core and rocky mantle and crust was slow, the metal core-forming elements would equilibrate with the rocky elements in the crust and mantle, leaving the crust partially reduced at the expense of oxidizing some of the iron in its core.

If, per Urey's logic, the crust and mantle were reduced, then the ocean and atmosphere would be too. Even today, at age 4.5 billion years, the Earth is still geologically active enough to regularly cycle its volatile inventory into and out of the mantle. For example, the march of plate tectonics forces volatile-rich crustal plates down into the mantle at subduction zones, such as the Mariana Trench, where geothermal heat liberates the volatiles and sends them back up to belch out of volcanoes in a cycle that, were it not for the rapid production of oxygen by photosynthesis, would ensure that the Earth's atmosphere remains more or less in chemical equilibrium with its crust.

In contrast to Urey's speculations, however, the minerals present in the oldest rocks still to be found on Earth suggest that they formed in a relatively oxidized (though not *oxic*) environment. This, many geologists argue, implies that the dense metallic iron and nickel sank to form the core too quickly for chemical equilibration with the mantle to occur, and thus the Earth's mantle and crust could have been, as they are today, composed of relatively oxidized silicates. Were this the case, volcanic outgassing would have led to a more oxidized atmosphere than would be expected from the bulk composition of the planet. Although without a good rock record of the first half billion years, we cannot tell which of these two scenarios came to pass, as we discussed with respect to Venus in chapter 3, photolysis does tend to oxidize terrestrial planets. Given this, while the rock record suggests that the Earth's atmospheric composition was nearer the middle, neu-

Table 4.2
The more common Miller-Urey reaction products formed under reducing conditions

Compound	Yield (% total fixed carbon)	Compound	Yield (% total fixed carbon)
Formic acid	4.0	Succinic acid	0.27
Glycine	2.1	Sarcosine	0.25
Glycolic acid	1.9	Iminoacetic propionic acid	0.13
Alanine	1.7	*N*-methylalanine	0.07
Lactic acid	1.6	**Glutamic acid**	0.05
β-Alanine	0.76	*N*-methylurea	0.05
Propionic acid	0.66	Urea	0.03
Acetic acid	0.51	**Aspartic acid**	0.02
Iminodiacetic acid	0.37	α-Aminoisobutyric acid	0.01
α-Hydroxybutyric acid	0.34		
α-Amino-*n*-butyric acid	0.34	Total	15

Note: Proteinogenic amino acids shown in bold type.

tral part of the spectrum from reduced to oxic by the time the planet was a billion years old, this is perhaps not proof that, a half billion years earlier, the primordial atmosphere was similarly oxidized.

In the decades since Miller's initial work, his experiment has been repeated many times with many different atmospheric compositions and energy sources, including Miller's electric discharge (simulating lightning), hot rocks (simulating volcanism), UV light (simulating sunlight), and ionizing radiation (simulating radioactive minerals and cosmic rays). As long as the experimental atmosphere is, as was true for the one Miller employed, fairly reducing, all these variations produce about the same results: a soup of amino acids and other small, simple organic compounds corresponding to about 10% to 15% of the total carbon input into the experiment, with the remaining carbon primarily ending up in a complex, high molecular weight tar lining the apparatus (typical reaction products are listed in table 4.2).

This is a pretty exciting observation, no? But it is not without its detractors. Historically, it was found that unless methane and ammonia or methane, nitrogen, and molecular hydrogen are present, Miller-type experiments produce relatively little in the way of biologically relevant small molecules. And, as noted above, geochemists have argued that the early Earth's atmosphere was not as reduced as Urey had thought. Fortunately, however, Jeffrey Bada,* a geochemist at Scripps Institution of Oceanography near San Diego, California, has since demonstrated that, under some conditions, a relatively oxidized carbon dioxide and nitrogen atmosphere also supports Miller-Urey chemistry. Starting from this more oxidized atmosphere, Miller-Urey reactions create acids, which lower the pH of the "ocean" and reduce amino acid yields, and nitrites, which destroy what few amino acids are produced. These effects can be countered, however, by adding soluble reducing agents to remove the nitrites and carbonates to buffer the pH. For example, performing the Miller-Urey experiment using a carbon dioxide, nitrogen, and water vapor atmosphere in the presence of iron and calcium carbonate, Bada's students obtained amino acids in modest yields.

Take a reduced (or even neutral), water-rich atmosphere, pass some lightning through it, sit back, and what do you get? Most of the amino acids that form our proteins (as shown in fig. 4.2). Pretty neat, eh? And, come to think of it, really all you need to know. But if you're willing to hitch up your trousers and slog through a bit of chemistry, the story gets a little richer—and more complex—than we've let on so far.

Miller-Urey Mechanisms

Amino acids are less stable than a mixture of methane, ammonia, and water, and thus energy must be invested in order to synthesize them. In Miller's experiment, the energy source was the lightning-like electric discharge, though, as we've noted, many other energy sources are likewise sufficient. In the case of the electric spark, high-energy electrons in the discharge tear other electrons out of the gas molecules, creating molecular species such as methyl and hydroxyl "radicals" that

*As a former student of Miller's, Bada is the intellectual "grandchild" of Urey.

contain an odd number of electrons (and thus electrons not paired up with partners). As the main driving force of chemical reactions is pairing up electrons to fill up orbitals, radicals are usually unstable and extremely reactive. For example, the hydrogen radical ($^{\bullet}H$, where the dot represents the unpaired electron) can attack methane to produce hydrogen gas (H_2) and the more stable methyl radical ($^{\bullet}CH_3$). The methyl radical, in turn, reacts with water to produce another molecule of hydrogen and the still more stable methoxy radical ($CH_3O^{\bullet}$). Of course, though, it doesn't stop there: the methoxy radical can undergo a number of other reactions, including joining with a hydrogen radical to produce methanol (CH_3OH) or reacting with a hydrogen radical to produce hydrogen gas and formaldehyde ($H_2C{=}O$). Alternatively, the methyl radical can react with another carbon radical to produce larger, more complex carbon compounds, or it can bond with ammonia in the first step of a process that ultimately produces hydrogen cyanide ($HC{\equiv}N$). The theme here is that reaction of a radical with a stable "even-number-of-electrons" molecule, such as water, always produces another radical (with its odd number of electrons), and thus radical chemistry is characterized by chain reactions that build progressively more stable radicals until a second radical is encountered, ending the cascade. Moreover, as long as they remain in the gaseous atmosphere, the larger molecules produced in these cascades also fall prey to the electric discharge (or photolysis), forming still more complex radicals and still larger, more complex compounds. A smattering of some of the radical reactions thought to occur in the discharge is shown in fig. 4.3; similar "radical chain reactions" occur in flames and are a central feature of the chemistry of combustion (see fig. 1.1).

Hints as to the mechanism by which Miller-Urey chemistry forms amino acids from simpler precursors can be obtained by monitoring the concentration of various molecular species as the reaction proceeds (fig. 4.4). For example, the first 24 hours see a rise in the concentrations of both the simplest nitrogen-containing organic molecule, hydrogen cyanide, and the simple oxidized carbon compounds called aldehydes (containing an $-HC{=}O$ grouping of atoms). After a few more days, however, the concentrations of these species fall to near

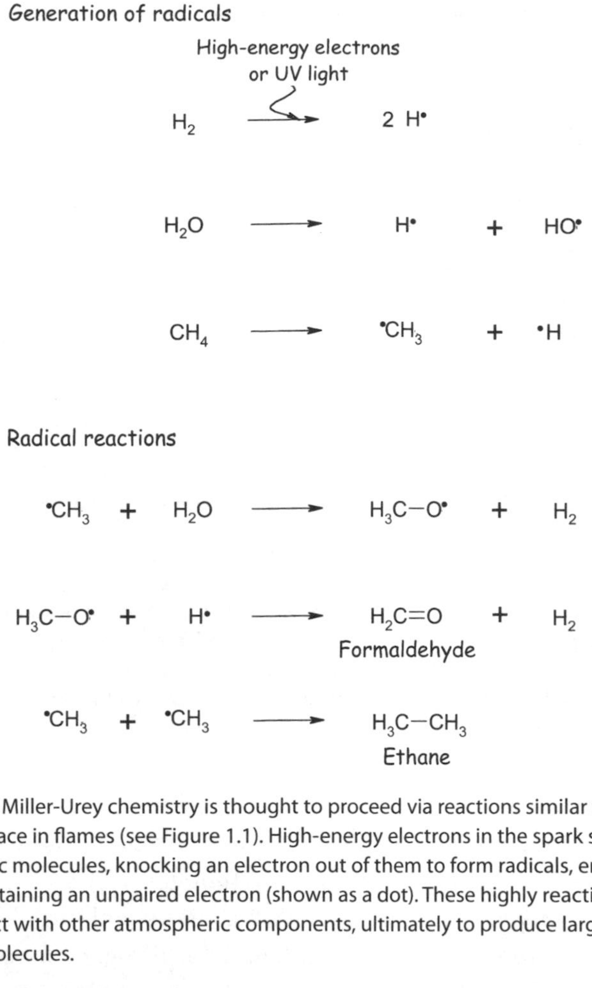

Generation of radicals

High-energy electrons or UV light

$$H_2 \longrightarrow 2\ H^\bullet$$

$$H_2O \longrightarrow H^\bullet + HO^\bullet$$

$$CH_4 \longrightarrow {}^\bullet CH_3 + {}^\bullet H$$

Radical reactions

$${}^\bullet CH_3 + H_2O \longrightarrow H_3C{-}O^\bullet + H_2$$

$$H_3C{-}O^\bullet + H^\bullet \longrightarrow H_2C{=}O + H_2$$

Formaldehyde

$${}^\bullet CH_3 + {}^\bullet CH_3 \longrightarrow H_3C{-}CH_3$$

Ethane

Figure 4.3 Miller-Urey chemistry is thought to proceed via reactions similar to those that take place in flames (see Figure 1.1). High-energy electrons in the spark strike atmospheric molecules, knocking an electron out of them to form radicals, energetic species containing an unpaired electron (shown as a dot). These highly reactive radicals quickly react with other atmospheric components, ultimately to produce larger, more complex molecules.

zero. In contrast to this rise-from-zero-then-fall-to-zero behavior, the concentration of ammonia—which, as one of the starting materials, is initially at a very high concentration—falls steadily as the reaction proceeds, and the concentration of amino acids rises steadily for several days before reaching a plateau. These trends suggest that hydrogen cyanide and simple aldehydes are intermediates in a reaction that, in net, "fixes" ammonia into amino acids. A century before Miller's experiment, the German organic chemist Adolph Strecker (1822–1871) had reported a synthetic route to the formation of amino acids

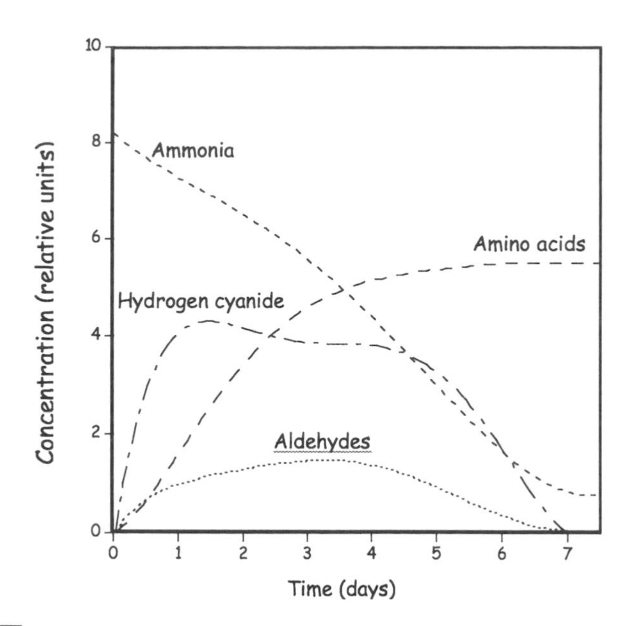

Figure 4.4 Insight into the mechanisms of Miller-Urey chemistry is provided by the time course of the consumption of ammonia (NH_3) and formation of hydrogen cyanide (HC≡N), aldehydes (R—HC·O), and amino acids. The peak in the concentration of hydrogen cyanide and aldehydes at intermediate times suggests that these molecules are waypoints in a multistep reaction that converts small-molecule precursors into larger and more biologically relevant amino acids.

that similarly employed hydrogen cyanide, ammonia, and aldehydes (fig. 4.5). Miller-Urey chemistry is now thought to be a variant of this "Strecker synthesis," in which ammonia reacts with an aldehyde to produce an imine, a molecule containing a nitrogen-carbon double bond. The carbon in this bond is relatively susceptible to attacks from reactive species such as cyanide ($^{-}$CN), with which it forms an α-aminocyanonitrile. The cyanonitrile group (—C≡N) is prone to hydrolysis, which is the breaking of a bond through its reaction with one or more water molecules. Specifically, the addition of two molecules of water (and the subsequent loss of a molecule of ammonia) causes conversion of cyanonitrile into carboxylic acid (—CO_2H), forming, in this case, an amino acid in the process.

Around 4% of the total carbon input (as methane) into a typical Miller-Urey experiment is converted into amino acids. Miller, using

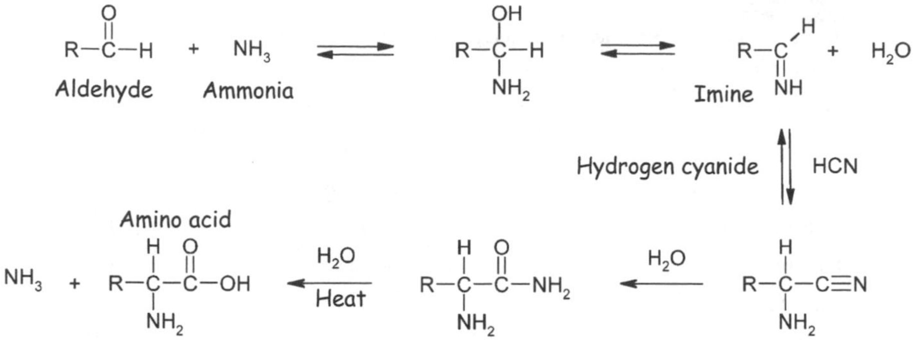

Figure 4.5 The Strecker synthesis is thought to parallel the mechanism by which amino acids are synthesized in the Miller-Urey experiment and perhaps to be the source of the amino acids delivered to Earth by the carbonaceous chondrite meteorites (discussed later in the chapter).

the analytical techniques of the time, identified five of these, three of which are among the 20 amino acids employed by life on Earth to make proteins. Follow-on experiments conducted decades later on Miller's original samples identified more; we now know that 10 protein-forming (or "proteinogenic") amino acids are produced under Miller's conditions (listed in table 4.3) and that four more can be synthesized using slight "tweaks" on Miller's original conditions. The 10 proteinogenic amino acids not found under Miller's conditions include the aromatic amino acids phenylalanine, tyrosine, and tryptophan, which are characterized by their large, bulky, ring-containing side chains; the positively charged amino acids histidine, lysine, and arginine; the sulfur-containing amino acids cysteine and methionine; and the amide-containing amino acids asparagine and glutamine (amide group $-CO-NH_2$). The latter four omissions, however, are technical rather than fundamental. For example, because Miller did not include any sulfur compounds in his reaction mixture, the lack of the two sulfur-containing amino acids is not surprising; when hydrogen sulfide (H_2S) is included in the reaction mixture, both of these amino acids are produced. Likewise, the boiling "ocean" in Miller's experiment destroyed the two amide-containing amino acids. The prebiotic synthesis of the other six proteinogenic amino acids, in contrast, seems a more fundamental problem. Those amino acids were almost

Table 4.3
Yields of the α-amino acids in a typical, hydrogen sulfide–free Miller-Urey experiment

Amino acid	Yield (μM)	Amino acid	Yield (μM)
Glycine	440	Norleucine	6
Alanine	790	**Isoleucine**	5
α-Aminobutyric acid	270	**Serine**	5
Norvaline	61	Alloisoleucine	5
Aspartate	34	Isovaline	5
α-Aminoisobutyric acid	30	**Proline**	2
Valine	20	**Threonine**	1
Leucine	11	Allothreonine	1
Glutamate	8	*Tert*-Leucine	0.02

Note: Proteinogenic amino acids shown in bold type.

certainly introduced by *biochemistry* after the origins of life. For the aromatic amino acids, the reason may be the difficulty in synthesizing the aromatic ring using gas-phase chemistry such as that seen in the Miller-Urey experiment.

The Prebiotic Synthesis of Sugars

Sugars are called carbohydrates because their elemental composition is that of a carbon atom combined with a water molecule: $C_n(H_2O)_n$, where n is in the range of three to seven (fig. 4.6). The pentoses, a family of sugars for which $n = 5$ (e.g., ribose), and the hexoses, for which $n = 6$ (e.g., glucose), are particularly important in biology on Earth. In defiance of the historical name, however, the hydrogen and oxygen atoms in carbohydrates do not take the form of water molecules (i.e., sugars are not "hydrates" of carbon). Instead, sugars are polyalcohols in which half of the hydrogens (plus one) are bound directly to a carbon and the other half (minus one) are bound to oxygen atoms to form alcohol groups (—OH). The "minus one" results from the fact that one carbon in every sugar molecule is bound to an oxygen by a double bond (C=O, where the oxygen has no binding

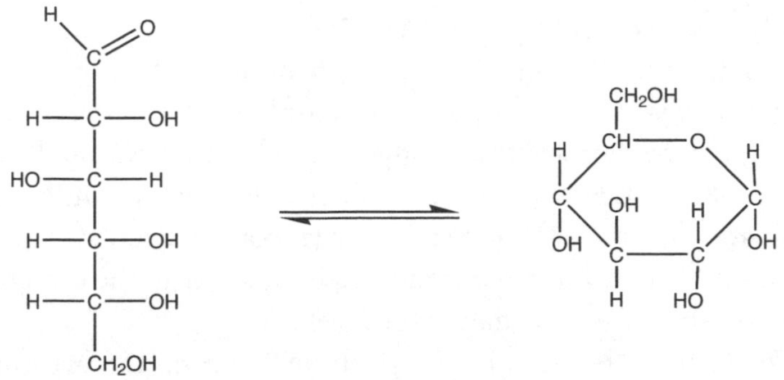

Figure 4.6 Sugars are polyalcohols of the general formula $C_n(H_2O)_n$. Shown here is glucose, an aldohexose (from *aldo*, designating the double-bonded oxygen at the end of the chain, and *hexose*, designating that $n = 6$). Sugars such as glucose typically exist in two or more interconverting forms. On the left is the linear form, and on the right is a cyclized conformation produced when one of the alcohol groups attacks the aldol carbon. For glucose, the cyclic form predominates in solution.

capacity left to accommodate a hydrogen) to form an aldehyde or ketone functional group, which is located at the end of or within the chain of carbon atoms, respectively.

The trouble with sugars is that, even the simplest, those containing just three carbons and three oxygen atoms, are complex enough that Miller-Urey chemistry does not produce them. That is, although Miller-Urey chemistry forms simple alcohols, aldehydes, and ketones, even the smallest sugar is simply too large to be synthesized in the atmosphere; long before the simplest sugar could be produced, the precursor molecule would rain out of the sky.* Sugar formation, then, must have occurred in the oceans or on land through polymerization of the smaller precursors that Miller-Urey chemistry can produce. How might that have happened?

In 1861, the Russian chemist Alexander Butlerov (1828–1886) described the formose reaction, which would later be hailed as the most plausible prebiotic route to the sugars. This starts with formaldehyde

*Consistent with this, the *Galileo* entry probe failed to detect molecules containing more than three carbon atoms when it sampled Jupiter's highly reduced, lightning-filled atmosphere. Thus our earlier argument that hydrogen-filled "blimpoids" are unlikely to be found living in the atmospheres of gas giant planets.

dissolved at a concentration of 1% to 2% in water under alkaline (basic) conditions. (Although formaldehyde itself is not a sugar, with its formula $H_2C{=}O$, it has the same "carbon-plus-water" composition as sugars.) After a short incubation, the reaction accelerates rapidly until, at its peak, up to 50% of the formaldehyde is converted into sugars. On further incubation, however, the sugar content decreases because the sugars that are formed are, ultimately, unstable in the presence of the strong base used to catalyze the reaction.

The formose reaction is thought to proceed via the two-carbon intermediate glycolaldehyde, which is formed from two molecules of formaldehyde (fig. 4.7). Glycolaldehyde is also a catalyst in the formose

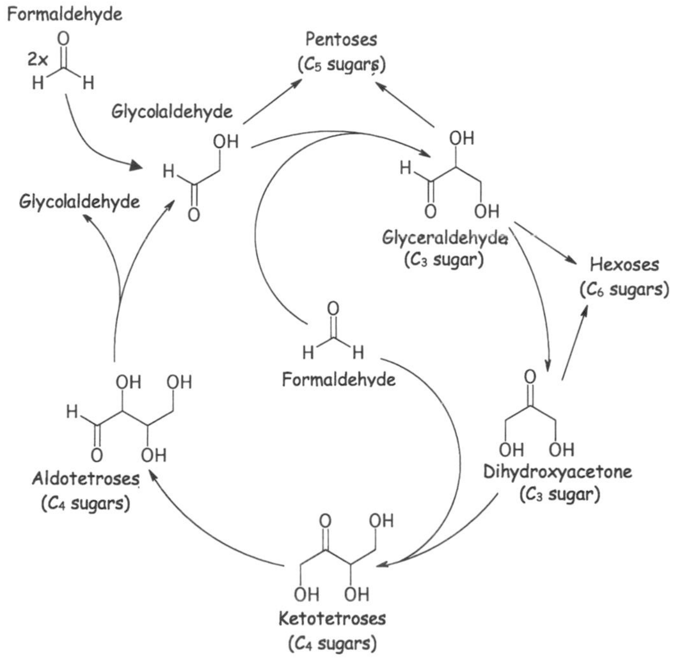

Figure 4.7 The base-catalyzed formose reaction converts one-carbon formaldehyde into higher molecular weight sugars. After the slow formation of the first molecule of glycolaldehyde from two molecules of formaldehyde (upper left), a reaction cycle is started that synthesizes both sugars and, with each turn of the cycle, *more glycolaldehyde*, which serves to accelerate the reaction in the next round. The reaction is catalyzed by reasonably plausible prebiotic catalysts, such as calcium hydroxide, but it is fairly nonspecific, producing a complex mixture of high molecular weight sugars.

reaction, and this autocatalysis accounts for the short delay observed before the reaction takes off; once a few molecules of glycolaldehyde have formed, they speed up the formation of yet more molecules of glycolaldehyde. Under alkaline conditions, the glycolaldehyde in turn reacts with a third molecule of formaldehyde to form the three-carbon sugar glyceraldehyde. Glyceraldehyde, in turn, can undergo a base-catalyzed isomerization reaction (a reaction that moves atoms around in a molecule but does not change its overall formula) that converts it into the other three-carbon sugar dihydroxyacetone. These two sugars, with their C=O functionality on an end carbon and middle carbon, respectively, are the only possible three-carbon sugars.

From glyceraldehyde and dihydroxyacetone, each only three carbons long, the formose reaction can produce many other larger and more complex sugars. For example, two of these three-carbon sugars can combine by an aldol condensation reaction (so named because it results in an *ald*ehyde-alcoh*ol*) to form most of the six-carbon hexoses. Alternatively, dihydroxyacetone can condense with another molecule of formaldehyde to form the four-carbon ketotetrose (four carbons, with C=O within the chain) erythrulose. Erythrulose, in turn, isomerizes to form the aldotetroses (also four carbons, but with the C=O at the end of the chain), which can undergo a reverse version of the aldol condensation, splitting the molecule into two glycolaldehydes and starting the cycle anew. Of note, the isomerizations and aldol condensations thought to occur in the formose reaction mimic the chemistry by which present-day Earth life produces sugars from smaller precursors. Whether this coincidence represents a historical artifact of prebiotic chemistry that was later hijacked by life or simply represents the most convenient chemistry by which sugars can be made (and thus the most likely chemistry for evolution to independently discover) remains an open question.

Although the formose reaction has some of the traits we are looking for—it is spontaneous and can produce high yields—it is not without its problems as an explanation for the prebiotic synthesis of sugars. The first problem is the formaldehyde. Not that it can't be made prebiotically; formaldehyde can be synthesized by the action of UV light on carbon dioxide and water in the atmosphere. Light with a wavelength of less than 240 nm (relatively far into the UV spectrum) splits water into the hydroxyl radical (•OH) and a free hydrogen atom. Similarly, short wavelength (and thus high-energy) UV light splits carbon dioxide into carbon monoxide and atomic oxygen, itself a potent radical generator. In the presence of carbon monoxide and water, the radicals formed in these processes generate formaldehyde, although ever so slowly. Current estimates suggest that, by this mechanism, it would take around 10 million years for formaldehyde to accumulate to the levels required for the formose reaction to proceed. Formaldehyde, though, is a fairly reactive species, and thus it seems unlikely that such a slow process could, at the end of the day, generate a high enough formaldehyde concentration to drive the reaction forward. Moreover, even if sufficient formaldehyde were available, the problem remains that the formose reaction is nonspecific and produces only small quantities of each of the many different types of sugars. In particular, the formose reaction produces very little ribose, which in many regards is the most fundamental sugar for life on Earth and, as we will see in the next chapter, is intimately coupled to the most promising theories of life's origins. Worse still, what little ribose is produced by the formose reaction is relatively unstable. Even at modest temperatures, the half-life of ribose in water is just decades. Given these issues, it is hard to imagine that much ribose could have built up simply from the formose-induced polymerization of formaldehyde in a primordial ocean.

Fortunately, a potential solution to at least this last "ribose" problem has been demonstrated in the laboratory. As a catalyst for the formose reaction, researchers have traditionally used calcium hydroxide, which is a fairly powerful base. Very little ribose accumulates under strongly basic conditions because the sugars produced further polymerize into a complex, brown goo. Steven Benner at the University of Florida has shown, however, that calcium borate, which occurs in nature as the minerals borax and colemanite, stabilizes sugars, such as ribose, in which two neighboring hydroxyls point in the same direction. It does so by forming a cyclic boron "ester"—a ring-shaped molecule with boron bridging the oxygens of the former alcohol groups. As this renders the sugar unreactive, formation of the ester terminates the formose reaction. Of course, while this pushes up the yield of ribose and a few other, structurally similar sugars, it does nothing to stabilize the ribose once the free sugar is released from the boron (the reaction is easy to reverse—just add water). And the chemical half-life of ribose remains quite short relative to the required lengths of time we'd expect for prebiotic chemistry to build up usable quantities of life's precursors.

The Prebiotic Synthesis of Nucleic Acids

Miller-Urey chemistry also fails when it comes to producing significant amounts of purine (two-ring) and pyrimidine (one-ring) nucleobases (fig. 4.8), important ingredients in life on Earth that serve as key components of DNA and RNA. In 1961, however, John "Juan" Oro (1923–2004) found that amino acids and small amounts of the purine base adenine form spontaneously from mixtures of hydrogen cyanide and ammonia in water. Following this, other researchers have explored Oro's chemistry in more detail and found that small amounts of several different nucleobases are detectable when solutions of ammonium

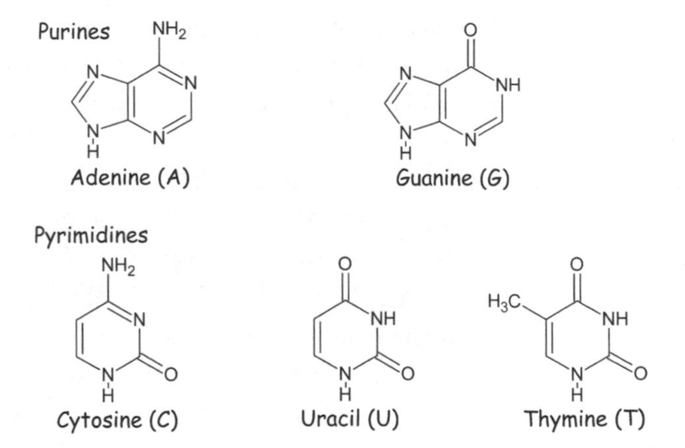

Figure 4.8 Five different nucleobases (often simply referred to as bases) are employed throughout Terrestrial biochemistry. Three of these, adenine, guanine, and cytosine, appear in both RNA and DNA. Rounding out the set we have uracil as the fourth nucleobase in RNA and the closely similar thymine as the fourth nucleobase in DNA.

cyanide, the compound formed when ammonia and hydrogen cyanide combine, are allowed to react and then boiled with acid.

Key to the chemistry of these reactions is the fact that hydrogen cyanide is a fairly reactive molecule and thus contains the energy necessary to drive a multistep prebiotic chemical pathway. Nevertheless, because four molecules of the compound are condensed in the first steps of the proposed adenine synthetic reaction, relatively high concentrations of hydrogen cyanide are required. Indeed, the concentrations typically employed in the laboratory experiments were hundreds of thousands of times higher than the hydrogen cyanide concentration calculated to exist on the surface of the early Earth. Worse still, because hydrogen cyanide is more volatile than water, it cannot be concentrated by, for example, the evaporation of seawater in a tidal pool. A year after Oro's experiment, however, Leslie Orgel (1927–2007) suggested a possible solution to this problem when he proposed that the requisite high concentrations can be achieved by freezing a solution of hydrogen cyanide, which concentrates it in the voids between ice crystals. To test this, in 1975 Miller placed a fairly dilute sample of ammonium cyanide in cold storage at −78°C. Some 27 years later (tenure provides the breathing room required for professors to take the long view), Miller thawed his samples for analysis and was able to detect small quantities of several nucleobases, including adenine. This suggests that their prebiotic synthesis may have happened only in the Earth's polar regions or in icy planetesimals in the outer Solar System.

The abiological synthesis of the purine nucleobases has been explored in some detail. The process is thought to involve the multistep condensation of four molecules of hydrogen cyanide to form diaminomaleonitrile (fig. 4.9). Under the influence of UV light (sunlight is sufficient), this compound rearranges and reacts with yet another molecule of hydrogen cyanide to produce the nucleobase adenine in about 7% overall yield. That is, 7% of the starting cyanide is converted into adenine, and the other 93% either does not react or reacts to produce other molecules. Alternatively, four molecules of hydrogen cyanide can react with the salt ammonium formate to produce adenine with a yield of better than 90%. To achieve this yield, however,

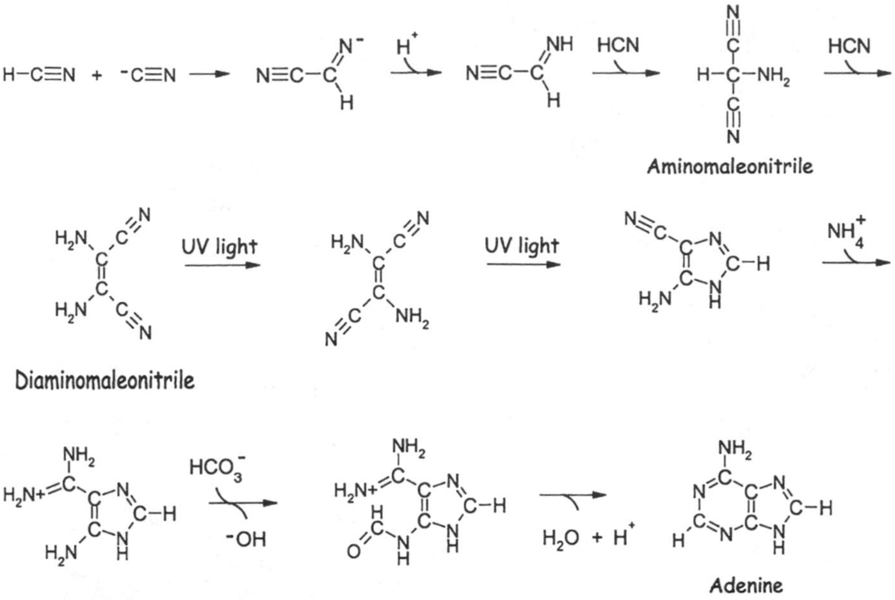

Figure 4.9 One of several pathways proposed for the prebiotic synthesis of the nucleobase adenine (a purine) from bicarbonate (HCO_3^-), ammonium (NH_4^+), and hydrogen cyanide (HCN), the latter of which is produced in Miller-Urey reactions.

the reaction mixture must be heated to dryness: the reaction occurs with the liberation of two water molecules, and thus driving off the water with heat favors formation of the nucleobase.

The other purine nucleobase guanine can also be synthesized by variations on the above synthesis of adenine. Guanine can be synthesized, for example, by the polymerization of ammonium cyanide, which, again, is formed by dissolving both ammonia and hydrogen cyanide in water (fig. 4.10). The synthesis works effectively over a broad range of temperatures, but even at ammonium cyanide concentrations far higher than those plausible on the prebiotic Earth, the overall yield of this nucleobase is less than 1%. Likewise, when ammonium cyanide polymerizes in ice, guanine is produced in parallel to adenine, albeit at yields 10 times lower. In contrast, an interesting gas-phase route has been described that may achieve more of the desired nucleobase. Specifically, Svatopluk Civis of the Czech Academy

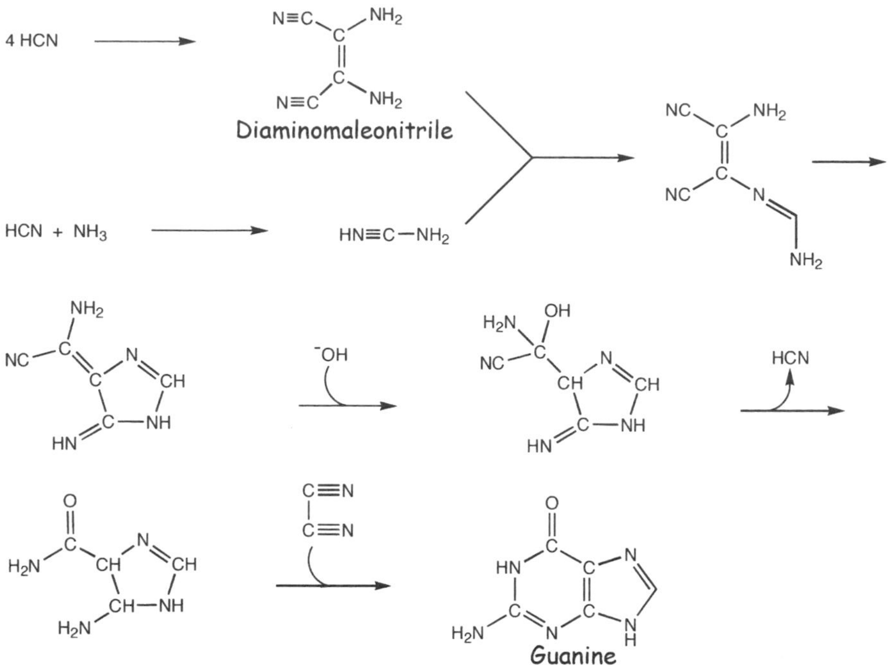

Figure 4.10 A proposed prebiotic synthetic route to the nucleobase guanine from ammonia (NH_3) and hydrogen cyanide (HCN). Even under optimized laboratory conditions, however, the yield of this reaction is low.

of Sciences used an ultra-high-power laser to simulate the effects of a meteorite impact on a primordial Earth with the Miller-Urey product formamide. From this, he recovered guanine at levels of a few milligrams per liter.

The prebiotic synthesis of the pyrimidine nucleobases has also been explored, with many of the proposed routes starting with the precursor cyanoacetylene, a minor product of the reactions induced in methane-nitrogen mixtures under the influence of a spark discharge (fig. 4.11). This precursor reacts with water to form cyanoacetaldehyde, and this in turn can react with the small, nitrogen-containing organic compound urea, which is also formed via Miller-Urey chemistry. The intermediate so formed spontaneously rearranges to form the more stable cytosine, one of the three pyrimidine bases used by biology on Earth. As nucleobases go, however, cytosine is relatively

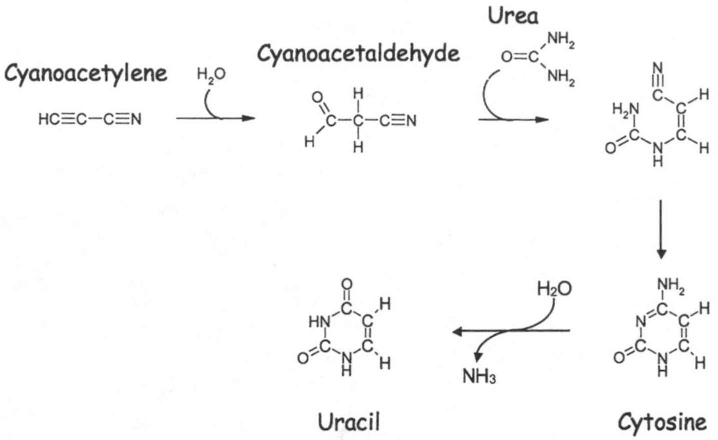

Figure 4.11 The proposed prebiotic synthesis of the nucleobase cytosine (a pyrimidine) from cyanoacetylene and urea, both of which are produced in Miller-Urey reactions. Cytosine, in turn, reacts with water and loses ammonia to form uracil.

unstable; its half-life in water is estimated to be around a century at room temperature. The good news about this reactivity is that the hydrolysis product of cytosine is uracil, another of the three pyrimidines found in our biochemistry, thus accounting for the prebiotic synthesis of this critical component. The bad news, however, is that the short half-life of cytosine means that it would have been difficult for significant quantities of this nucleobase to build up over a geological time frame.

The above is all fine and dandy but, by and large, Terrestrial biology doesn't employ free nucleobases so much as three-part structures consisting of a nucleobase covalently attached to a ribose (or deoxyribose) sugar that, in turn, is linked to one or more phosphate groups (fig. 4.12). Specifically, nucleobase-sugar combinations come in two general forms. The nucleo*sides* consist of a nucleobase covalently linked to a sugar by a glycosidic bond, and nucleo*tides* consist of nucleosides plus phosphates. Nucleosides and nucleotides are synthesized from free sugars, nucleobases, and phosphates by *dehydration* reactions, which are the reverse of the hydrolysis reactions described above. That is, this linkage chemistry proceeds with the removal of a water from the molecule. Because of this, these reactions are gener-

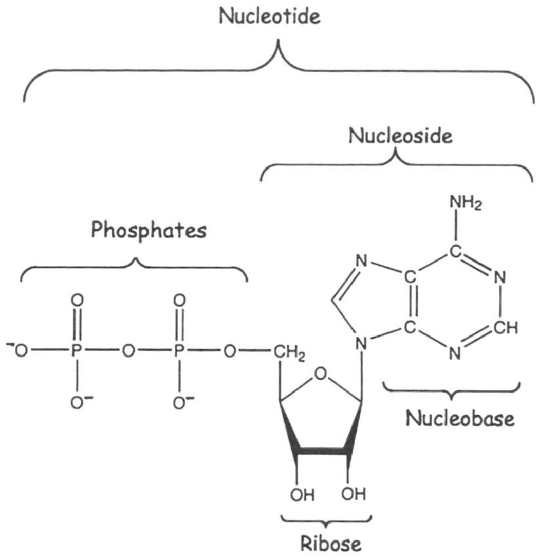

Figure 4.12 Nucleosides are a covalent combination of a sugar (either ribose, as shown here, or, in DNA, deoxyribose, which we discuss in the next chapter) and a nucleobase. A nucleo*tide* is a nucleo*side* plus one or more phosphate groups. Shown here is the ribonucleotide adenosine diphosphate (abbreviated ADP).

ally unfavorable when they take place in liquid water; you'll remember from chapter 1 that there are a whopping 55.5 mol of water in each liter of water, which is an amazingly high density of molecules. With all this water around in aqueous environments, chemical reactions tend to want to go in the direction that consumes water molecules rather than the direction that liberates them. This, in part, is why the formation of nucleosides and nucleotides, both of which occur by dehydration reactions, is energetically unfavorable in water and must be "driven" with some energy input in order to move forward efficiently. Today, our bodies use the energy obtained from oxidizing carbohydrates to drive these reactions. But this, of course, was not available before the advent of life. How, then, might the prebiotic synthesis of nucleotides have been achieved?

Leslie Orgel, who spent several decades at the Salk Institute, demonstrated a relatively straightforward route to the synthesis of nucleosides of the nucleobase hypoxanthine. If a mixture of hypoxanthine, ribose, and magnesium ions is heated to dryness, the nucleoside inosine is formed (fig. 4.13). Even better, the nucleoside produced is in the so-called β configuration, in which the nucleobase is on the opposite side of the sugar ring from the two remaining hydroxyls, which is the structure observed biologically on Earth. How does this reaction work? The magnesium ion binds to two of the hydroxyls on the ribose, thus "activating" the carbon to which the nucleobase will link. A reactive nitrogen atom in the hypoxanthine attacks this activated carbon to form a nucleoside. Heating to dryness pushes the linkage reaction forward by driving off the water that is liberated as the reaction progresses.

Unfortunately—isn't it starting to appear as though there's always an "unfortunately" in this business?—this reaction is neither general enough nor specific enough to plausibly account for the prebiotic synthesis of nucleosides. It is insufficiently general because hypoxanthine is the only nucleobase that is reactive enough to form nucleosides by this mechanism, and of the four nucleobases for which we have plausible prebiotic syntheses, it's the one that is least relevant, at least for contemporary Earth life. It is insufficiently specific because many of the possible five-carbon sugars react equally well, which would waste precious nucleobases without producing the desired ribose-containing nucleoside.

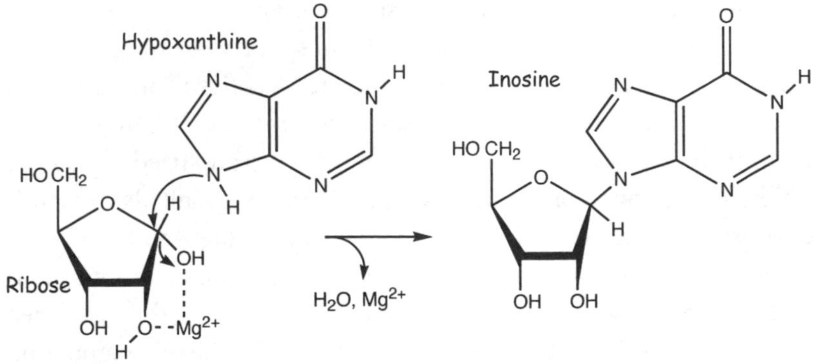

Figure 4.13 If a magnesium ion is present to act as a catalyst, the nucleoside inosine spontaneously forms from ribose and the nucleobase hypoxanthine. No other nucleobase, however, has been shown to participate in this coupling reaction.

And that's not all. Orgel's reaction is an example of what synthetic organic chemists pejoratively call a "relay synthesis." That is, Orgel's group never succeeded in going from the basic precursors, cyanide and ribose, to a fully formed nucleoside. Instead, they started with pure hypoxanthine and ribose (which they purchased, commercially, taking the "baton" only at this late stage in the game), rationalizing that previous work had demonstrated that these molecules can be synthesized under prebiotic conditions. Moving forward, they showed that hypoxanthine can be linked to ribose under other plausible prebiotic conditions. The problem here, though, is that the conditions that support the formation of the ribose and of the nucleobase are very different from one another and from the conditions that support their linkage to form the nucleoside. The experimenters thus had to hypothesize that one part of the prebiotic Earth provided the right conditions to synthesize the sugar, a second part provided the base, and, after they were transported there, a third place would foster their linkage to form the nuleoside. And while this is plausible (the Earth's a big and diverse place, after all), it does suggest that the yield of the nucleoside would have been rather low at best. Fortunately, however, there may be a solution to this problem: in 2009, John Sutherland and his team, then of the University of Manchester, demonstrated a route to the formation of nucleosides that avoids this problem. And, as an added bonus, it produces a nucleotide—a nucleoside with a phosphate attached, like the ones used in Terrestrial biology.

Like those he followed, Sutherland had spent more than a decade thinking about the stepwise—and separate—synthesis of sugars and nucleobases, without ever demonstrating a route that could lead to them from start to finish. Finally, Sutherland's group tried a radically different tack, one that scrambles together the cyanonitrile (−C≡N) chemistry that leads to the nucleobase and the aldehyde (−HC=O) chemistry that generates the sugar so as to avoid the problem associated with creating a bond between a preformed sugar and a preformed base (fig. 4.14). To do this, they added cyanimide, the first compound in the synthesis of the nucleobase, to glycolaldehyde, the first compound in the synthesis of the sugar, to form a molecule that was half a sugar bonded to half a nucleobase. Combining this with glyceralde-

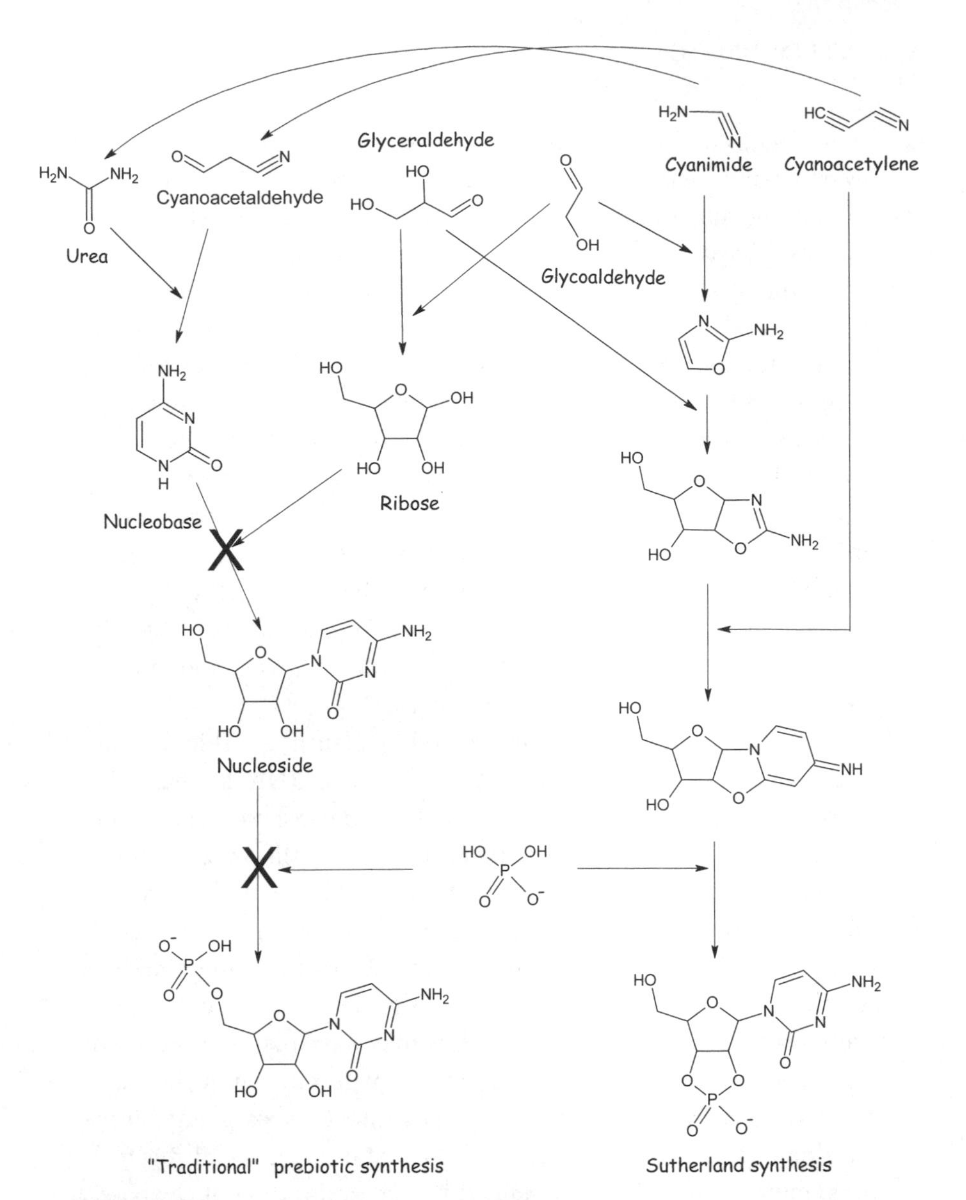

Figure 4.14 According to traditional thinking (illustrated on the left), ribose and the nucleobases were synthesized independently before being linked together with phosphate to form the nucleotides. Spontaneous, prebiotic bond formation between ribose and nucleobases, however, has not been demonstrated for any of the four nucleobases found in RNA. The addition of "unactivated" phosphate (last step) has likewise proved problematic. Fortunately, however, John Sutherland and coworkers demonstrated an alternative synthesis of cytidine phosphate that occurs spontaneously without utilizing either ribose or the nucleobase as intermediates. The addition and subtraction of water molecules, which occurs in several steps of the reaction, has been omitted for clarity.

hyde, the second component of the sugar, and cyanoacetylene, the second component of the nucleobase, they produced an intermediate containing, albeit hidden within it, both a ribose and a cytosine nucleobase. Treating this with phosphate (which serves as both a catalyst of the reaction and a reactant in it), they liberated the nucleoside of ribose and cytosine connected by the same glycosidic bond used in Earth's biology. Better still, the phosphate remains attached to the ribose—thus the product is a nucleotide rather than the less useful nucleoside. A caveat of Sutherland's postulated prebiotic synthesis, however, is that the various steps in the reaction pathway require varying conditions (e.g., changing pH), and thus he had to postulate that the environment that would have supported such a prebiotic synthesis would have had to fluctuate significantly with time. Still, the chemical principles that this work illustrates may yet prove more general, such that similar synthetic routes that rely less on the invocation of complex environmental fluctuations may exist.

Given the difficulty in identifying solidly plausible prebiotic routes to the formation of the nucleotides in DNA and RNA, an increasing number of researchers have argued that the nucleobases, ribose, and phosphate that are the components of contemporary life on Earth might not have served as the foundation of its origins. The American chemist Nick Hud, for example, systematically surveyed more than 80 candidate pyrimidines and purines to show that 2,4,6-triaminopyrimidine, which, although it does not play any role in contemporary biology, is the most reactive with ribose and thus may have played a precursor role in our biology, being later supplanted by the current nucleobases. Likewise, the Swiss chemist Albert Eschenmoser has argued that the triazines, six-membered aromatic rings containing three nitrogen atoms, rather than the two contained in our pyrimidine nucleobases, are a perhaps intriguing alternative to the current nucleobases as they are readily synthesized under plausible prebiotic conditions. In parallel, alternative nucleic acids constructed from sugars other than ribose, including a number of four-, five-, and six-carbon sugars, have been demonstrated in the laboratory. This said, what, if any, roles these many alternatives might have played in the origins of life on

Earth and, for that matter, whether they might support the creation of life elsewhere, remain open questions.

So far, so good. But free nucleosides and nucleotides are not the stuff upon which life is founded. Instead, life is built around *polymers* of nucleotides, long chains of nucleobases bound to ribose that are in turn linked, hydroxyl group to hydroxyl group, by intervening phosphate groups in what are known as phosphodiester bonds. These polymeric nucleic acids, called RNA (ribonucleic acid) or DNA (deoxyribonucleic acid), are higher in energy than monomers because the bonds that link the monomers together in these compounds are themselves unstable; they too are the products of dehydration reactions. A major question is thus: how do we generate "activated," high-energy nucleotides whose chemical energy can be harnessed to drive polymerization? (Now might be a good time to check out sidebar 4.2, though, which touches on the inherent rareness of high-energy molecules.) Perhaps the most compelling answer to this is the use of high-energy polyphosphates, which is in fact the system Terrestrial biology uses today to store and shuttle energy.

The simplest nucleotides are the monophosphates, in which a single phosphate is linked to a hydroxyl group on the nucleoside's sugar. And while Sutherland's synthesis produces these by default, they can also be synthesized, in up to 60% yield, simply by heating a solution of nucleosides and phosphoric acid with urea, which acts as a catalyst. Unfortunately, however, nucleoside monophosphates are not the sort of high-energy compound we need to polymerize nucleotides. In contrast, a string of phosphates linked by oxygen bridges is high in energy because, at neutral pH, each phosphate takes on a negative charge, and the negative charges on adjacent phosphates repel one another. Contemporary Earth life utilizes the energy available from this repulsion to drive the polymerization of RNA (and DNA), not to mention countless other biochemical reactions within our cells. And it seems it just might be possible for nucleoside polyphosphates to be synthesized under reasonably plausible prebiotic conditions.

The postulated reaction involves the triphosphate molecule trimetaphosphate, which consists of three molecules of phosphoric acid

Sidebar 4.2

Weighing the Probabilities

How likely is it that a thick, rich soup of amino acids and activated nucleotides can form on a primordial, terrestrial planet? If we limit ourselves to currently well-understood chemistry, the answer is, unfortunately, "not very." We'll give just two examples.

The serious mismatch between the rates at which cytosine and ribose might be produced and the rates at which they are destroyed (and thus, in net, their equilibrium concentrations) suggests that these critical compounds would be quite rare. The problem of the stability of likely biopolymers is similarly acute. The polymerization of RNA, for example, is an energy-consuming process, and thus the formation of longer and longer RNA polymers becomes less and less likely unless the polymerization reaction can be coupled with some other, energy-liberating reaction. This is why searches for chemistry that might "activate" nucleosides has been so intense; without such activation, it is difficult to see how the relevant reactions could proceed to generate even trace amounts of polymers of sufficient length.

Of course, even uphill chemical reactions can proceed, they just do so at lower and lower yield as the energy consumed in the reaction increases. The ratio of product to starting material for the conversion of molecule A into product molecule B is proportional to $e^{-\Delta G/kT}$, where ΔG is the energy difference between molecules A and B, and kT is the mean energy of molecular collisions at a given temperature, T. The k is a constant, termed the Boltzmann constant, which relates temperature to thermal energy. Because the fraction of molecules in a population with a given amount of thermal energy drops off *exponentially*, the polymerization of unactivated nucleosides to form even short polymers is astronomically unlikely. But it is not impossible. And thus, once again, the origins of life are tied up in the anthropic principle (i.e., our existence proves only that the probability of life forming was not zero and says nothing about how close to zero it might have been). Alternatively, of course, the solution to this conundrum may lie in our incomplete knowledge. Specifically, while it is fairly advanced, our knowledge of chemistry in general and of the chemistry of primordial, rocky planets in particular is hardly exhaustive. There may yet be high-yield synthetic routes by which all the components of the putative primordial soup could have formed. In short, the jury is still out.

that have become linked together by dehydration reactions. On heating a dry mixture of nucleotide monophosphates with trimetaphosphate (a cyclic molecule comprised of three phosphate groups linked into a circle) in the presence of the catalyst magnesium (again, the heat drives off water, propelling the reaction forward), nucleoside polyphosphates are formed in good yield. When water is added, these compounds rapidly hydrolyze to form nucleoside triphosphates, which then only slowly hydrolyze to form di- and monophosphates. Unfortunately, however, the relevant phosphate compounds are quite rare on Earth; phosphate is usually the limiting mineral, for example, in freshwater ecosystems, and its high energy forms are not stable.

Of course, just because polyphosphates play the role of activator on contemporary Earth, this doesn't mean that it's the only way to activate nucleosides or, indeed, that it was the activation chemistry employed in the origins of life. For example, in the 1990s, Leslie Orgel proposed that the small, nitrogen-containing ring compound imidazole could react with phosphate to form a phosphorimidazolide containing a high-energy nitrogen-phosphorus bond that activates the phosphate for further reactions. Unfortunately, however, while phosphorimidazolides can be synthesized in the laboratory, no plausible prebiotic route to their synthesis has been described. And if some alternative activation chemistries were in play during the origins of life, they have not yet been identified.

The Missing Ingredient: Fats

Life on Earth is not built on amino acids and nucleotides alone. The dry weight of a bacterial cell, for example, is about 6% lipids, a word stemming from the Greek *lipos*, meaning "fat." To a biochemist, though, lipids are anything that can be extracted from tissues by an organic solvent, and thus this family of molecules includes both fats and oils. Unfortunately, fats and oils are not readily produced by Miller-Urey chemistry, yet it seems likely that they played key roles in the origins of life. How, then, were appreciable amounts of these materials synthesized under prebiotic conditions? The answer to this question also remains unknown, but several seemingly plausible theories have been put forth.

Most lipids are long-chain molecules, and all consist almost entirely of carbon and hydrogen. Miller-Urey chemistry, in contrast, tends to produce only oxygen- and nitrogen-containing species, such as alcohols and amino acids. The only way long carbon chains can form via Miller-Urey chemistry is through radical-radical reactions in which methyl radicals add to a growing carbon chain. As the chances

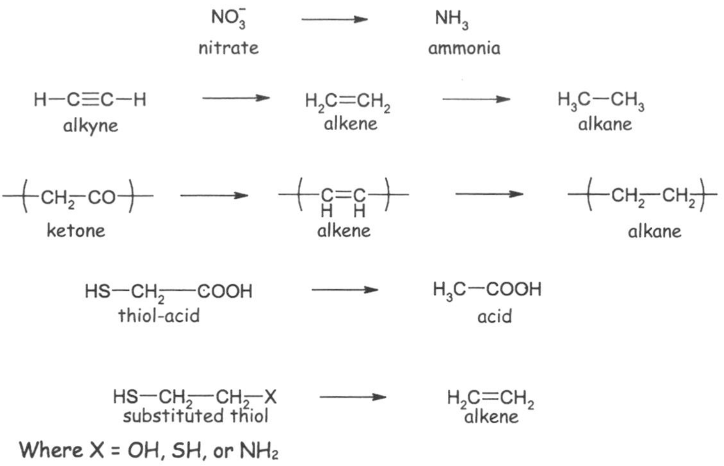

NO_3^- (nitrate) $\longrightarrow$ NH_3 (ammonia)

$H-C\equiv C-H$ (alkyne) $\longrightarrow$ $H_2C=CH_2$ (alkene) $\longrightarrow$ H_3C-CH_3 (alkane)

$-(CH_2-CO)-$ (ketone) $\longrightarrow$ $-(CH=CH)-$ (alkene) $\longrightarrow$ $-(CH_2-CH_2)-$ (alkane)

$HS-CH_2-COOH$ (thiol-acid) $\longrightarrow$ $H_3C-COOH$ (acid)

$HS-CH_2-CH_2-X$ (substituted thiol) $\longrightarrow$ $H_2C=CH_2$ (alkene)

Where X = OH, SH, or NH_2

Figure 4.15 Shown are the wide variety of reduction reactions that can occur on the mineral troilite (FeS). Such reactions, which would probably have occurred as ocean water filtered past highly reduced rocks deep in the Earth's primordial crust, could be a prebiotic source of lipids, which are critical for biology and are not produced by Miller-Urey chemistry.

of the radical reacting with a water or ammonia molecule (to stop the growing chain) are reasonably large, purely carbon-hydrogen compounds become rarer with increasing chain length simply due to the improbability of one carbon radical reaction following another. Likewise, long-chain carbon compounds are not very volatile. Miller-Urey chemistry is gas-phase chemistry and, as we've mentioned, molecules with more than about three or four carbons fall out of the atmosphere as liquids or solids.

Perhaps the most realistic theory described to date for the prebiotic formation of lipids is that they were synthesized by the reduction of that other class of large carbon-containing compounds: sugars. But again, these compounds are not volatile and thus the reduction cannot occur in the atmosphere. So where can this happen? Deep in the planetary crust, where reduction can be catalyzed by the iron mineral troilite, which is composed of iron sulfide (FeS). In the presence of the reducing gas hydrogen sulfide (remember: the crust is reduced, at least in this model of prebiotic chemistry), troilite is a strong reducing agent. In fact, it is a strong enough reductant to produce hydrogen; reduce alkenes (compounds with carbon-carbon double bonds), alkynes (with carbon-carbon triple bonds), and thiols (with —SH groups) to saturated hydrocarbons; and replace ketones (C=O groups) with thiols (fig. 4.15). And even today, more than 4 billion years after our planet formed, the Earth's geology is still so active that deep-sea vents in the mid-ocean ridges recycle an entire ocean's volume of water every 8 million years, and thus catalytic reduction deep within a planet's crust is certainly a possibility.

Non-Miller-Urey Sources of Life's Building Blocks

Given the possibility that primordial atmospheres are too oxidized to promote Miller-Urey chemistry, we should note that a small but vocal group within the astrobiology community has argued that the prebiotic precursors to life are not formed in situ but are delivered to terrestrial planets from space, where they can be synthesized under more reducing conditions.

Comets are known to contain ammonia, methane, and water (as solids, all of which are referred to as "ices" by planetary scientists), as are the asteroids that formed out beyond the snow line (discussed in chapter 3). Both types of objects are also subjected to significant radiation in the form of solar UV light and cosmic rays and therefore may contain Miller-Urey-type small molecules. Do they? The half dozen spacecraft that have visited comets have flown by at such high velocities that delicate molecules such as we are interested in here have difficulty surviving their encounter with the craft's instruments. Nevertheless, in 2009 NASA announced the discovery of glycine by its *Stardust* spacecraft, which swept up dust particles while flying by comet Wild-2 at 6 km/s in 2004 and carried them back to Earth for analysis two years later. More detailed analysis of the organic inventory of a comet, however, had to wait until 2014, when the European Space Agency's *Rosetta* mission, which dropped into orbit around comet 67P/Churyumov-Gerasimenko in 2014 (fig. 4.16), where it, too, found glycine in the dust the comet was spraying into space.

Figure 4.16 In 2014, the European Space Agency's *Rosetta* spacecraft dropped into orbit around the comet 67P/Churyumov-Gerasimenko, a 4 km wide chunk of primordial ice and rock that was once a resident of the Kuiper belt. The spacecraft detected glycine in the jets of gas and dust seen here erupting from the comet's surface. (Courtesy of the European Space Agency)

In contrast to comets, which we've had to sample via robotic spacecraft, meteoritic material is delivered to the Earth with some frequency, and thus it can be analyzed in the laboratory. Among the many classes of meteorites are the carbonaceous chondrites, which are water- and organic-containing stones thought to arise beyond the snow line, in the outer half of the asteroid belt. Because they consist of clays and other highly hydrated minerals, carbonaceous chondrites typically disintegrate soon after landing on our warm, wet planet, and thus they have been difficult to study. An important counterexample, however, arose in the fall of 1969 when a large carbonaceous chondrite exploded in the air over Murchison, Australia. The fragments of the Murchison meteorite were quickly collected under dry conditions, allowing their leisurely study in the laboratory.

Hot water extractions of the water-soluble organic material in the Murchison meteorite indicated that it contains a wide range of amino acids. In fact, some 70 different amino acids have been identified in the Murchison meteorite so far, including at least six of the 20 proteinogenic amino acids found in life on Earth (table 4.4). Probably not coincidentally, many of the amino acids most readily formed in the Miller-Urey chemistries are found in abundance in the Murchison material, suggesting that interstellar Strecker synthesis was occurring in the prestellar nebula. Consistent with these arguments, radio astronomers have identified the unambiguous spectral signatures of a number of small organic compounds, the most complex being the two-carbon, sugar-like molecule glycolaldehyde, in nebulae thousands of light-years from Earth.

Of interest to us, the organic inventory of carbonaceous chondrites isn't just limited to amino acids. Dave Deamer at the University of California, Santa Cruz, for example, has extracted lipids from Murchison samples. When added to fresh water, these lipids spontaneously form hollow spheres, called vesicles, reminiscent of cell membranes. But Deamer's experiments have led him to question the

Table 4.4
Amino acids observed in the Murchison meteorite

Isovaline	β-Aminoisobutyric acid
α-Aminoisobutyric acid	Pipecolic acid
Valine	**Glycine**
N-methylalanine	β-Alanine
α-Amino-*n*-butyric acid	**Proline**
Alanine	γ-Aminobutyric acid
N-methylglycine	**Aspartic acid**
N-ethylglycine	**Glutamic acid**
Norvaline	

Note: Proteinogenic amino acids shown in bold type.

supposition that life originated in the ocean. It turns out that, in salt water, the lipids simply clump together without forming a hollow sphere, so Deamer has theorized that a freshwater pond would have been a more hospitable environment for life's origins. Finally, in 2019, Tomoki Nakamura and coworkers in Sendai, Japan, reanalyzed Murchison samples and identified a number of sugars, including ribose, suggesting that this class of molecule can also be synthesized in prestellar nebulae or protoplanetary disks.

So, it seems, meteorites and comets can deliver biologically relevant molecules from the outer reaches of a solar system to its rocky, inner planets. But the question remains: could they do so in sufficient quantity to be relevant to the origins of life? Jeffrey Bada, whom we met earlier in this chapter, long argued that the delivery of organic building blocks from space is insufficient. He arrived at this conclusion by searching for α-aminoisobutyric acid in places like Antarctica and Greenland, where meteoritic material falling on snow and ice can build up over significant time frames. This molecule is rare on Earth because, unlike the proteinogenic amino acids, it is not associated with life. But it is efficiently synthesized by Miller-Urey chemistry and is a major organic component of the Murchison meteorite. Bada found less than 0.05 mg of the compound per square centimeter of ice surface, leading him to argue that this extraterrestrial molecule—and by extrapolation, others—would not have built up to high enough concentrations to contribute much to the primordial soup.

More recently, however, Bada had to reevaluate this claim when he discovered that enormous quantities of "buckyballs" had arrived on Earth intact from outside the Solar System. Known more formally as fullerenes, buckyballs are large, spherical molecules of pure carbon that resemble geodesic domes, and thus were named after the inventor of said domes: Buckminster Fuller (1895–1983). But buckyballs are a common component of soot, so how did Bada know that the buckyballs he found were of extraterrestrial origin? The spherical structure of a buckyball allows it to trap atoms inside. With his associate Luann Becker, Bada explored a 2-billion-year-old impact site in Ontario, Canada, that contained about a million tons of carbon rich in buckyballs, many of which contained trapped helium. While helium is the second most common element in the Universe, it is rare on terrestrial planets (with its light weight and chemical inertness, it tends to be lost to space over geological time). It thus appears likely that these millions of tons of buckyballs, and presumably equally large amounts of other organic compounds, were indeed delivered from the outer Solar System by impacts. The consensus is now that both extraterrestrial delivery and in situ Miller-Urey chemistries contribute to the formation of the rich, prebiotic soup of organic materials necessary for life to form.

Prebiotic Polymerization

In our exploration of prebiotic chemistries, we still face a significant hurdle. While we understand where many of the *monomers*—individual amino acids and nucleobases—come from, monomers do not equal life. Even free nucleobases are only relevant to life in the context of *polymers* of *nucleotides*, the three-part, covalent combinations of sugars, bases, and phosphate we described above. The reactions that link such monomers into complex polymers are presumably critical steps in the origins of life, not only on Earth, but anywhere. But these reactions consume significant amounts of energy and thus are extremely unfavorable. How, then, might they have spontaneously occurred on a prebiotic planet? We don't really know, but several (admittedly rather speculative) mechanisms have been proposed.

Activated nucleotides, as we've noted, are presumably the starting materials from which RNA polymers are synthesized. How such polymerization might take place has been the subject of extensive study. One possibility is simply the spontaneous polymerization of phosphorimidazolides (or other, still unknown activated nucleotides) in solution. This reaction, however, tends to be extremely inefficient. The problem is that each nucleotide has several free hydroxyl groups, each of which can form a phosphoester linkage (fig. 4.17). In fact, for the activated base adenosine phosphorimidazolide, incorrect linkages are

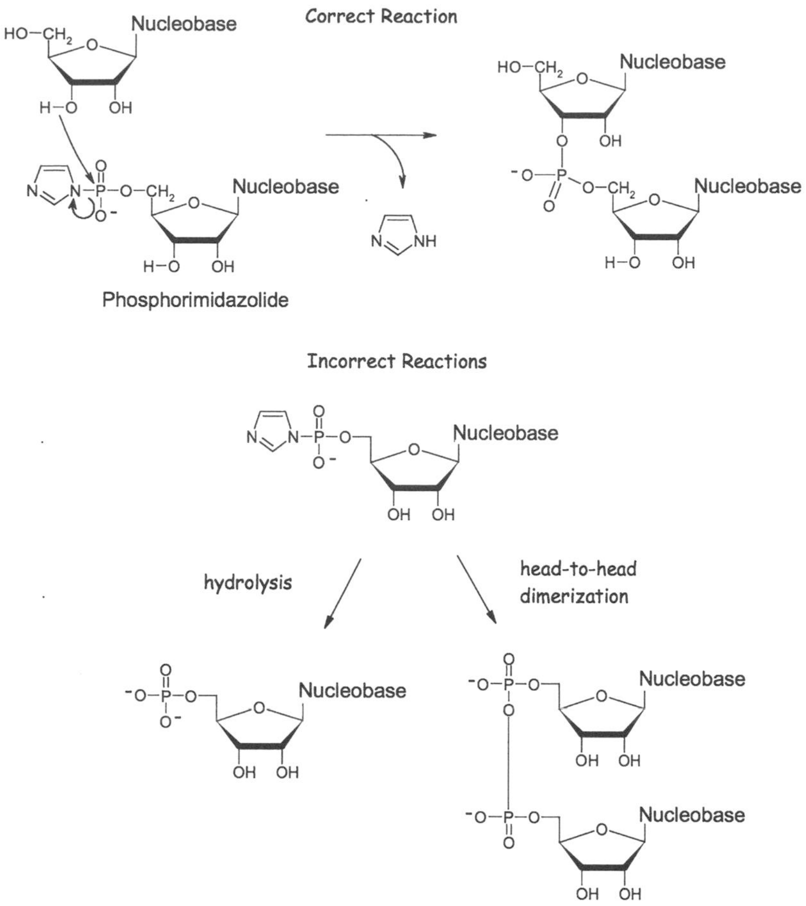

Figure 4.17 Activated nucleotides, such as the phosphorimidazolide shown here, readily polymerize in aqueous solution. Due to a multitude of side reactions, however, the yield of correctly linked RNA polymers produced by such polymerization is low.

six times more likely to form than the correct linkages found in the RNA polymers in life on Earth. Fortunately, mineral catalysts may provide a means of overcoming this problem. James Ferris (1932–2016) at Rensselaer Polytechnic Institute in New York State has shown, for example, that RNA polymers up to 55 monomers long can be synthesized (with mostly correct linkages) on the common clay montmorillonite; it seems that the surface of the clay binds the growing polymer and directs the addition of each new monomer. Alternatively, the polymerization could be *template directed*. RNA (like its relative DNA) can bind to a complementary sequence to form a double helix. This suggests that an existing RNA polymer could act as a template to direct the specific polymerization of a new strand of RNA. Using polycytidine (the polymer of the cytosine-containing nucleotide) as a template to direct the polymerization of activated guanosine (the guanine-containing nucleotide), Orgel found that the correct linkage is favored by 2:1 in the final polymerized product. The presence of zinc ions increases the yield of polymerized product without harming the yield of proper linkages. When the activator 2-methylimidazole is employed instead in the presence of magnesium, zinc, or iron ions, a template containing both uracil and cytosine can be efficiently copied as long as cytosine dominates the template's composition. If the cytosine and uracil contents approach equality, the yield of polymerized material plummets. This limitation presents a serious obstacle to the formation of a self-replicating system from the simple, template-directed polymerization of RNA; as we discuss again in the next chapter, a cytosine-rich sequence that serves as an efficient template will produce cytosine-poor products that cannot, in turn, serve as new templates.

With all these caveats and disclaimers and open questions, the prebiotic synthesis of RNA polymers is looking pretty bad, right? Sadly, the situation may be even worse than we've let on. Not only is the synthesis of nucleotide polymers an (energetically) uphill battle (see sidebar 4.2), but staying "uphill" is equally difficult. For example, the half-life of the phosphodiester linkages in unstructured, single-stranded RNAs is less than 1,000 years in water at the freezing point and less than 1 day at 35°C. And given that even a modest-length RNA polymer contains hundreds of phosphodiester bonds, the lifetime of an intact polymer is much, much less than the lifetime of any one linkage. This observation has, once again, led to the speculation that some as yet unknown polymer that shares RNA's abilities to catalyze reactions and encode information and yet is more stable (and perhaps easier to synthesize prebiotically) was instead involved in the origins of life, a point that we'll come back to a third time in the next chapter.

Conclusions

We've come a long way in the half century following the discoveries of the double helix and prebiotic chemistry, but the latter did not advance as fast as molecular biology, and perhaps not as far as Miller would have hoped while marveling over his results back in 1953. Thanks to his work and that of those who followed, we understand in detail how quite a few of the key molecular elements of life as we know it might have been created in the prebiotic world. But equally many questions about prebiotic chemistry remain.

And what happened to the pioneering protagonists of our story? In the early 1960s, Harold Urey went on to head the Chemistry Department at the newly founded University of California, San Diego, and he brought along his protégé Stanley Miller. Miller spent the next 40 years pursuing origins-of-life research, publishing several hundred papers and numerous books on the topic and spawning a whole community of scientists pursuing the same elusive goal, although progressing only slowly. Indeed, when Miller died in 2007, the unfulfilled promise of his 1953 breakthrough was a key theme of many of his obituaries. However, the legacy of his pioneering experiments lives on. For example, when his former student Jeffrey Bada inherited the contents of Miller's lab, he discovered samples of experiments conducted in 1958, which were clearly labeled and referenced in Miller's notebooks. The previously unreported 1958 experiment was the first Miller had conducted that included hydrogen sulfide (H_2S). Reanalyzing these samples using modern high-performance liquid chromatography and mass spectrometry, Bada found that they contained the oxidation products of the sulfur-containing amino acids cysteine and methionine, which Miller himself didn't report making until the 1970s. Give them a few million years more, and Miller's samples may come up with life after all.

Further Reading

Prebiotic Chemistry

Cleaves, H. James, II. "Prebiotic Chemistry: What We Know, What We Don't." *Evolution: Education and Outreach* 5 (2012): 342–60.

Miller, S. L. "A Production of Amino Acids under Possible Primitive Earth Conditions." *Science* 15, no. 3046 (1953): 528–29.

Orgel, Leslie E. "Prebiotic Chemistry and the Origin of the RNA World." *Critical Reviews in Biochemistry and Molecular Biology* 39, no. 2 (2004): 99–123.

Parker, Eric T., Henderson J. Cleaves, Jason P. Dworkin, Daniel P. Glavin, Michael Callahan, Andrew Aubrey, Antonio Lazcano, and Jeffrey L. Bada. "Primordial Synthesis of Amines and Amino Acids in a 1958 Miller H_2S-Rich Spark Discharge Experiment." *Proceedings of the National Academy of Sciences USA* 108, no. 14 (2011): 5526–31.

Sutherland, John D. "Studies on the Origin of Life — the End of the Beginning." *Nature Reviews Chemistry* 1 (2017): 0012.

Alternatives to RNA in the Origins of Life

Hud, Nicholas V., Brian J. Cafferty, Ramanarayanan Krishnamurthy, and Loren Dean Williams. "The Origin of RNA and 'My Grandfather's Axe.'" *Chemistry & Biochemistry* 20, no. 4 (2013): 466–74.

Chapter 5

The Spark of Life

Sometime in the third century BCE, the Greek philosopher Aristotle (382–322 BCE) enunciated the theory of spontaneous generation. Synthesizing the work of earlier scholars, Aristotle argued that life could arise spontaneously from nonliving matter provided the latter was suffused with *pneuma*, or "vital heat." After all, was it not obvious that maggots arose spontaneously in spoiled meat? This idea, as with so many of Aristotle's claims, held fast through the Middle Ages, when it was still widely held that old rags mixed with wheat could give rise to fully formed adult mice.

One of the first to question spontaneous generation was the Tuscan (this being before Italy was "a thing") naturalist Francesco Redi (1626–1697). Often called "the father of experimental biology," Redi showed that maggots only appear in meat that has been exposed to adult flies; cover the opening of a meat-containing jar with fine gauze, and maggots no longer "form." A hundred years later, in 1765, Redi's studies were extended by the Italian Lazzaro Spallanzani (1729–1799), who used a microscope, unavailable in Redi's time, to show that even microscopic organisms do not spontaneously form in broth that has been sterilized by boiling and left sealed and out of contact with air.

Redi's observations, however, were not the final word on spontaneous generation. In his experiments, for example, he kept his flasks sealed—what if some component in the air was necessary for spontaneous generation to occur? Interest in this question grew over the next century until, in 1859, the French Academy of Sciences founded a prize of 2,500 francs for the scientist who could finally lay the question to rest by conclusively proving or disproving spontaneous generation. The prize was quickly won by Louis Pasteur (1822–1895), who would later go on to invent immunization, pasteurization, and, more or less, the germ theory of disease.

To clinch the prize, Pasteur placed sugar solutions, urine, and beet juice in swan-shaped flasks whose long, curved necks allowed communication with the air but would prevent, he postulated, any small, living particles from dropping into the liquids with the settling dust. In flasks that he sterilized by boiling, nothing happened. In contrast, the liquid in the unboiled flasks rapidly spoiled. Of course, that could simply have meant that boiling had somehow made the liquid unpalatable for the hypothesized organisms. Decisive proof, though, was provided when Pasteur snapped off the swan-shaped necks, allowing dust to settle into the previously boiled organic brews. The liquid then spoiled, demonstrating that the organisms responsible had settled out of the air and proving that the liquid's prior sterility reflected the absence of spontaneous generation rather than ruined media that could no longer support life. Pasteur proclaimed, "Never will the doctrine of spontaneous generation recover from the mortal blow of this simple experiment."*

Pasteur's conclusions, though, were in conflict with a theory published just three years earlier by a scientist on the other side of the English Channel: Darwin's theory on the origins of species. Charles Darwin (1809–1882) had obviated the need for God in the creation of species by showing how one species could slowly transmute into another. But Darwin's theory left open the question of how the very first species arose. At the very least, the *first* species could not have arisen by the transmutation of some earlier species and thus could not have arisen without some form of spontaneous generation. Darwin seems to have been troubled by the subject, noting in 1866 that "as for myself I cannot believe in spontaneous generation, and though I expect that at some future time the principle of life will be rendered intelligible, at present it seems to me beyond the confines of science."** Five years later, however, in a letter to his close friend and

*In a lecture at the Sorbonne in Paris in 1864; quoted by his son-in-law, René Vallery-Radot, in his book *Life of Pasteur* (New York: Doubleday, Page and Co., 1919), 109.

**As quoted in Charles Darwin, Francis Darwin, and A. C. Seward, *More Letters of Charles Darwin: Volume 1* (London: J. Murray, 1903), 273.

fellow naturalist Joseph Hooker (1817–1911), Darwin went on to speculate how life might have first arisen: "But if (and oh what a big if) we could conceive in some warm little pond, with all sorts of ammonia and phosphoric salts, light, heat, electricity etcetera present, that a protein compound was chemically formed, ready to undergo still more complex changes."* More than a century and a half later, though, the details of life's origins remain one of science's greatest mysteries.

Panspermia

Within a decade of Pasteur's publication of his results, several prominent European scientists—most notably the English physicist Lord Kelvin (1824–1907), after whom the absolute temperature scale is named, and the Prussian physicist Hermann von Helmholtz (1821–1894) of, for example, the Gibbs-Helmholtz equation in chemical thermodynamics—suggested a possible work-around. Could we not avoid the difficulty of spontaneous generation if life had originated elsewhere and been transported to the Earth through space? Their early speculations on this, the panspermia hypothesis, from the Greek *pan sperma*, meaning "all seed," were fleshed out in great detail in a widely discussed body of work by the Swedish chemist and 1903 Nobel laureate (another example of how winning the prize provides a lot of leeway to pursue wacky ideas) Svante Arrhenius (1859–1927), who was already famous for relating the temperature dependence of chemical reaction rates. Originally in a 1903 journal article, then expanded in a popular book some five years later, Arrhenius argued that life, in the form of hardy, dormant spores, could survive in the cold dark vacuum of space for long enough to be transferred between the stars.

In Arrhenius's version of events, bacterial spores could escape from the upper atmosphere of their home planet and be launched into interstellar space by the pressure of light; photons have momentum and

*Quoted in F. Darwin, ed., *The Life and Letters of Charles Darwin* (New York: Basic Books, 1898), II: 202–3.

thus can accelerate something as small as a bacterial spore to quite high speeds. Eventually, some of the spores would fall upon another planet (Arrhenius estimated that sunlight could push a bacterial spore from the Earth to Mars in 20 days, and from the Sun to the nearest star in as little as 9,000 years) and inoculate the virgin world.

In the century after he first published his work, others have explored Arrhenius's hypothesis in quantitative detail. The well-known astronomer Carl Sagan, for example, calculated that, to leave the Solar System, a spaceborne spore must be less than 0.5 μm (a micrometer is a millionth of a meter) in diameter—any larger and light pressure, which depends on cross section (and thus goes with the square of radius), will lose out against the Sun's gravitational pull, which depends on mass (and thus goes with the cube of radius). Only the very smallest of Earth's bacteria, however, are this small. Sagan also described the various types of stars that spores could be blown from and to—from bright stars, which endow the spores with significant speed, to the solar systems of dimmer stars. He could not, however, come up with a solution to the vexing problem of radiation. Most Terrestrial organisms are rather quickly destroyed if exposed directly to the hazards of space; while the lichen species *Rhizocarpon geographicum* and *Xanthoria elegans* were once revived after a couple of weeks above the Earth's atmosphere (in an experimental holder strapped to the outside of a Russian spacecraft), most of the organisms they shared their vessel with were quickly killed by the Sun's raw UV light. That said, if protected from this by a layer of clay, spores of the bacterium *Bacillus subtilis* have been revived even after several years in space.

Putting aside our concerns regarding the viability of spores after millennia-long interstellar trips *sans* spacecraft, there is a deeper problem with the panspermia hypothesis: while it might explain the origins of life on Earth, it does not answer the more fundamental question of how life arose from inanimate matter in the first place. And thus, even if life could survive a trip between the stars, we astrobiologists can't sweep the mystery of life's origins under the rug simply by saying it took place elsewhere. Arrhenius himself sidestepped the issue of how life arose in the first place by suggesting that it might be eternal; after all, at the time, it was assumed that the Universe was immortal, so why couldn't life have always existed as well? This argument was raised again half a century later by Fred Hoyle, the disparager of the big bang and champion of his alternative cosmology of continuous creation, which postulated that the Universe had no beginning (or end). But alas, continuous creation has since been thoroughly debunked. It is now well established that the Universe is "only" 13.8 billion years old. At least once during the past 13.8 billion years, life must have arisen spontaneously; if not here, then somewhere else. And, so, the question remains: how did it do so?

Theories of the Origins of Life

Modern, scientific consideration of the origins of life began, as mentioned in chapter 4, with the Soviet and Scottish scientists Aleksandr Oparin and J. B. S. Haldane, the latter of whom seems to have coined the phrase *primordial soup*. Oparin proposed a "cells-first" theory of origins. Impressed by the cell-like appearance that oily organic materials can adopt in water, he proposed that the physical structure of the cell came first in the form of a suspension of oily droplets and hollow, water-filled, oily "vesicles," together called coacervates, from the Latin *coacervātus*, meaning "heaped together." Oparin noted that, in particular, the water-filled vesicles could serve as a vessel, providing a sequestered environment isolated from potentially disruptive influences. Moreover, he noted, under some laboratory conditions such vesicles *grow and divide* in a manner reminiscent of cell division.

Oparin's speculations aside, life is more than just swelling, water-filled vesicles of lipids.* Life, at least as we have defined it, requires genes and metabolism. The dispute over which of the two came first has divided researchers in the field, fueling a debate that many observers have likened to the classic chicken-or-egg argument, with both the metabolism-first and genes-first camps simultaneously claiming they have the upper hand.

*Though sometimes, midway through a dull work week, it may not seem so.

Metabolism First

The metabolism-first camp argues that the first life was formed from a network of self-sustaining chemical reactions of monomeric organic molecules catalyzed by either organic or inorganic catalysts. This can be a bit hard to visualize, but we've seen examples of it. In chapter 1, we discussed fire as a self-replicating, albeit uncatalyzed, network of free radical reactions (see fig. 1.1). And in chapter 4, we discussed the formose reaction, a calcium hydroxide–catalyzed reaction network that consumes formaldehyde to produce sugars (see fig. 4.7). In both cases, if provided with sufficient input materials (fuel and oxygen for the former, formaldehyde for the latter), these reaction networks replicate. This replication is obvious in the case of fire, but it happens in the formose reaction too; the reaction starts off very slowly as the spontaneous formation of a molecule of glycolaldehyde is rare. But as the reaction progresses, one glycolaldehyde is converted into two with each turn of the cycle, exponentially accelerating (replicating) the reaction over time.

Günter Wächtershäuser, a German chemist turned patent lawyer who started dabbling in origins-of-life chemistry in his spare time, proposed just such a metabolic network in the late 1980s. Wächtershäuser postulated that an assembly-line network of chemical reactions, which he dryly described as "two-dimensional chemi-autotrophic surface metabolism in an iron-sulfur world," was set up on the surface of catalytic minerals, such as iron sulfide, in hydrothermal vents deep beneath the sea. The hydrothermal vent environment can drive the reactions by using the chemical disequilibria set up when the hot, briny water emerges into the colder ocean.

Wächtershäuser's hypothetical "pioneer organism" was composed of a mineral substrate on which transition metal atoms, such as iron or nickel, catalyzed the reduction of carbon dioxide to other small, organic molecules using the reducing power of hydrogen sulfide. These molecules remained on the mineral surface and served as ligands—small molecules that bind to and thus modify the chemical activity of the transition metals. By improving a metal's catalytic activity, some of these new ligands formed an autocatalytic "surface metabolism" in

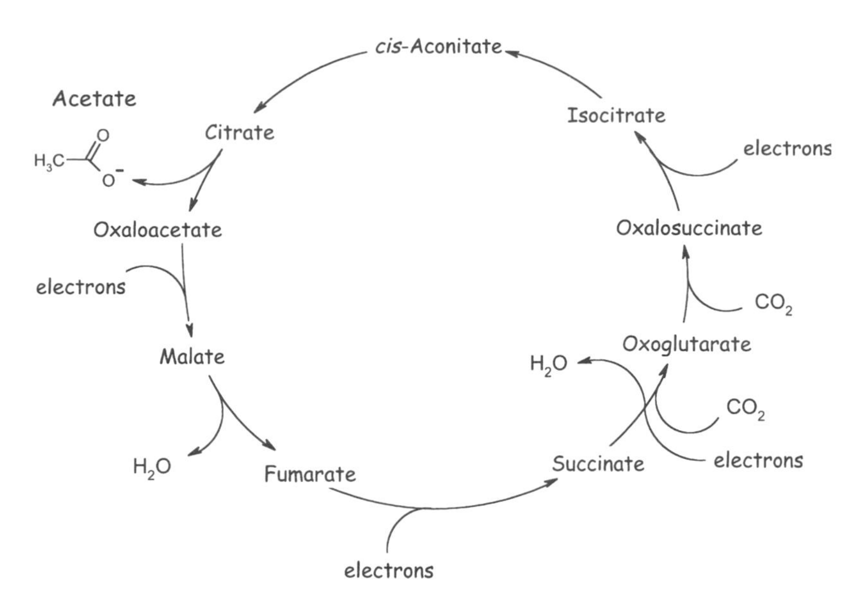

Figure 5.1 Günter Wächtershäuser argues that life could start as purely metabolic networks spontaneously arising on the surface of iron sulfide minerals. For example, he postulates that a reverse (reductive) Krebs (or citric acid) cycle can be catalyzed on the mineral troilite (FeS), which (in net) takes the reducing potential of hydrogen sulfide and uses it to reduce carbon dioxide to larger organic molecules. To date, however, few if any of the steps in this postulated reaction network have been demonstrated in the laboratory.

which the formation of new reactive centers was accelerated by the very same small molecules that they produced. Specifically, Wächtershäuser proposed that the key set of reactions (fig. 5.1) in his "metabolic life" formed a backward, reducing version of the Krebs (citric acid) cycle—the central *oxidative* biochemical pathway in aerobic organisms. As the mineral substrate for this activity, Wächtershäuser proposed troilite (FeS), a now rare form of iron sulfide that was probably more plentiful on the early Earth (under current conditions, the more oxidized iron-sulfur mineral pyrite, FeS_2, is far more common). The troilite, in turn, can catalytically reduce various organic molecules, converting itself back into pyrite. Wächtershäuser also suggested that the surface of iron sulfide would constrain the distribution and orientation of the products of each reduction in such a way as to support a complex, self-sustaining sequence of metabolic reactions that, ultimately, would lead to the formation of new catalysts and new metabolic pathways. Thus, argued Wächtershäuser, once an autocatalytic metabolism is established, it will go on to produce increasingly complex catalysts, increasingly complex product molecules, and increasingly numerous and diverse metabolic pathways.

In support of his hypothesis, Wächtershäuser pointed to at least a few potentially relevant reduction reactions that can take place on iron sulfides. For example, in collaboration with Claudia Huber at the Technical University of Munich, Wächtershäuser showed that, in the presence of iron and nickel sulfides, hydrogen sulfide reduces carbon *mon*oxide to acetic acid—carbon monoxide, though, not the carbon *di*oxide that is essential to his postulated reductive citric acid cycle. The two researchers have similarly shown that, in the presence of ammonia and hydrogen sulfide, iron and nickel sulfides can reduce a specific class of molecules called α-ketoacids to the α-amino acids used in proteins. So, it seems that biochemistry-like reductive reactions can take place in the presence of mineral sulfides. It remains very much an open question, though, whether a complete metabolic network could spontaneously self-organize (and operate autonomously) on an iron sulfide surface under plausible prebiotic conditions. What, after all, would drive a collection of seemingly dissimilar chemical reactions to spontaneously self-organize?

Moving forward from Wächtershäuser, William Martin of the University of Düsseldorf and Michael Russell of NASA's Jet Propulsion Laboratory in California have argued that similar metabolic pathways formed within small, iron sulfide–lined cavities in the rocks of hydrothermal vents are a better candidate for the first life. In part, this is because the cavities would tend to retain any newly formed metabolic products at higher concentrations than would occur on a mineral surface in the open ocean. Moreover, steep temperature gradients inside the walls of the vent would increase the range of reactions that are feasible. Martin and Russell argue, for example, that the synthesis of small, monomeric molecules such as amino acids, which requires the input of substantial energy, would occur in the hotter regions and their polymerization, which is reversed at high temperatures, would progress more readily in cooler regions.

Critical to the work of Wächtershäuser and of Martin and Russell is the argument that the self-organization inherent in their putative metabolic pathways stems from the constraint of molecules being adsorbed onto (and synthesized on) the surface of a mineral. The organization imposed by the mineral surface, they argue, would foster the formation of reaction networks and increase their specificity. To date, however, no such mineral-driven guidance has been observed for any set of chemical reactions even remotely approaching the complexity that would be required to form a self-sustaining network that fed on simple precursors and synthesized larger molecules. Perhaps this is to be expected: any one mineral is unlikely to specifically catalyze more than one or two of the many distinctly different chemical reactions required for a metabolic network as complex as the Krebs cycle, much less catalyze *only* those reactions that are productive and not competing side reactions that would disrupt the network. Moreover, even if such multicatalytic minerals did exist, there is no laboratory evidence that complex catalytic networks could spontaneously self-organize on their surfaces. Thus, while the iron-sulfur theory has stimulated discussion and research, it is, at least currently, not based on firm laboratory verification. Nor are there any laboratory-based hints that something like this might be possible under some as yet to be defined set of conditions.

Of course, we might still feel confident about the possibility of iron-sulfur life, even in the absence of laboratory examples, if we could come up with an argument in favor of evolution in the iron-sulfur world and, likewise, could compellingly explain how we could have evolved from such a start. That is, even though our interest is in defining the range of possible forms that life can adopt anywhere, not just on Earth, if we could argue that we had evolved from iron-sulfur life, we'd feel more confident about the possibility of iron-sulfur-based metabolism as a precursor to life on both the early Earth and elsewhere in the cosmos. And while there are small "iron-sulfur" clusters (consisting of up to four irons and four sulfurs) in some of the proteins that catalyze oxidative biochemistry, no clear route from "there" to "here" has yet been described. For example, why would nucleic acids, so central to biology on Earth, provide a selective advantage for these systems? For that matter, given that the FeS itself is not replicating, what, if anything, would provide a selective advantage—which is first and foremost about improving the ability to reproduce—for such a system? In the absence of compelling answers to these questions, we consider it unlikely that our very first ancestor was a gene-free metabolic network quietly chugging away on the surface of some mineral, and in the absence of a compelling demonstration of such chemistry in the laboratory, we must question whether any life anywhere could arise from such a start.

An alternative twist on the metabolism-first idea is that the first living organisms were not made up of the sort of complex, multistep networks that we typically envision when we say "metabolism" but instead were built from metabolic-like pathways involving lipid aggregates. Lipids are "water-hating" molecules (termed "hydrophobic" in biochemistry) and thus, in solution, lipids tend to spontaneously organize into compact spheres called micelles or into hollow, water-filled, membrane-bound balls called vesicles (soap is a lipid that spontaneously forms micelles around grease, solubilizing it and allowing it to wash down the drain). Even today, most biological membranes are not created from scratch but rather through a growth-and-division process that is at least somewhat analogous to life. And the coacervates Oparin studied in the 1920s could sometimes be coaxed to adsorb smaller precursors, and even to divide. But replication alone does not life make. Is there a plausible chemical scenario in which a blob of membranes could be said to evolve? Doron Lancet of the Weizmann Institute in Israel and Dave Deamer at the University of California, Santa Cruz, have argued that there might be (fig. 5.2).

The "lipid-world" argument starts with the observation that lipids are an extremely diverse set of compounds. For example, modern eukaryotic cells contain three broad classes of lipids in their cell membranes: phospholipids, sphingolipids, and sterols. The first two differ in terms of the polar, water-loving "head group" attached to the water-hating fatty acid tail of the lipid. Sterols, however, represent a different design, the most common example of which is cholesterol. Each of these three classes, in turn, can be subdivided by chemical details, such as the chemistry of the water-hating tail, chemistry of the water-loving head group, and so on. There are dozens of distinct types of lipids in the

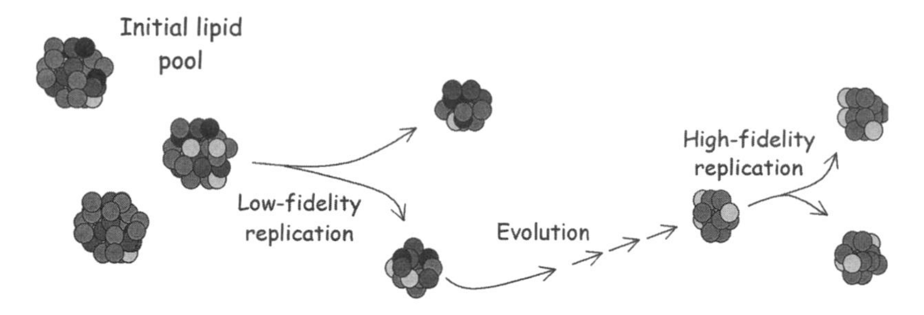

Figure 5.2 The lipid-world hypothesis is predicated on the thought that there are so many different types of lipid that some rare collections of them might exhibit the property of adsorbing new lipid components in the same ratio as the existing ratio in the aggregate, thus swelling and eventually splitting to produce offspring similar to themselves. If this "breeding" were not quite perfectly "true," any mistakes—"mutations"—that improved the aggregate's rate of growth and division would cause the system to evolve.

membranes of our cells, dozens and dozens of others in the membranes of prokaryotes, and countless more that can be synthesized in the laboratory and may have arisen prebiotically. With such tremendous diversity, who can say what range of chemical and physical properties is within reach? More specifically, notes Lancet, is it not possible that among the myriad of possible lipid compositions in a simple vesicle there might be some particular mixture that encourages the adsorption of other lipids in exactly the right ratios to grow and produce offspring *with almost exactly the same composition as, and thus very similar physical and chemical properties to, the parent vesicle*? That is, might there exist some "special" composition of lipids that has the ability to "breed true" and thus accurately pass down its traits to the generations that follow, such that lipid "A" attracts lipid "B" into the blob, and then lipid "B" attracts lipid "C," and so on until, finally, lipid "Z" completes the cycle (network) by drawing in lipid "A"? And, if this is so, is it much more of a stretch to postulate that these lipid aggregates could sometimes accidentally take on some new lipid component that would improve the efficiency with which they reproduced or improve their ability to survive and grow in some new environment *and* that also would allow them to breed true and pass these properties to their offspring? Were such lipid aggregates possible, they could be said to be both reproducing and evolving. They could be said to be living things.

The lipid-world hypothesis nicely explains how a presumably very dilute primordial soup could spontaneously organize into complex structures; to minimize the extent to which their hydrophobic elements are exposed to water, lipids readily self-assemble into micelles or vesicles, even when at extremely low concentrations. That said, the lipid-world hypothesis is still just that, a hypothesis. For example, to date, the micelle and vesicle chemistry we have observed in the laboratory just isn't selective; no one has yet come up with a mixture of lipids that even comes close to breeding true, much less one that starts there and slowly evolves. One problem is that the aggregation of lipids to form micelles and vesicles is generally nonspecific; only relatively poor differentiation is observed. And, thus, there isn't much in the way of laboratory evidence supporting the ability of lipid vesicles to breed true, a key element of the lipid-world hypothesis. A second problem is shared with the iron-sulfur hypothesis: how would this evolve into us? Even though our cells are enclosed within membranes (and, indeed, membranes may have played an absolutely critical role in the evolution of life—just not a solo role!), lipid-world chemistry is hardly reminiscent of our most fundamental metabolic and genetic chemistry. To quote its proponents, "A complex chain of evolutionary events, yet to be deciphered, could then have led to the common ancestors of today's free-living cells, and to the appearance of DNA, RNA and protein enzymes."* And thus, the lipid-world hypothesis remains an interesting speculation, backed perhaps by theoretical models but lacking much in the way of supporting empirical observations.

Metabolism-first theories, such as the iron-sulfur world or the network of interactions inherent in the lipid-world model, face yet another, perhaps even more fundamental, difficulty: gene-free networks are generally resistant to evolutionary change because such change would require that multiple mutations occur simultaneously.** Let's look at a network in which product A catalyzes the formation of

*Daniel Segré and Doron Lancet, "Composing Life," *EMBO Reports* 1, no. 3 (2000): 217–22.

**We should note that, over a beer or two, Lancet and Deamer have argued to us that their lipid-world hypothesis is a "genes-first" (discussed next) rather than "metabolism-first" hypothesis, but to our thinking, their model envisions a cooperative, complex network of interactions between lipids and not the one-to-one sort of templating we think of when we say *gene*.

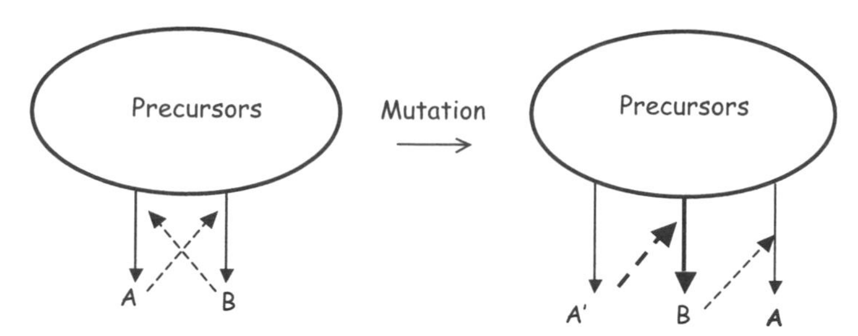

Figure 5.3 A generic problem with metabolism-first theories of the origins of life is that gene-free networks are generally resistant to change. Let's consider a network in which product A catalyzes the formation of product B, which catalyzes the formation of more product A. This hypothetical network would be a self-replicating chemical system, but it does not meet our definition of life as it is exceedingly difficult for such a system to *evolve*. Even though there may be an A-like molecule, A′, that is a better catalyst of the formation of B, the B so produced *catalyzes the formation of A, not A′*. Only in the extremely unlikely event that A′ catalyzes the formation of B′ (which catalyzes the formation not of A but A′) will this system have evolved to a more fit state. For more complex networks (e.g., A catalyzes the formation of B, which catalyzes the formation of C, which catalyzes the formation of more A), the problem compounds exponentially with the number of nodes in the network.

product B, and vice versa (fig. 5.3). It's always possible that there is some "mutant" version of A, say A′, that is a *better* catalyst of the formation of B. Thus, the mutant A′ would provide a selective advantage for the putative metabolism-only organism. But while A′ is a better catalyst, it is still likely to catalyze the formation of B, and B catalyzes the formation of A, not A′. To provide an "inheritable" selective advantage, the mutant A′ must instead catalyze the formation of a mutant B′ that, in turn, catalyzes the formation of A′. That is, for networks to evolve, they require *multiple, simultaneous mutations.** And mutation probability does not change algebraically with the number of mutations but changes geometrically: the formation of two simultaneous mutations is not *twice* as improbable as the formation of one mutation but is, instead, the *square of the improbability* of forming one. Given

*The internet, with its complex network of servers, is similarly robust. If any single node is knocked out, others are there to take its place. And thus, changing one node does not change the network itself. In fact, the internet was originally built under funding from the US Department of Defense as a communications network that could survive nuclear war.

that even single mutations must be relatively rare events—if mutations occur too readily, too few of the offspring will be viable—if we square the probability (or cube it, if it is a three-part network, and so on), the ease with which mutations produce a selective advantage diminishes very rapidly. It is thus difficult for complex networks to evolve via the slow accumulation of stepwise mutations; at a very fundamental level, it seems unlikely that networks could give rise to life before the emergence of genes because, in contrast to networks, *genes allow for the stepwise, additive accumulation of favorable mutations.*

Genes First

The genes-first camp argues that the first living organisms were likely to be genes, that is, information-containing single molecules *that could catalyze their own replication*. These simple, self-replicating molecules would then, under the influence of selective pressures, evolve into increasingly complex organisms and, on Earth at least, could eventually have evolved into organisms with complex biochemistry such as ours. The genes-first hypothesis has a significant advantage over the metabolism-first camp. Namely, while networks are fundamentally resistant to evolutionary change, a single molecule that catalyzes its own formation can evolve more readily *if modifications of the catalyst breed true* and can be passed down through the generations (fig. 5.4). The

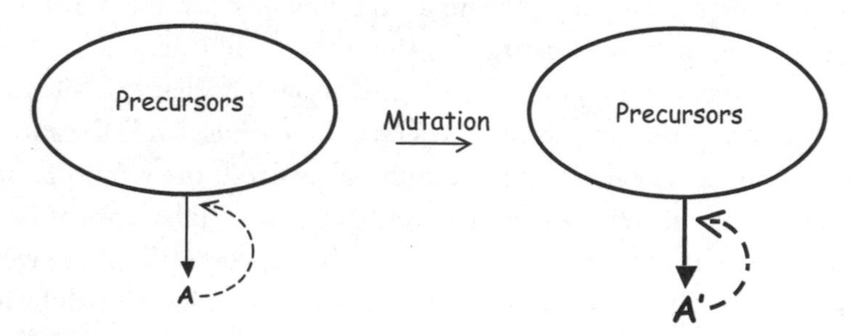

Figure 5.4 A solution to the network problem is possible if there is a single molecule, A, *that can catalyze the template-directed formation of copies of itself.* In this case, a mutation producing a more fit molecule, A′, say with better catalytic activity, would be a heritable change.

question, then, is, Can we think up chemically plausible catalysts for which this holds? Excitingly, several possibilities have been suggested.

Even the simplest self-replicating chemical systems, crystals, have been seen to exhibit the property of modifications that breed true. For example, irregularities in the surface of a crystal, such as the so-called screw dislocations that cause the surface of the growing crystal to spiral upward rather than form discrete layers, can continue to propagate as a crystal grows and, in some circumstances, can breed true if the crystal shatters and nucleates the formation of more crystals. Once again, though, we must remember that even with mutations, replication alone is not sufficient to meet our definition of life. Even self-propagating modifications do not constitute evolution unless they provide a selective advantage for the "organism."

A. G. Cairns-Smith (1931–2016), while at the University of Glasgow, suggested that some minerals have the right stuff to form "mineral genes" that can evolve and thus could have been the first living things. Clay is made up of sheets of charged alumina and silicates packed together in alternating layers (fig. 5.5); the slickness of wet clay arises when water intercalated between the layers lubricates them and allows them to slide past one another. One common clay, montmorillonite, has the formula $(Na, Ca)_{1/3}(Al, Mg)_2(Si_4O_{10})(OH)_2 \cdot nH_2O$. Now, you may have noticed several odd things about this formula. For one, there's the "nH_2O," which simply reflects the fact that different amounts of water can be inserted between the layers of clay. For another, have you ever seen a chemical formula that contains commas? This one does because positively charged sodium (Na) and calcium (Ca) ions can replace one another at some of the negatively charged "sites" on the surfaces of the alumina or silica layers and, thus, while the sum of the sodium plus calcium in the mineral is fixed, their ratio is not. Likewise, aluminum (Al) and magnesium (Mg) ions are similarly sized and thus can replace one another in the clay's alumina layers, leading to an effectively infinite number of molecules each differing from the others in its ratios of sodium to calcium and magnesium to aluminum. These clays form two-dimensional, water-separated sheets in which one sheet might or might not look exactly like the sheet above or below it.

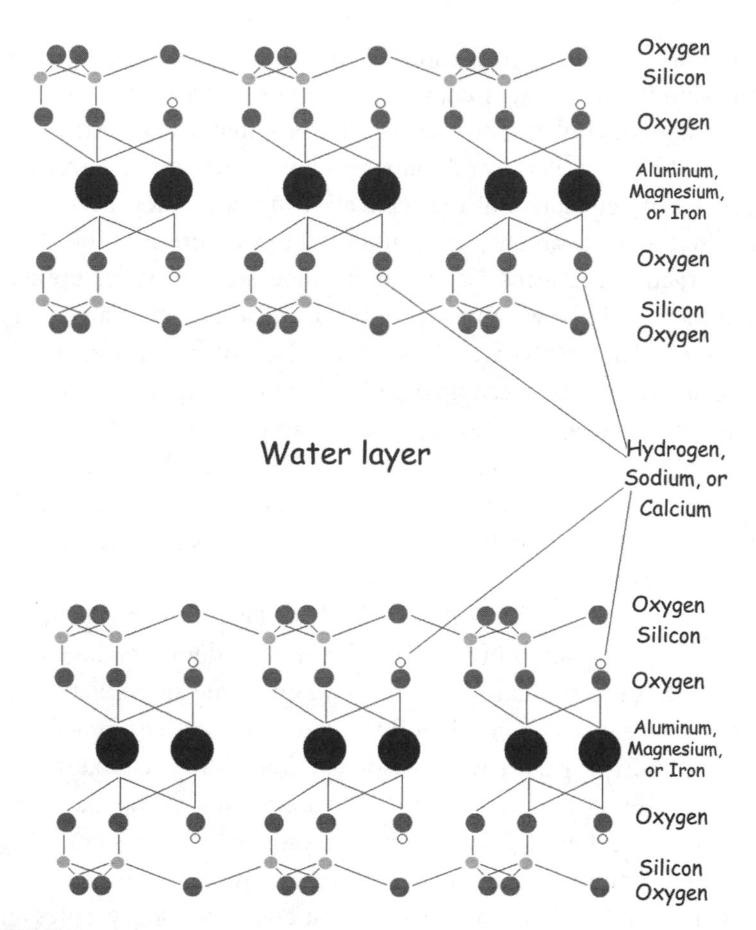

Figure 5.5 Clays are aluminosilicate minerals consisting of alternating sheets of alumina and silica. Oxygen atoms at the faces of these sheets take on a negative charge and bind positively charged ions, such as sodium or calcium. Other ions, such as iron or magnesium, sit within the mineral lattice itself. Shown here is the structure of the montmorillonite family of clays, in which an alumina layer is sandwiched between two silica layers, and this triple-layer structure is linked to others by weak, ionic interactions. Water disrupts these ionic interactions, allowing adjacent layers to slide against one another, which accounts for the slickness of wet clay.

Cairns-Smith theorized that the sheet-to-sheet variability in clays might allow them to act as "genes." That is, the specific array of charged ions in one layer of clay could act as a template and catalyze the formation of a complementary new layer of clay—the "next-generation" layer. Any "mutations"—that is, mistakes in the packing of the alumina or silica, or changes in the positive ions that associate with them to neutralize their negative charge—that occur during copying would be inherited by all follow-on layers, and if these mistakes improved the efficiency of the replication process, they would provide a selective advantage, thus fulfilling our definition of life. These first *in*organic organisms, Cairns-Smith suggested, provided the scaffold on which life as we know it—built of sugar and spice and everything nice—later evolved. Cairns-Smith also noted that ions in clay can act as catalysts to speed up organic chemical reactions, including, as we noted in chapter 4, the polymerization of RNA, and thus over the course of millions of years could have been a significant source of biologically relevant polymers in the primordial soup. Despite this potential strength, though, Cairns-Smith's theory is not without serious weaknesses.

Perhaps the dominant weakness in this "clay-world" hypothesis is that no chemistry anything like it has ever been demonstrated in the laboratory. It thus remains very much an open question whether defects and irregularities in mineral sheets can be made to breed true, such that they propagate with reasonable fidelity to the next layer of clay. Likewise, it may be telling that no remnants of the clay-based metabolism exist in modern metabolism on Earth: there is nothing in current biochemistry that looks remotely like silicate minerals. Cairns-Smith argued that this simply means that the original clay genes and catalysts have been completely and utterly replaced, leaving not the slightest vestigial traces. This is, of course, quite possible: clays are not particularly good catalysts and thus we might expect them to be completely abandoned after evolution invented better catalysts based on organic materials. Still, were such vestiges present, they would be nice, tangible evidence supporting such origins. In their absence, we are left only with thought-provoking, but as yet experimentally unsupported, speculations.

What we really want is a "replicase." That is, some relatively simple molecule that, unlike the hypothetical clay chemistry of Cairns-Smith, has been shown in the laboratory to be able, if not to make more copies of itself (which is our goal), at least to make molecules similar to itself. The first simple examples of such molecules were provided by two independent research groups at the Scripps Research Institute in La Jolla, California. In the first, Reza Ghadiri and his research team designed a polypeptide (a short protein) that autocatalytically directs the synthesis of copies of itself from two smaller, chemically activated polypeptide fragments, each consisting of half of the template sequence. Likewise, Gerald Joyce has demonstrated not only molecular replication but even rudimentary evolution in the test tube using RNA molecules that catalyze the fusion of smaller RNA molecules into new molecules that support the linkage reaction.

Nevertheless, systems such as these are probably poor analogs for the origins of life itself: as Ghadiri argues, these examples simply show how molecules that copy themselves from (slightly) simpler precursors are physically possible. But they are poor candidates for the origins of life because plausible prebiotic reactions are very unlikely to generate the short polymers from which these reactions proceed. A similar problem has arisen in efforts to invoke DNA as the original self-replicating molecule. In 1968, Leslie Orgel began to explore the idea that a single strand of DNA could serve as a template on which activated nucleotides could spontaneously polymerize to form the complementary sequence. After decades of work, the conclusions are mixed. Polymerization can proceed reasonably efficiently starting from phosphorimidazolide monomers (an activated nucleotide that we discussed in chapter 4), but only on templates that are rich in the nucleobase cytosine. For example, the sequence GGCGG is obtained in 18% yield from the template CCGCC. But cytosine-poor polymers make very poor templates, and thus the GGCGG product cannot be used to synthesize more of the CCGCC starting material; the reaction grinds to a halt after just one round and is therefore not self-replicating. Orgel's idea also fails our second test—there is no evidence of such origins remaining in our biochemistry. That is, the high-energy nitrogen-phosphorous bonds that drive the polymerization of phosphorimidazolide-activated monomers are not seen anywhere in current biology, dealing this idea another blow with respect to its having played a role in the origins of life on Earth.

Orgel's work with DNA-templated DNA polymerization, Ghadiri's work with self-replicating peptides, and Joyce's work with self-linking RNA raise another chicken-or-egg conundrum. Namely, in current Terrestrial biochemistry, DNA encodes the information necessary to make proteins, and proteins are required in order to copy DNA. It is thus not at all clear whether either of these two species can replicate without the other, and thus it is not at all clear that either of these two could have been the first, self-replicating molecule. A more promising candidate for the chemistry of the origins of life would be a molecule that has been shown to have the capacity, at least in a small way, to copy itself from simpler, monomeric precursors. Just such a molecule may be at hand in the form of RNA.

The RNA World

As we hinted above, by the late 1960s, Francis Crick (1916–2004; co-discoverer of the structure of DNA), Leslie Orgel (we've met him before), and Carl Woese (1928–2012; we'll get to know him better in chapter 7) had each independently developed the hypothesis that nucleic acids could have formed the basis of the first living things, a hypothesis that motivated much of Orgel's later work on DNA-templated DNA polymerization. But underlying that work was the assumption that some inorganic catalyst would be responsible for the polymerization reaction. While the ability of RNA molecules to fold up into the sort of intricate, three-dimensional structures associated with protein-based catalysts was well established by the early 1970s, the intellectual stance that proteins are *the* biological catalysts was so firmly entrenched—indeed, it was considered a major element of biology's *central dogma*—that the idea of nucleic acids as catalysts seemed far-fetched. Molecular biologists had relegated RNA to secondary, or more precisely, tertiary importance. In cellular organisms like us, for example, RNA is neither the genetic material (a job carried out by DNA) nor the catalytic machinery by which most of our metabolism is conducted (a job carried out by proteins). Instead, RNA was long thought to serve such lowly (if essential) tasks as messenger RNA's role in transporting genetic information from the DNA, where it resides, to the protein-synthesizing ribosomes. That said, RNA was known to serve as the genetic material in a few small viruses (HIV and the coronaviruses, for example, encode their genetic information in RNA). If only RNA could catalyze reactions as well! Then we'd have what we are looking for: a molecule that could form the basis of both genetics and metabolism. And in 1981, just such catalytic activity was discovered, like many truly revolutionary observations, essentially by accident.

The catalytic possibilities of RNA were discovered in the laboratory of Thomas Cech at the University of Colorado. Cech and his students were studying the mechanism by which ribosomal RNAs, key components of the ribosome, are "processed." Two years earlier, it had been discovered that the information contained in many RNA molecules is not continuous; the coding regions of freshly synthesized RNAs are interrupted by apparently meaningless sequences called introns that must be spliced out of the immature RNA before it is used. Cech found that cellular extracts from his favorite study organism, the single-celled paramecium *Tetrahymena*, could carry out the splicing reaction in the test tube. When faced with such an observation, the natural thing for a biochemist to do is to try to "fractionate" the extract into its component parts in order to discover which protein in the system is responsible for the catalytic activity. The first fractionation experiment they attempted, using extracts from the cell nucleus, carried out the splicing reaction perfectly, suggesting that the investigators were on the right track. But there was a fly in the ointment. The test tube "next door" to the nuclear extract was a negative control that contained the RNA but lacked the nuclear extract, and yet splicing seemed to have occurred there as well. At first, they thought they had just mixed up the tubes, but repeating the experiment another *five* times produced the same puzzling result. Was the problem that the RNA was contaminated with a small amount of *Tetrahymena* proteins, including, perhaps, the splicing protein? After all, the RNA had also been purified from *Tetrahymena* extracts, so some of the relevant proteins could have come along for the ride. Treatment of the RNA with proteases, though, which should destroy any leftover proteins, did not stop the reaction, suggesting that contaminants were not responsible for the observed splicing. Finally, in desperation, the research group synthesized the RNA in *Escherichia coli*, which is quite unrelated to *Tetrahymena* and thus produces RNA samples entirely free of *Tetrahymena* proteins. And still the supposed control RNA spliced. With such compelling and undeniable evidence in hand showing that the reaction could occur even in the absence of *Tetrahymena* proteins, Cech finally went public with a positively revolutionary idea: an RNA molecule could splice itself *without the help of proteins*. And while, in the strictest sense, the RNA splicing reaction is not catalysis—a true catalyst remains above the fray and emerges unchanged from the chemical reaction it fosters—this result was the first hint that RNA could accelerate biochemical reactions in a manner previously thought solely the realm of proteins.

It quickly became apparent that Cech's autocatalytic RNA was not a one-off example. While Cech's group was conducting these experiments, Sidney Altman at Yale University had been studying another catalytic RNA-processing step—the maturation of transfer RNA (more about this type of small RNA in chapter 6). One of the last steps in the process is cleavage of the RNA to remove several nucleotides from one end. The purified cellular component that carries out this process is a two-molecule complex consisting of a protein tightly bound to an RNA. Altman had been assuming that, as the central dogma stated, the *protein* part of the complex is the catalytic agent and the RNA plays some structural role or serves to bind the enzyme's target. But when Altman's students attempted to run the reaction using the purified RNA component alone, they found that, in the presence of a large quantity of magnesium ions (presumably to take the place of the protein in its role of neutralizing the negative charges on the RNA and allowing it to fold), the RNA was the catalyst. Indeed, it was the protein that serves a merely structural role! For these discoveries, Cech and Altman shared the 1989 Nobel Prize in Chemistry.

In the years since the first ribozymes, short for *ribo*nucleic acid (RNA) en*zymes*, were discovered, the list of catalytic roles that RNA can play has grown (fig. 5.6). For example, it has since been shown that the ribosome, the massive protein and RNA "factory" in which proteins are synthesized, is a ribozyme: proteins, it turns out, are syn-

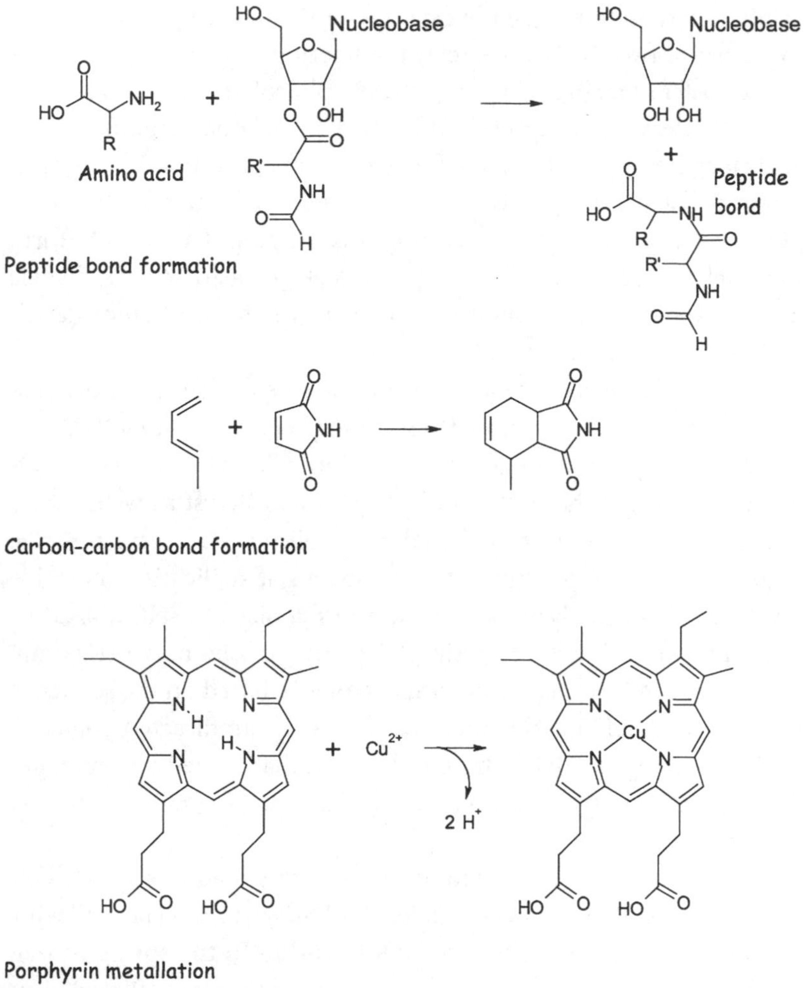
HO
Nucleobase
O
NH2
HO
+
O
OH
R
Amino acid
R'
NH
O
H
Peptide bond formation
OH OH
+
Peptide
bond
HO
NH
R
R'
NH
+
NH
Carbon-carbon bond formation
N
H
N
H
N
N
+ Cu2+
2 H+
Cu
HO
O
Porphyrin metallation

Figure 5.6 While only a handful of naturally occurring ribozymes have been identified, a number of artificial ribozymes have been generated in the laboratory. Shown here are three examples that illustrate the broad range of reactions within reach of ribozyme-based catalysis.

thesized by the catalytic activity of RNA, an important point to which we'll be returning in the next chapter. Meanwhile, the catalytic activity of RNA is far more diverse than indicated by the few examples that natural selection provides; many more catalytic functions have been produced by artificial selection in the laboratory. These additional catalytic activities include the cleavage of DNA, the ligation (linking together) of two RNA molecules, the ligation of DNA, acyl transfer (the transfer of an organic group from a phosphate to another molecule), the cleavage of peptide bonds (the type of bond that links amino acids together in a protein), the formation of peptide bonds from activated, acylated compounds, and the insertion of a metal into a porphyrin (a large, nitrogen-containing organic ring structure) to form the catalytic organometallic compounds at the heart of such diverse metabolic processes as photosynthesis (in the form of chlorophyll) and oxygen transport (in hemoglobin).

Did the discovery of ribozymes provide a solution to the origins-of-life question? As we've noted, it had long been known that RNA can serve as the repository of genetic information, and by the early 1980s it was established that RNA could serve as a catalyst as well. Could RNA, then, have been the original self-replicator? To serve this role requires a very special kind of catalysis: a self-replicating molecule must be able to catalyze the formation of copies of itself. RNA, like other polymers, forms from simpler precursors by polymerization. The formation of a specific polymer sequence based on the sequence of another, templating polymer is called template-directed polymerization. Was the first living thing a ribozyme that carried out template-directed RNA polymerization? Is such a ribozyme even physically plausible?

One of the first scientists to head down this road was Walter Gilbert, the Nobel Prize–winning molecular biologist from Harvard who, in 1986, first coined the phrase "RNA world." In the spring of that same year, when on a tour of potential grad schools, I (Plaxco) was sitting in Professor Gilbert's office when a student interrupted, popping her head in the door and simply saying, "nothing's come up yet." Professor Gilbert looked at me thoughtfully for a moment, said "You look like you can keep a secret," and then proceeded to describe the experiment his student was conducting. She had synthesized a test tube full of random RNA sequences, fed them with activated nucleotides, and then, once a week, was checking to see whether any one sequence was replicating and thus increasing in number. Given that, in the many years that have since passed, the results of said experiment have not appeared in print, it's a safe guess that none of the sequences ever started to dominate the mix (it likewise seems that the statute of limitations has run out on the promise not to tell this story). The experiment's failure, though, might have been quantitative rather than fundamental. That is, while a test tube full of completely random, 70-base RNA polymers contains 10^{14} different sequences, this represents only a miniscule (one in 10^{28}) fraction of the more than 10^{42} possible sequences that can be encoded by a 70-base RNA polymer. Perhaps sequences with the properties we want are less common than one in 10^{28}?

The question then becomes how to improve the odds. One approach would be to start with a larger test tube (more random sequences), but the most RNA that can reasonably be synthesized and handled is only a few hundred times the amount that Gilbert's team used. Realizing this, David Bartel's group at the Massachusetts Institute of Technology took a different approach. Specifically, they applied two new tricks. First, to avoid having to search through a huge morass of unfolded RNA polymers (no well-defined three-dimensional structure equals no complex chemical activities), they started with a folded, catalytically active ribozyme sequence. Specifically, they employed a sequence that Bartel had discovered when he worked in Jack Szostak's lab and that already performed chemistry similar to the kind they wanted: the ribozyme was a ligase, a type of enzyme that catalyzes the splicing together of two longer RNA polymers and thus creates the kind of phosphodiester bonds that are also formed during polymerization. Indeed, the ligase ribozyme is already capable of polymerizing RNA, but only up to three nucleotides and not in the necessary template-directed fashion. Bartel's group then mutagenized the ligase to create a "random" starting pool of some 10^{15} different RNA sequences, all of which had significant similarity to the parent ligase. Second, they employed a clever trick to ensure they could pick out sequences that had *any* template-directed polymerase activity, even if the activity was insufficient to actually support full-fledged self-replication. To do so, they added to every element in their mutated library a short "priming sequence" that acted as the initiation site for the polymerization reaction, along with a short RNA that would act as an internal template to guide the sequence-specific polymerization reaction.

To fish for the (presumably) rare sequences with the desired activity, Bartel's group added some activated nucleoside triphosphates to the pool of molecules, including the modified, sulfur-containing nucleotide 4-thiouracil triphosphate (4-thioUTP). Any RNA molecule in the pool that could polymerize a short stretch of RNA on its template would incorporate this 4-thioUTP into its structure as 4-thiouracil. Bartel's students then separated out any such 4-thiouracil-containing RNA molecules by forcing the RNA products of their reaction to migrate through a mercury-containing gel; mercury binds to the sulfur, impeding the motion of thio-containing RNAs. They then isolated the slowly migrating RNA molecules, converted these into their complementary DNA sequences (using an enzyme called reverse transcriptase), copied and amplified this pool of DNA sequences (using the enzyme DNA polymerase), and, finally, converted the amplified pool of DNA sequences back into RNA (using the protein RNA polymerase)—phew!—before starting the cycle over again. During the first few cycles, they uncovered mostly mutant RNA that had "evolved" the property of covalently binding 4-thioUTP without catalyzing template-directed polymerization. But with each additional round, the population of RNA molecules that exhibited polymerase activity (and thus incorporated more than one 4-thiouracil) increased. Ultimately, after *18* rounds of selection, copying, reengineering, reamplification, and reselection, Bartel group member Wendy Johnsen recovered a 189-nucleotide-long RNA, the "18.12 polymerase," that could use a template to polymerize a specific, complementary sequence of RNA. That is, they'd found molecules that could catalyze the copying of molecules of their own kind.

But is the 18.12 polymerase ribozyme the self-replicating "replicase" we are looking for? It's promisingly close, but it's not there. Even this "best" RNA polymerase extends its primer by only 14 nucleotides before the ribozyme, which itself is relatively unstable, breaks down. Moreover, it only works if given an unstructured (i.e., unfolded) template; it is incapable of unwinding the sort of structure it, itself, adopts, and thus it cannot use something as complex as itself as a template.

Building on the Bartel group's pioneering work, a number of other researchers have since "evolved" more efficient ribozyme polymerases. A promising example is the work of David Horning and Jerry Joyce, both then at the Scripps Institute in San Diego, California. Horning started with the 18.12 polymerase, mutagenized it, and then selected from among a pool of a hundred trillion variants those best able to copy two structurally complex RNA templates. Specifically, Horning selected for the ability to generate RNA sequences that fold into the precise three-dimensional shapes required to bind the small molecules guanosine triphosphate and vitamin B12 and then used a combination of binding to these molecules and size separation as a means of fishing out those sequences that had succeeded in copying these longer templates. Repeating this cycle a total of 24 times, each time increasing the selective pressure by increasing the length of the required polymerization and decreasing the time allowed to generate it, Horning identified a polymerase that not only exhibited a dramatically improved ability to copy complex templates, but that also polymerized RNA a hundred times faster than the best rate previously achieved. The catalytic efficiency of the new ribozyme is so great that it can synthesize a highly structured, 76-base yeast transfer RNA (more on this type of RNA in the next chapter), albeit at a yield of only 0.07%.

Moving forward, in 2020, Katrina Tjhung, also working with Joyce, reported the best ribozyme RNA polymerase described to date. To do so, she devised a new selection scheme in which successful polymerases would generate the hammerhead ribozyme, which is an RNA sequence that performs RNA cleavage at a specific RNA sequence. She then added the complement of the hammerhead ribozyme as a template to Horning's best polymerase, mutagenized the polymerase, and attached the library of mutant sequences to a solid support. Any ribozyme that copied the hammerhead complement to form a hammerhead ribozyme would, by doing so, cleave itself off the support such that it could be collected, amplified, and mutagenized to set up the next round of selection. Tjhung performed this cycle 14 times (coupled with Horning's work, this pushed her 38 generations past the 18.12 polymerase) while progressively reducing the time allowed for polymerization from two hours to five minutes in order to select for more rapid catalysis. Using this approach, the best ribozyme polymerase Tjhung derived was able to synthesize the above-described transfer RNA in 2.4% yield; while this is still low, it is more than 300 times better than the yield its immediate ancestor was able to achieve.

This 179-base ribozyme efficiently copies sequences more than a fifth of its length and can, with low yield, even copy something more than a third of its length. Thus, if the catalytic abilities of the ribozyme were increased just a few fold, it would be efficient enough to create copies of itself. Product length, however, is not the only metric we need to concern ourselves with: the fidelity of replication is also crucial. Specifically, self-replicating things must breed true enough that a fair fraction of their offspring are functional enough to reproduce. Since deleterious mutations are more common than neutral or helpful mutations, this means replication fidelity must be high. Conversely, too high a reproductive fidelity inhibits evolution; if all your offspring are perfect carbon copies of you, then none of them will have any selective advantage over any of the others. These two contravening effects together produce an optimal level of fidelity that is somewhere between "high enough that sufficient numbers of offspring survive and reproduce" and "low enough that evolution has at least the occasional mutation to work with." This trade-off is seen in the fact that, even today, evolution *optimizes*, rather than maximizes, the fidelity of replication: the mutation rates in today's organisms are tuned such that, irrespective of the size of the genome, there is, on average, one mutation per genome per generation.

The above arguments suggest that an optimal level of fidelity for the RNA polymerase ribozyme would be about one mutation per generation, which corresponds to an accuracy of about 99.4% for ribozymes of the length we're discussing. In making three mistakes per hundred nucleotides polymerized, however, Bartel's original ribozyme missed this mark by a factor of five. The fidelity of the catalytically improved polymerase of Horning and Joyce likewise clocks in at 97%, and the improved polymerase of Tjhung is just 96%. The poor fidelity of these ribozymes, however, likely reflects a technical hurdle rather than some fundamental limitation of their chemistry. The problem is that, while in vitro selections for improved catalytic efficiency are conceptually easy (if perhaps difficult in practice)—one can select for longer products by separating larger molecules from smaller molecules—how one would select for improved fidelity is not so clear. Or at least it is not clear how one would do this before one has a self-replicating system that can evolve itself.

The fidelity and catalytic efficiency problems aside, the best RNA polymerase ribozymes identified to date do have many of the characteristics that we are looking for in an RNA-based replicase. They are, for example, quite *un*selective in their choice of templates and thus, were they more efficient, they most likely could produce copies of themselves and any other genes that might be of use. And they do show the right type of activity—they synthesize RNA polymers with linkages between the same set of hydroxyl groups seen in naturally occurring RNAs. It thus appears likely that an appropriately catalytic, self-replicating ribozyme could exist, an observation that provides a major boost for the idea at the core of the RNA-world hypothesis: that the first living thing on Earth used RNA for both genetic material and catalytic activity, including the replication of the genetic material.

RNA World Redux

In the RNA-world scenario, the first living thing, your very, very, very first ancestor, consisted of three parts: a ribozyme with RNA polymerase activity, a template RNA to direct the polymerization, and a membrane or some other type of physical container. The need for two RNA molecules, and not just one, stems from how ribozymes catalyze reactions. Because ribozymes need to fold into specific, complex structures to perform their functions, it is exceedingly unlikely that any "self"-replicating molecule can actually *serve as its own template*. In order to serve as a template on which a new RNA molecule can be synthesized, a molecule must be unfolded and exposed to the monomers that will polymerize on it. And unfolded molecules are not catalytic; catalysis is intimately related to the precise, three-dimensional placement of atoms. Thus, in the RNA world, the spark that separates life from nonlife is the creation of not one ribozyme sequence, but two. The need for compartmentalization (e.g., a surrounding membrane) stems from the need that genes remain physically associated with the products they encode. Without such association, the metabolic products of this primitive earliest life-form would simply diffuse away and thus would not provide a selective advantage for the genes that created them. Putting this all together, the RNA world predicts that our very earliest ancestor was two RNA molecules floating together inside a membrane sac, taking up monomers from the surrounding environment and polymerizing them into more copies of the RNA until the RNA load grew so great that the membrane split. Grow, split, grow. The first life.

There we have it: the RNA-world hypothesis in a nutshell. We wrap a leaky lipid membrane around a self-replicating ribozyme, allow nucleotide monomers to diffuse in, and the monomers are polymerized into new copies of the ribozyme. The high molecular weight of the ribozyme traps it within the lipid membrane, perhaps driving its growth and subsequent division. Neat idea, right? It's easy to see, for example, how the lipid membrane could have evolved into today's cell membranes, which have abolished the putative leakiness, presumably by employing longer, better "sealing" lipids and (now protein-based) pores that are highly selective and let in only those small molecules needed by the cell. Pursuing this idea in the mid-2000s, the research team of Jack Szostak (2009 Nobel laureate in Medicine and Physiology) at Harvard Medical School in Boston demonstrated, if not quite this exact system, at least some interesting analogs. Abandoning the long-chain lipids used in our cells, which produce virtually impenetrable membranes, the Szostak lab fabricated vesicles from simpler, less ideal lipids that allowed the transport of small molecules. In 2004, for example, Szostak team member Irene Chen showed that vesicles containing high concentrations of RNA "steal" lipids from vesicles lacking RNA, presumably to relieve some of the osmotic pressure created by their contents. And in 2008, Szostak team member Sheref Mansy trapped a synthetic, modified DNA that could act as a template for the polymerization of a DNA analog inside these leaky vesicles and supply the requisite phosphorimidazolide-activated nucleosides to the solution surrounding the vesicles to form "protocells." Not surprisingly, if the vesicles had just the right leakiness—if they trapped the DNA polymer but allowed in the nucleotide monomers to flow in and out—the polymer grew within the vesicles. As the (so far) culmination of this work, in 2013, Szostak group member Katarzyna Adamala showed that, with a magnesium citrate complex as catalyst, activated RNA monomers could diffuse into template-containing fatty acid vesicles and drive templated-directed RNA polymerization in the fashion of Orgel's work described above, albeit, as with Orgel's work, this produced only rather short sequences.

Of course, the membranes that surround our cells are more than just passive sacks: they also have to expand and divide as the cell grows. Exploring this, in 2009, Szostak team member Ting Zhu showed that vesicle growth under some conditions produces long, thread-like tubes that easily rupture under mechanical agitation, producing multiple daughter "cells." And, in 2011, Szostak and Itay Budin described ways of getting cycles of growth and division driven by increasing the concentration of lipids followed by their redilution under turbulent conditions, as might happen in a tide pool that evaporates at low tide only to be filled again at the next high tide.

All told, the RNA-world hypothesis remains the most chemically plausible and best experimentally supported theory of the origins of life. In RNA, we have a molecule whose credentials as both a genetic material *and* as a catalyst are well established. And not just any type of catalysis, but RNA-templated RNA polymerase activity, the very type of activity that would be fundamental to the simplest living thing, were that living thing built of RNA. And while, admittedly, the very best ribozyme RNA polymerase described so far can't copy itself, or even copy a copy of itself, this is presumably a quantitative issue and does not reflect a fundamental inability of RNA to perform such action. Moreover, while it's not clear how RNA polymers might have been synthesized under prebiotic conditions, at least we understand how most of the precursors of RNA were probably synthesized in the primordial soup. Finally, as we'll explore in great depth in the next chapter, the RNA-world hypothesis is also the most *biologically* plausible theory we have of the origins of life. That is, as biologists we can spin a compelling narrative explaining how a single, self-replicating RNA polymer could have evolved, ultimately, into us through plausible, stepwise evolution

Sidebar 5.1

Weighing the Probabilities

Ribozyme sequences capable of catalyzing the template-dependent polymerization of new RNA seem to be quite rare. To create the first (and not a very efficient one at that), Bartel and coworkers had to start from an already folded, catalytic RNA (minimizing the number of unfolded sequences investigated helped to cut down on the total number of sequences that had to be searched). And still they required 18 rounds of selection and optimization—which couldn't happen on a prebiotic world (selection in the natural world requires an ability to *replicate*, and these molecules aren't there yet)—to make a sequence that *still couldn't copy itself*. The longest polymer synthesized by the most efficient of Bartel's ribozymes was just 14 nucleotides long, less than one-tenth the length of the ribozyme itself. And still Bartel's students had to test (in parallel, fortunately!) well in excess of 10^{15} sequences to find one that achieved even that level of activity. Follow-on studies using Bartel's sequence as a starting point to explore still vastly more sequences have improved on the efficiency of Bartel's best efforts quite significantly, but even these still aren't "good enough" to replicate themselves or in terms of catalytic efficiency or fidelity; ribozyme sequences able to copy themselves presumably exist, but they seem to be rare indeed.

How rare? The Japanese cosmologist Tomonari Totani at the University of Tokyo has tried to estimate this based on the supposition that two sequences (a polymerase and its template) of some minimum length are required and that, of all the possible sequences of minimum length, only 10,000 exhibit sufficiently high-efficiency, high-fidelity polymerase activity. Starting from these constraints (the second admittedly arbitrary in its precise value) and using seemingly liberal estimates regarding the efficiency with which activated nucleotides are made on prebiotic terrestrial planets, Totani still finds that, if the minimum length is more than 35 bases, the probability of achieving self-replication *even once in the entire observable Universe* is small. And given the physical chemistry of RNA, it is a good bet that the minimum length of a high-efficiency RNA polymerase is probably longer than this (the best reported to date is 179 bases). Conversely, however, and as Totani notes, inflationary cosmology (see sidebar 2.1) suggests that the true Universe is at least 10^{78} times larger (in volume and mass) than the observable Universe. Given this, the chances of self-replication arising somewhere across the entire Universe are quite respectable. Which, due to the anthropic principle (see sidebar 2.3), is all we need.

Totani, Tomonori. "Emergence of Life in an Inflationary Universe." *Scientific Reports* 10 (2020): Article Number 1671.

in which each stepwise increase in complexity provided a selective advantage over the last. The RNA world has a great deal going for it; infinitely more, really, than any other theory as to the origins of Terrestrial life. But is the whole package now neatly wrapped up? In a word: no. (But what would be the fun of that, anyway?)

Several serious questions remain regarding the plausibility of the RNA world. One is the sheer improbability of spontaneously generating one of the rare sequences that could copy itself through the random, prebiotic polymerization of RNA monomers. This becomes even more of a hurdle when one considers that, as noted above, the first living thing had to consist of two such sequences together. If the first living chemical system did not arise until a ribozyme template-dependent RNA polymerase sequence was spontaneously generated in the presence of a second sequence that encoded the same catalytic function (such as a copy of itself, although any sequence that worked as a polymerase would be suitable) and could serve as a template, then it was a rare event indeed: the need for two sequences *squares* the improbability of the random synthesis event (see sidebar 5.1). Given this, Szostak and others continue to search for plausible, prebiotic routes to templated-directed polymerization that don't require a ribozyme. But while our mechanistic understanding of these reactions has improved significantly in recent years, polymerization efficient enough and general enough to support self-replication remains, so far, out of reach.

Mirror Molecules

Beyond the debunking of spontaneous generation, Louis Pasteur's career had many other highlights, including the invention of pasteurization. He was also the first person to note, some 15 years before his work on "corpuscles that exist in the atmosphere" (i.e., germs), that the component chemicals of life are *chiral*, from the Greek *cheir*, meaning "hand." This coinage, which came from Lord Kelvin, stems from the fact that your hands are a convenient example of chirality; that is, they are mirror images of each other that are not superimposable. This is illustrated by the fact that, no matter how many ways you twist and turn it, your left glove will not fit on your right hand. Amino acids and sugars are similarly chiral, coming in left- and right-handed versions that, while similar in many aspects, are mirror images of one another that are not interchangeable.

On Earth, proteins are composed exclusively of L-amino acids (the L being taken from the Latin *laevus*, or "left"). The equivalent D-amino acids (from *dexter*, meaning "right") do show up now and then, such as in the cell walls of some bacteria, but they are far from common. The nucleic acids DNA and RNA likewise contain only "right-handed" sugars, termed D-deoxyribose and D-ribose, respectively. (Note, however, that the historical designations "right" and "left" were assigned arbitrarily and do not imply that L-amino acids are in some way opposite in structure to D-sugars.) Because the chemistries of mirror-image molecular pairs (i.e., pairs of *enantiomers*) are indistinguishable, which chirality was selected does not appear to be at all important. For example, a protease (a protein that catalytically cleaves other proteins) synthesized in the laboratory using only D-amino acids folds fine, but as expected, it folds *into the mirror image of the naturally occurring protein.* This mirror-image protein, in turn, is just as catalytically active as its naturally occurring counterpart, but, consistent with its mirrored structure, it cleaves only protein chains composed of only D-amino acids. Just as your right-hand glove fits only your right hand, this right-handed protein bound to and cleaved only right-handed substrates. And yet, while polymer chains of either handedness appear to work equally well, polymers of *mixed* handedness do not. If the chirality of the monomers changes randomly from left to right along the length of a polymer, this tends to preclude folding into nice, regular structures. And without folding, there is little in the way of catalysis. Homochirality thus appears to be critical.

The requirement for homochirality represents another potentially serious problem for the RNA world. The problem is that biological processes, such as the prebiotic chemistry of the early Earth, produce equal amounts of left- and right-handed molecules, rendering it improbable in the extreme that random chemistry would produce a polymer containing only one. To be precise, the probability of Bartel's 189-nucleotide ribozyme being polymerized from an equal-molar mixture of left- and right-handed monomers is one in 2^{189} (one in 8×10^{56})! This number of 189-nucleotide RNA sequences, taken together, would have the mass of 40,000 Suns, which nicely illuminates the extent to which the requirement for homochirality increases the improbability of a self-replicating RNA sequence arising spontaneously from an activated mixture of L- and D-ribonucleotides.

Like the poor efficiency with which RNA is synthesized under plausible prebiotic conditions, the requirement for homochirality reduces the probability that RNA-world-type chemistry is at the root of the origins of life. But does this render the RNA-world hypothesis invalid? It does not. For one, the need for homochirality does not push the probability of spontaneously generating a self-copying ribozyme to zero, and the anthropic principle says that even super-astronomically improbable mechanisms may lie at the heart of the origins of life—our existence only says that the probability of life arising in our Universe is not zero, but it could be infinitesimally close to zero. Disliking this explanation, though, a good number of researchers have spent much effort over the last few decades trying to come up with approaches that avoid or solve the problem of homochirality. Four broad classes of theories have been suggested in an attempt to address or circumvent this problem. The first is that there are abiological processes that preferentially degrade or create one enantiomer. Alternatively, there may be abiological processes that can amplify small, random deviations in the ratio of L and D monomers or chiral catalysts that polymerize only one monomer out of a mixture of two. Finally, it is possible that life arose starting with polymers lacking chirality and only later evolved to include chiral polymers like RNA and proteins.

Two of the abiological processes postulated to produce enantiomeric excesses are the weak nuclear force and the action of circularly polarized light. Neither idea, however, appears all that compelling. In the 1950s, it was discovered that β-particles, energetic electrons emitted from atomic nuclei during some forms of radioactive decay, are preferentially emitted with a specific handedness (in terms of their direction of travel and direction of spin). This implies that the weak nuclear force, the force that holds nuclei together, is itself asymmetric, an effect that could, potentially, have implications in the origins of homochirality. Despite intensive investigations, however, no one has demonstrated the selective degradation of one enantiomer over another through the effects of β-radiation, so its role in prebiotic chemistry must be seriously questioned. The theory suggesting that circularly polarized light produced enantiomeric excesses on the early Earth is on perhaps better footing, but even this one is only very weakly supported by laboratory experiments. As a charged object moving through a magnetic field, an electron traveling through the intense magnetic fields of a neutron star will spiral, causing it, per Maxwell's equations, to emit light, typically, under these conditions, as UV light. Due to the spiral motion of the electrons, the light will be circularly polarized. That is, the electric field vector that describes it oscillates in a circular fashion, as opposed to the more well-known plane polarization caused by your sunglasses, in which the electric field vector of the polarized light is constrained within a plane. The two enantiomers of a chiral molecule absorb circularly polarized light to different extents. Given this, it has been postulated that if (1) there were a nearby neutron star producing copious amounts of circularly polarized UV light, and (2) the prebiotic precursors to life arose in the pre-solar nebula (where they would feel the full brunt of any UV light), then this might account for, for example, the selection of L-amino acids and D-ribose on Earth. Laboratory studies of this mechanism, however, indicate that it is a very weak effect. When amino acids were blasted with so much UV light that 99% of all the molecules (of both enantiomers) were destroyed, the remaining dregs of amino acids showed enantiomeric enrichment of at most only a few percent. If such an effect is a mandatory step in the formation of life, then the origins of life were a fortuitous event indeed.

Given the lack of plausible mechanisms for generating homochiral monomers without biology,* researchers are now focusing on mechanisms that might amplify small enantiomeric excesses (caused, for example, by small, random fluctuations) into larger excesses. These include autocatalytic reactions in which the product of a reaction catalyzes its own production or inhibits the formation of the other enantiomer. The "Soai reaction," for example, produces enantiomerically enriched products even when starting from a seemingly completely

*This said, a few percent excesses of one handedness have been identified among some of the amino acids found in the Murchison meteorite, including in some of the nonproteinogenic amino acids, suggesting that this asymmetry is not due to biological contamination that occurred on Earth. The origin of these small, if tantalizing, asymmetries remains completely unknown.

racemic mixture of materials. This said, this reaction is of no specific relevance to prebiotic chemistry, and no plausibly prebiotic reaction has yet been reported that achieves similar autocatalytic production of near homochirality. Alternatively, Donna Blackmond and her team at the Scripps Institute in California have shown that, if two enantiomers that can interconvert in solution but preferentially form only single-enantiomer crystals, then a mixture of crystals of both enantiomers will spontaneously resolve into a single handedness given some energetic input (such as grinding) via a process called "Viedma ripening," in which one crystal form, and thus one enantiomer, eventually wins over time.

Still another mechanism potentially supporting chiral amplification would be catalysts that could selectively polymerize only L or pure D monomers out of a mixture of the two. To this end, Kenso Soai of Tokyo University has shown that, when added to various non-polymerization reactions that normally produce 50:50 mixtures of L and D products, pure, powdered D- or pure, powdered L-quartz* can produce significant excesses of one handedness over the other. As yet, however, this has not been shown to hold for RNA polymerization on mineral surfaces.

Given the lack of any clear solution to the "chirality problem," several researchers have proposed that there was a pre-RNA world in which a nonchiral, RNA-like polymer was the basis of the first life, and chiral RNA came into the picture only later, under the influence of selective pressures. Several potential pre-RNA polymers have been suggested. For example, Albert Eschenmoser, then at the Scripps Research Institute in La Jolla, California, demonstrated that nucleotides containing the sugar threose can be strung together into RNA-like polymers that form complementary duplexes, and he has argued that threose, being a four-carbon sugar, is likely to be synthesized in greater yield than the five-carbon ribose under prebiotic conditions. Threose, however, is also chiral, so Eschenmoser's TNAs (threose nu-

*The silica (SiO_2) units that make up quartz are not, themselves, chiral, but the crystal forms that quartz adopts are. When molten silica crystallizes, it forms a 50:50 mixture of left- and right-handed crystals. Next time you're shopping in a New Age store, be sure to check out its collection of quartz crystals and see whether you can identify the two mirror-image crystal forms.

cleic acids) do not solve the homochirality problem. In contrast, Pernilla Wittung, while working in Peter Nielsen's laboratory in Copenhagen, demonstrated in the early 1990s that a nucleic acid composed of a nonchiral, polypeptide-like backbone—peptide nucleic acid, or PNA, composed of nucleobases linked via the molecule *N*-(2-aminoethyl)glycine, which is produced in the Miller-Urey experiment—can couple with itself to form stable double helices. Other than RNA, however, no polymer has ever been demonstrated in the laboratory to support template-directed polymerization of like polymers in the absence of protein catalysts, and so speculation that there was a pre-RNA world remains just that: speculation.

Conclusions

In 1859, Louis Pasteur won the French Academy's prize for showing us how life does *not* start. And now, more than a century and a half on, we still do not know how it does. Self-replicating metabolic pathways? Self-replicating clays? An RNA (or RNA-like) replicase that can copy a copy of itself? We simply do not know, because every one of these theories faces major, unsolved hurdles. But, at least in the RNA-world hypothesis, we have a chemically and biologically plausible—if perhaps quantitatively improbable—theory concerning our origins.

Further Reading

Metabolism-First Hypothesis

Orgel, Leslie E. "Self-Organizing Biochemical Cycles." *Proceedings of the National Academy of Sciences USA* 97, no. 23 (2000): 12503–07.

The FeS World

Cody, George D. "Transition Metal Sulfides and the Origins of Metabolism." *Annual Review of Earth and Planetary Sciences* 32 (2004): 569–99.

The Lipid World

Lancet, Doron, Daniel Segre, and Amit Kahana. "Twenty Years of 'Lipid World': A Fertile Partnership with David Deamer." *Life (Basel)* 9, no 4 (2019): 77–88.

The RNA World

Atkins, John F., Raymond F. Gesteland, and Thomas R. Cech, eds. *RNA Worlds: From Life's Origins to Diversity in Gene Regulation.* Woodbury, NY: Cold Spring Harbor Laboratory Press, 2011.

RNA Polymerase Ribozymes

Horning, David P., and Gerald F. Joyce. "Amplification of RNA by an RNA Polymerase Ribozyme." *Proceedings of the National Academy of Sciences USA* 113, no. 35 (2016): 9786–91.

Johnston, Wendy K., Peter J. Unrau, Michael S. Lawrence, Margaret E. Glasner, and David P. Bartel. "RNA-Catalyzed RNA Polymerization: Accurate and General RNA-Templated Primer Extension." *Science* 292, no. 5520 (2001): 1319–25.

Protocells and RNA Self-Replication

Joyce, Gerald F., and Jack W. Szostak. "Protocells and RNA self-Replication." *Cold Spring Harbor Perspectives in Biology* 10, no. 9 (2018): a034801.

Szostak, Jack W. "The Narrow Road to the Deep Past: In Search of the Chemistry of the Origin of Life." *Angewandte Chemie* 56, no. 37 (2017): 11037–43.

The Origins of Chirality

Blackmond, Donna G. "The Origin of Biological Homochirality." *Cold Spring Harbor Perspectives in Biology* 2, no. 5 (2010): a002147.

Chapter 6

From Molecules to Cells

After taking part in the unraveling of the DNA double helix, Francis Crick moved to the logical next step: investigating the expression of genetic information—how the sequence of nucleotides so neatly lined up in the double helix structure is eventually translated into amino acid sequences. With his wildly creative thinking, he contributed some insights that turned out to be true (such as the adaptor hypothesis, which essentially predicted the role of transfer RNAs) and others that were perhaps just that little bit too imaginative, such as the "comma-less" genetic code, which restricted the use of three-letter codons to those that would be safe against one-letter shifts of the reading frame (we describe the genetic code later in this chapter). He was much intrigued by both the complexity of the protein biosynthesis machinery and the universality of its language.

In 1972, Crick resorted to yet another wild idea to explain the puzzling phenomenon that all the millions of species on our planet use essentially the same, incredibly complex protein synthesis machinery, with the same genetic code. In a paper he coauthored with Leslie Orgel, which appeared in the planetary science journal *Icarus* the following year, the two argued that, in particular, the protein synthesis machinery is so complex that it was very hard to understand how it could have been created by the slow, stepwise progression of evolution. This quandary was so serious, they proposed, that we must consider even the wildest alternative hypotheses. For example, they argued, a possible alternative solution would be that life on Earth, with its seemingly impossibly complex biochemistry, may have been artificially designed by intelligent aliens from some other planetary system in our galaxy who then purposefully sent it here by "space mail."

The aliens in question solved the problem of the difficulties associated with the origins of our complex biochemistry by themselves having a much simpler biochemistry. Simple enough, Crick and Orgel argued, that their biochemistry could have arisen spontaneously from nonlife more easily than our rather complicated biochemistry could have. The two further decorated their proposal—which built on earlier panspermia ideas discussed in chapter 5—with plenty of detail, including the design of the spaceship that might have delivered the first spores to Earth. It is a remarkable tribute to the complexity of contemporary Terrestrial biochemistry that, after decades spent deciphering some of its greatest mysteries, these two highly distinguished researchers still found many aspects of biochemistry so puzzling that they would seriously consider such a radical alternative (or at least semiseriously: Orgel admits that the paper was for him rather tongue-in-cheek, but he suspects "Francis took it somewhat more seriously"*).

Today, more than four decades later, few researchers working on the translation of DNA into protein would resort to such speculative explanations, but the evolution of protein biosynthesis still poses a serious conundrum. Protein synthesis as it occurs today in all cellular organisms (and, independently, in certain subcellular compartments, such as the mitochondria) requires an incredibly large tool kit, including ribosomes consisting of more than 50 proteins and at least three large strands of RNA, multiple tRNAs (the *t* stands for "transfer" because this RNA's job is to transfer an amino acid to the growing polypeptide chain), and several dozen non-ribosomal proteins, including the tRNA synthetases (the enzymes that covalently attach the appropriate amino acid to each tRNA), as well as protein factors that initiate translation, ensure that elongation proceeds efficiently, and, ultimately, drive the release of the full-length protein. If any one of these components is removed, the whole process fails. How could such an "irreducibly complex" set of interacting pieces (i.e., a set in which removing any one piece eliminates all function) have arisen by the stepwise, random workings of evolution? How could the RNA world transform itself into a DNA-RNA-protein world? (And if we

*Personal communication with the author (Plaxco).

rewound and ran the film again, would the story unfold the same way? See sidebar 6.1.) Luckily, clues to this mystery are contained within the very complexity of the protein synthesis machinery itself.

My Name Is LUCA

Judging from the extraordinary degree to which the biochemistry of all Terrestrial life is similar, it seems that everything living on Earth today is related through some long-lost great-to-the-*n*th-power grandmother: the last universal common ancestor (LUCA) of all living things on Earth. Her descendants branched out into two of the major domains of life, the Archaea and the Bacteria, some of which later recombined to form the third, the Eukarya, which comprises the plants, animals, fungi, and protists.

Extrapolating backward from today's biodiversity, we can look across all life on Earth to identify those things we share in common (that are *homologous*, i.e., have arisen from a common ancestor), under the economical assumption that any traits that are common across all life are shared because we all inherited them from LUCA, and thus she shared them too. Using this comparative approach, biochemists have been able to infer quite a bit about LUCA and her biochemistry. For example, we know that LUCA stored genetic information in DNA, produced a couple of hundred proteins working as enzymes, receptors, or transporters, and made use of the 20 amino acids that are now the standard components of proteins. Similarly, LUCA used the same genetic code that we do to translate DNA sequences into the amino acid sequences of proteins and performed this translation using more or less the same RNA-based protein synthesis machinery that all life on Earth employs.*

LUCA's biochemistry was thus already quite complex; by the time LUCA was around, life had come a long way since the birth of the very first simple, self-replicating molecules. But because so much was

*This said, somewhere along the way the eukaryotic lineage added further complexity to its ribosomes, which contain far more components of archaeal or bacterial ribosomes.

Sidebar 6.1

Was There Another Way?

All life on Earth today employs proteins for the vast majority of its catalytic functions. Similarly, save for a few small, simple viruses, all Terrestrial life employs DNA as its genetic material. But did this have to be? That is, is there some physical or chemical imperative that forced evolution's hand to make these "choices"? Or was the choice to adopt protein catalysts and DNA information storage simply a historical accident? Can we even meaningfully speculate on these issues?

As we describe in this chapter, the "invention" of DNA was encouraged by the selective advantage that DNA provides as an information storage agent in terms of its chemical stability. But stability is not the sole requirement of a genetic material. It must also be able to mutate (sometimes, but not too often!) to create new information (i.e., to allow evolution to proceed), without changing its physical properties so much that the replication and translation machineries no longer recognize it. This principle, which Steve Benner of the University of Florida termed COSMIC-LOPER (capable of searching mutation-space independent of concern over loss of properties essential for replication), states that an ideal genetic material will be one whose physical properties remain largely unchanged by changes in sequence. And in addition to being chemically stable, DNA is COSMIC-LOPER; the shape and chemistry of the double helix is, to a first approximation, independent of its nucleotide sequence.

But was DNA the only way for evolution to meet the demanding, simultaneous requirements of chemical stability and COSMIC-LOPER? There are other nucleobase-containing polymers that form complementary helices (such as the PNAs and TNAs we discussed in chapter 5), which suggests that there are other ways of storing genetic information in molecules. But do these other polymers have the right set of properties? Quite simply, we do not know.

And what about proteins? As we described in chapter 5, the improved catalytic ability of protein-based enzymes relative to RNA-based ribozymes provides a strong selective advantage for any organism that evolves the ability to make proteins. But did these newfound catalysts have to be polymers of α-amino acids? The answer is a qualified "perhaps so." There are several supporting arguments. First, α-amino acids are formed in relatively high yield in the Miller-Urey experiment and are a common component of the organic material in carbonaceous chondrites. It seems reasonable to assume that the ready availability of α-amino acid precursors would provide a significant boost for any organism using proteins, as opposed to some other polymer, to accelerate its metabolism.

A second issue is the physical chemistry of polypeptides, which seems to suit them for folding into well-defined—and thus functional—structures. The peptide bond between amino acids in a protein contains a hydrogen atom attached to a nitrogen atom (see sidebar 4.1). Because the nitrogen is strongly electronegative (meaning that electrons engaged in bonds between nitrogen and a less electronegative atom spend more time with the former atom than with the

(*continued*)

Sidebar 6.1 *(continued)*

latter), the hydrogen atom takes on a small (partial) positive charge. The peptide unit also contains a strongly electronegative oxygen atom, which takes on a partial negative charge. These two atoms can participate in a hydrogen bond, in which a hydrogen with a small positive charge is attracted to an oxygen with a partial negative charge. When a protein is unfolded in water, the hydrogen of the peptide bond can hydrogen-bond to oxygen atoms in the surrounding solvent molecules, and the oxygen can hydrogen-bond to hydrogens in these same solvent molecules. When a protein folds, however, the solvent is excluded and thus all the hydrogen bonds need to be formed internally. As Linus Pauling (1901–1994) realized back in 1948 (while sick in bed and doodling on some paper), polypeptides can satisfy all their hydrogen-bonding needs internally. For example, they can fold into what he called an α helix, in which the hydrogen of one peptide linkage is hydrogen bonded to the oxygen of a peptide linkage some three amino acids farther along the chain. Moreover, all the bond lengths and bond angles in the amino acids are extremely stable in this configuration; there are no nasty clashes of one atom against another, and the interatomic bonds are all quite satisfied with the angles they have to adopt—and thus researchers were not surprised to find that α helices are indeed a common feature of folded proteins. More recently, Jayanth Banavar, then at Penn State University, and Amos Maritan, of Padova University in Italy, have pointed out that the packing density of this helix is effectively perfect. If you want to twist a linear tube into a helix shape leaving as little empty space as possible, the radius of curvature of the helix has to equal the tube radius, and the pitch of the helix has to be precisely 2.512 times its radius. And for the α helices found in proteins, these two measurements are within a few percent of their optimal values. This perfect packing also encourages the formation of a stable, folded state—because nature abhors voids.

While the existence of ribozymes proves that proteins are not the only polymeric catalysts, the one-two-three punch of proteins' greatly improved catalytic properties, formation from readily available precursors, and folding-friendly physical chemistry render them extremely well suited for the role of metabolic catalysts. Pondering these questions during a seminar Professor Banavar presented on the topic of helix packing, one of us raised our hand and asked, "When we finally travel to Alpha Centauri and meet up with the Centaurians, will they, too, use polymers of α-amino acids as their dominant catalysts?" His reply? "Let us endeavor to do so and find out."

Benner, Steven A. "Chance and necessity in biomolecular chemistry: is life as we know it universal?" In *Signs of Life: A Report Based on the April 2000 Workshop on Life Detection Techniques*. Committee on the Origins and Evolution of Life, National Research Council. Washington, DC: National Academies Press, 2002.

Maritan, Amos, Cristian Micheletti, Antonio Trovato, and Jayanth R. Banavar. "Optimal Shapes of Compact Strings." *Nature* 406 (2000): 287–90.

lost in going through the bottleneck that was LUCA and, presumably, many earlier bottlenecks (we must remember that there may have been thousands of other species, contemporary with LUCA, whose descendants just weren't lucky enough or fit enough to survive in competition with her and her relatives), we can only speculate about how the first cells came into being and which of the many compounds and functions that we now consider essential may have been absent in LUCA's progenitors. But we do have constraints that can guide our speculations: we assume that the primordial soup was the starting point, and LUCA was the final product.

Back to the RNA World

One of the most widely accepted theories concerning the phase between primordial soup and ancestral cells is the RNA-world hypothesis, which we introduced in chapter 5. The discovery of ribozymes as potential relics of the RNA world was a major inspiration for RNA biochemists as it suggested that RNA could carry out the various functions considered essential for life. But the first ribozymes to be discovered were limited to the processing of other RNAs and, perhaps worse, to the processing of themselves (i.e., as we mentioned in chapter 5, they aren't really catalysts in the true sense of the word). This is hardly the sort of diverse chemistry around which complex life can be built; for that, we'd expect at the very least to need reactions that convert small molecules that look nothing like RNA into the true precursors of RNA.

The first ribozyme-catalyzed reaction that does *not* involve the processing of RNA was observed by Thomas Cech's group in 1992 with a genetically modified *Tetrahymena* ribozyme. The altered ribozyme could cleave the bond between the amino acid methionine and its matching tRNA; this is the reverse of the reaction catalyzed by the tRNA synthetases, which add amino acids to tRNAs as the first step in the synthesis of proteins. Three years later, Michael Yarus, a colleague of Cech's at the University of Colorado in Boulder, and his team screened a random mixture of 10^{14} different RNA sequences and managed to select a sequence that catalyzed the attachment of an activated amino acid to itself. Following this, Peter Lohse and Jack Szostak demonstrated an RNA sequence that is able to transfer aminoacyl groups to a free amino acid to form a dipeptide—which is just what the peptidyltransferase center of the ribosome does when it elongates the polypeptide chain by one unit.

It thus seems that RNA can bind substrates and catalyze a range of reactions above and beyond the postulated ability of some RNA molecules to copy or edit themselves (as they might have done if the RNA world was not just an intermediate stage but the beginning of life). This chemical promiscuity might have made all the difference to the earliest organisms. The first self-replicating, evolving, and hence living thing would have been the only organism in the history of the planet that was not in competition with other organisms. That situation would have changed just a soon as it produced daughters. Immediately, there was competition for the precursors with which these organisms reproduced, and all too soon some molecule would have become hard to obtain. Then what? Then death. Unless, of course, one of these replicating RNA strands was error prone (or, put more positively, could evolve) and started to copy some different RNA sequence that catalyzed some reaction not directly tied up in replication. An example might be a reaction that converted some not-quite-right precursor into the correct, but now rare, precursor. This would provide a *tremendous* selective advantage for this organism: it could now outcompete its sisters by employing an alternative molecular source to replace the now rare precursor. It would also mark the invention of *metabolism*—the set of chemical reactions by which the various molecules that are available to an organism from its environment are converted into the precise molecules that the organism is built of.

It is possible that complex metabolisms arose during the era of RNA-based organisms. The present-day widespread use of ribonucleotides as cofactors in key metabolic pathways suggests that these pathways may have originated during the RNA-world era. Indeed, so compelling is this idea that, in 1976, a half dozen years before the discovery of ribozymes, Harold White used the widespread existence of ribonucleotide-containing cofactors to argue that "a metabolic system composed of nucleic acid enzymes . . . existed prior to the evolution of ribosomal protein synthesis."* Examples of these cofactors include the ribonucleotide ATP (adenosine triphosphate), which is used as the major energy "currency" of the cell, the adenine-containing coenzyme A (CoA), used as a "handle" with which to carry acyl groups such as acetate ($CH_3C{=}O$), and the ribonucleotides FAD (flavin adenine dinucleotide) and NAD (nicotine adenine dinucleotide), which accept high-energy electrons from reactions that liberate them and donate high-energy electrons to reactions that consume them. These potential remnants of the RNA world suggest that RNA-based organisms might have had a rich metabolism that included complex oxidation and reduction chemistries. And because the organism was enclosed by something, all this metabolic machinery, like the replicative machinery that evolved before it, was physically associated with the genes that encoded it, providing them a selective advantage.

But if our ancient ancestors had such rich RNA-based metabolisms, why don't we see RNA-based metabolisms today? The presumed answer is that any RNA-based organism would have been driven to extinction, remorselessly outcompeted by its more fit cousins who had the good fortune to invent or inherit protein-based catalysts. This idea is not so far-fetched given that the catalytic efficiency of enzymes is typically orders of magnitude higher than that of comparable ribozymes. For example, while the so-called hammerhead ribozyme, which catalyzes the cleavage of RNA, accelerates the degradation reaction a millionfold over the rate of the uncatalyzed reaction, a similar-sized protein, the enzyme ribonuclease, is a hundred thousand times more catalytically active still. Proteins, made up of 20 different types of amino acids, are much more complex than RNA, with its four nucleobases, and with this increased complexity comes increased catalytical flexibility and efficiency. Thus, any RNA-based organism that evolved the ability to synthesize proteins gained a tremendous advantage over its protein-free competitors.

*Harold B. White III, "Coenzymes as Fossils of an Earlier Metabolic State," *Journal of Molecular Evolution* 7, no. 2 (1976): 101–104.

Of course, seeing a comparison between the exceptional catalytic ability of proteins and the rather paltry skills of RNA-based catalysts, skeptics might question whether the RNA world ever existed. There is always the possibility of an alternative explanation: the relatively rare ribozymes might be fairly recent, not very successful experiments of evolution rather than survivors of a principle that was once more generally applicable. This ribozymes-first versus ribozymes-later argument raged through much of the 1990s (among those of us who cared) until it was finally resolved in favor of the primordial role of ribozymes. The resolution came with the unraveling of the mysteries of the ribosome, the subcellular factory that churns out the cell's proteins.

Even the bacterial ribosome (which is somewhat simpler than the version in the cytoplasm of eukaryotes, such as ourselves) consists of 57 different proteins and three RNA molecules folded together into an enormous macromolecular machine of several hundred thousand atoms in all (fig. 6.1). But which of these many components is responsible for the fundamental chemical reaction that the ribosome catalyzes: the synthesis of proteins? Long after the initial characterization of the ribosome, it was assumed that the catalyst must be one or more of the proteins (remember: until 1982, it was thought that *all* biocatalysts were proteins). Research throughout the 1980s and 1990s, though, had shown that each individual ribosomal protein was expendable—that is, after removal, in turn, of each of the 57 ribosomal proteins, the ribosome continued to function. But despite several well-publicized—and quite possibly successful—attempts, nobody had unequivocally succeeded in constructing a completely protein-free version of a ribosome that remained active. The RNA-world community had to wait for an atomic-resolution structure of the ribosome to find out whether the ribosome, at heart, is a protein-based enzyme or an RNA-based ribozyme.

When a string of high-resolution structures of ribosomal subunits and complete ribosomes started to appear, the answer was perhaps even clearer than expected. Reported by the research groups of crystallographer Tom Steitz (1940–2018) and "ribosomologist" Peter Moore, the structure of the 50S subunit from *Haloarcula marismortui* ribosomes made the cover of *Science* magazine in 2000 and won Steitz,

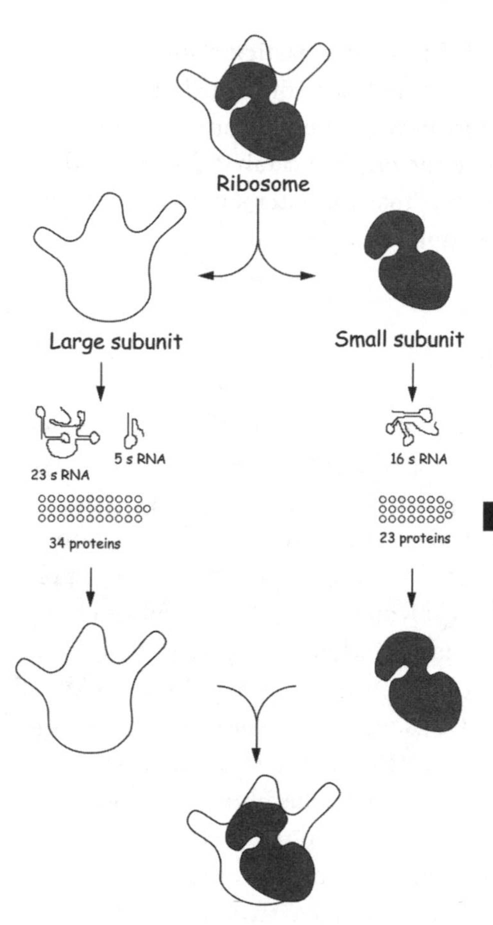

Figure 6.1 The bacterial protein-manufacturing machinery, the bacterial ribosome, can be separated in the test tube into the three strands of RNA and 57 different proteins from which it is made, and then put back together again. Using this approach, researchers have found that the resulting ribosomes are still active if any one of the proteins—no matter which one—is left out, showing that none of them are crucial for ribosome function.

along with Ada Yonath of the Weizmann Institute in Israel and Venkatraman Ramakrishnan of the MRC Laboratory of Molecular Biology in Cambridge, England, the 2009 Nobel Prize in Chemistry. The structure they published finally pulled back the curtain on the peptidyltransferase site of the ribosome, the spot at which new amino acids are added to the growing polypeptide chain. What must have delighted the RNA-world community more than any specific, mechanistic detail, however, was the observation that the peptidyltransferase active site is made up entirely of RNA: the nearest protein is 1.8 nm away (about 15 times the length of, say, a carbon-carbon bond) and thus clearly excluded from any participation in the catalytic activity. As RNA champion Tom Cech concluded in a commentary accompanying the structure paper, "The ribosome is a ribozyme." It became abundantly clear that the original catalyst for the synthesis of proteins was an RNA molecule that was later surrounded by proteins that serve various supporting roles.

Polypeptides Join the Fold

Proteins are long polymers each of a single, specific sequence, and thus, not surprisingly, the cellular machinery with which they are synthesized is quite complex. The question then naturally arises as to how such complex machinery could have arisen in one fell swoop. In short, it probably couldn't. Instead, it is thought that the first polypeptides were homopolymers (polymers consisting of multiple copies of a single amino acid) or short, random-sequence polymers. But homopolymers and random polypeptides generally do not fold and are not catalytically active, and thus they lack the selective advantages we generally associate with true proteins. Given this, why would some previously peptide-less life-form evolve the ability to polymerize amino acids? Several suggestions have been put forth. The chemical strength of the peptide bond implies, for example, that even simple, short chains of amino acids could form stable structural elements for an early cell. Likewise, simple polymers of positively charged amino acids could help counteract the negatively charged phosphates in RNA and improve its ability to fold into functional structures (remember: the ribosome has more than 50 proteins, which are mostly there to help the catalytic RNA fold). Alternatively, the first (simple) polypeptides may have provided a selective advantage by improving transport across the cell membrane. Because membranes are hydrophobic, it is hard to imagine that a highly charged RNA molecule could insert within a membrane to support the transport of some important nutrient into the cell. Many simple polypeptides, in contrast, are quite hydrophobic and could easily serve this role. Indeed, David Deamer of the University of California, Santa Cruz, has demonstrated just this:

short polypeptides consisting solely of alanine or leucine spontaneously insert into membranes and produce channels capable of transporting hydrogen ions (H^+). Moreover, these channels are specific: they do not transport other positively charged ions, such as sodium or potassium. So perhaps the first polypeptides were employed to regulate the properties of the membrane, to facilitate the entry of vital nutrients into the cell.

The fact that even short polymers of one or two amino acid types might have provided a selective advantage simplifies the problem at hand. That is, rather than needing to develop the entire, exceedingly complex protein synthesis machinery from scratch, evolution could have generated each part piecemeal, starting with the (presumably less complex) machinery required to make simple homopolymers, which would provide a modest selective advantage, before moving on to translation as we know it today.

The start of this evolutionary journey probably involved a close association between amino acids and RNA, which over many millions of years culminated in the complex and well-conserved mechanisms we still observe in today's protein biosynthesis. The first step, for example, might have been the use of amino acids to add functionality to RNA. Evolution could easily have invented an RNA molecule that "charged" itself—that is, autocatalytically attached an amino acid to itself to form a covalent complex. As described in chapter 5, such molecules have been created in the laboratory, so we know that RNA is capable of performing such chemistry. These mixed molecules could have provided a selective advantage for an originally RNA-only organism, say, by being more catalytic due to the addition of a new functional group. Similarly, even in contemporary Terrestrial biochemistry, we see examples of "aminoacylated" RNAs that are used to "donate" amino acids in metabolic processes. A glutamate attached to its tRNA is used, for example, in the synthesis of porphyrins, the type of organic carbon-and-nitrogen rings that hold the iron in hemoglobin and that date back to LUCA. Given this, it is quite possible that the first self-charging RNA resulted from a need to introduce amino acids (from Miller-Urey chemistry) into metabolic pathways aimed at making more RNA. Either way, it is easy to understand how the formation of the first RNA–amino acid complexes could have provided a selective advantage for RNA-based organisms. And if it existed, it seems likely that this amino acid–charged RNA might have been the progenitor of today's tRNA (but without the need for today's tRNA-charging enzymes).

The second step in the evolution of protein synthesis was probably the formation of a ribozyme that could use the RNA–amino acid complexes to synthesize homopolymers or random polymers of amino acids. Once again, ribozymes capable of synthesizing peptide bonds starting from activated amino acids (amino acids participating in high-energy bonds, such as those in aminoacylated RNAs) have been demonstrated in the laboratory, so we know that RNA is capable of performing this sort of chemistry. But is such a step a plausible intermediate in the origins and evolution of life? The products of this reaction, short random or homogeneous polymers of amino acids, may have provided a selective advantage due to their suitability for structural roles in the cell. Under the influence of this selective pressure, evolution could have quite readily produced a ribozyme capable of synthesizing polypeptides from charged tRNA precursors. This ribozyme would have been the ancestor of the modern ribosome.

Random-sequence polypeptides, though, are not proteins. How did sequence specificity enter the picture? Today the sequence specificity of protein synthesis arises because messenger RNA (mRNA) is used as a template to direct the incorporation of new amino acids into the growing polypeptide. Thus, the nascent ribosome must also have learned how to direct synthesis using such a template. Presumably this development occurred when the nascent ribosome found that it could bind its primitive tRNA substrates more tightly when they, in turn, had bound to yet another RNA, which would be the first messenger RNA. The precise sequence of this RNA would then determine the sequence of amino acids incorporated into newly synthesized polymer. The advent of sequence-specific synthesis would have provided a very significant selective advantage. Namely, it allowed the cell to make complex, highly efficient, protein-based catalysts and, because a protein's sequence is encoded in a gene, to pass this selective advantage on to its offspring. With this, we have the first true ribosome.

The Genetic Code

The ribosome translates the sequence of nucleotides in a messenger RNA intermediate, which was itself generated using the sequence of a DNA-encoded gene, into the corresponding sequence of amino acids that makes up a given protein, following the rules contained in the *genetic code*. The genetic code is simply the conversion table that links the $4^3 = 64$ possible ways of combining the four nucleotides into three-base codons to the 20 amino acids generally used in proteins, plus the chain termination or "stop" codons that tell the ribosome to halt synthesis when the job is done (table 6.1). When first deciphered in the late 1960s, the genetic code seemed to give no clues as to its early evolution. Indeed, when it was initially discovered that the same code applied in all the organisms studied so far, most researchers believed that it simply reflected a "frozen accident." The thought was that the universal code that assigns a given set of three bases to each amino acid was not fundamentally different from any of the other approximately 10^{20} possible patterns that could map the codons to the amino acids, but that once the existing code was set down randomly, the cost of changing it would have been so prohibitive that it was essentially frozen. Exchanging the assignment of one codon, for instance, would alter the corresponding amino acid in hundreds of proteins simultaneously, almost certainly with fatal consequences.

A quarter of a century later, however, a slightly more dynamic view of the code began to take hold. For one thing, it is now clear that a small number of exceptions to the code exist,* illustrating the fact that the code can evolve. Specifically, codon frequency was found to differ slightly among organisms, offering a viable, though rare, route to code evolution. A rare codon in a small enough genome might become so rare that it was used in only one or a few genes, and at that point it could be reassigned to a different amino acid if the change

*These occur mainly in mitochondria, the energy-producing subcellular organelles that power our cells and contain their own DNA. In the mitochondria of starfish, for example, the codons AGA/AGG and UGA specify serine and tryptophan, respectively, whereas in the standard code they represent arginine and a stop codon, respectively. These substitutions are possible because the mitochondrial genome is quite small: it encodes only 13 proteins. But altered genetic codes have also been found in some nuclear genomes, including in some species of yeast.

Table 6.1
The genetic code

First base in codon	U	C	A	G	Third base in codon
	Second base in codon				
U	Phenylalanine	Serine	Tyrosine	Cysteine	U
U	Phenylalanine	Serine	Tyrosine	Cysteine	C
U	Leucine	Serine	Stop	Stop	A
U	Leucine	Serine	Stop	Tryptophan	G
C	Leucine	Proline	Histidine	Arginine	U
C	Leucine	Proline	Histidine	Arginine	C
C	Leucine	Proline	Glutamine	Arginine	A
C	Leucine	Proline	Glutamine	Arginine	G
A	Isoleucine	Threonine	Asparagine	Serine	U
A	Isoleucine	Threonine	Asparagine	Serine	C
A	Isoleucine	Threonine	Lysine	Arginine	A
A	Methionine	Threonine	Lysine	Arginine	G
G	Valine	Alanine	Aspartate	Glycine	U
G	Valine	Alanine	Aspartate	Glycine	C
G	Valine	Alanine	Glutamate	Glycine	A
G	Valine	Alanine	Glutamate	Glycine	G

benefited that protein without producing fatal changes in other proteins. Despite these counterexamples, though, in the vast majority of cases it would seem to be effectively impossible for the codon assignments to vary on a reasonable evolutionary timescale—hence the suggestion that the current codon pattern, which presumably was originally randomly assigned, has been frozen in its current form by the nearly always fatal consequences of changing it.

But there is a problem with this frozen accident hypothesis: the current genetic code appears to be far more highly optimized than one would expect were it simply an accident. That is, the current genetic code is set up such that a large fraction of mutations at the level of nucleic acid sequence are silent, or at least chemically conservative, at the amino acid level. This is easy to see by looking at the genetic code. For example, look closely at each of the 16 *four-codon blocks* in the table of codon assignments (see table 6.1). The second block in the first row represents the four codons with the sequence UCN, where *N* indicates any of the four nucleobases U, C, A, or G. All four of these codons encode the amino acid serine, and so *any* mutation of the third position in any of these codons will be *silent*! And it's not just serine: this trait is shared by eight of the 20 amino acids, and partially by several others, so more than a quarter of all possible point (single-base) mutations are completely silent. That is, while they are changes in the DNA sequence, they do not produce changes in the sequence of amino acids that the DNA encodes. Moreover, many other mutations are *conservative*; mutation of the first base of many codons, for example, tends to swap chemically similar amino acids, such as leucine (CUN, where, again, *N* stands for any of the four nucleotides) for either isoleucine (AU followed by U, C, or A) or methionine (AUG).

The robustness against mutation that is captured in the current genetic code provides a significant advantage over alternative genetic codes that might not have this property. Keeping track of silent mutations and using various chemical scoring functions (such as hydrophobicity) to rank the impact of potentially conservative amino acid substitutions, theoreticians have found that the current genetic code is *highly* optimized in terms of suppressing the effects of mutation. In fact, the probability of picking by chance a genetic code as error proof as ours would be close to one in a billion, which suggests that the present-day code is, somehow, the product of intensive evolutionary optimization. But this finding seems highly paradoxical. If changes to the code are almost invariably fatal, how could the code evolve? Hints regarding the answer to this question are apparent in the structure of the code itself.

Look again at table 6.1. The amino acids in the first column, those with codons having the structure NUN, are the hydrophobic (water-hating) amino acids phenylalanine, leucine, isoleucine, methionine, and valine (see fig. 4.2 for their structures). Similarly, all the amino acids in the third column, those encoded by codons with the structure NAN, are hydrophilic. This has led to the suggestion that the first code was very simple; it may have used three nucleotides in a codon, but only the middle of the three *mattered*. If the middle nucleotide was U, a hydrophobic amino acid (perhaps the simplest, valine, but we don't know for sure) was encoded; if the middle nucleotide was A, a hydrophilic amino acid (again, perhaps the simple amino acid aspartate, but we don't know for sure) was encoded instead. The earliest code might thus have encoded only four different amino acids, and consistent with our earlier arguments about the origins of translation, the earliest proteins might have been of very simple composition.

How, then, did we go from this simple, four–amino acid, "only the middle position counts" code to the current 20–amino acid code? In time, selective pressures would ensure that new amino acids were recruited into the process of building proteins. When new amino acids were added, more complexity needed to be added to the genetic code to accommodate them. According to this hypothesis, this was done using the first position in the codon. Valine is differentiated from leucine by the first nucleotide in their GUN and CUN codon sets; the middle position encodes "hydrophobic" and the first distinguishes which of the several hydrophobic amino acids. Using just the first two codons, we can encode $4 \times 4 = 16$ amino acids. To enlarge the set beyond 16, we have to involve the third position. We see some examples of this in the current genetic code. For example, the amino acids glutamate and aspartate, which are chemically similar in that they are the only negatively charged amino acids (a glance back at fig 4.2 reveals how similar they are), are differentiated only by the third position in their codons. But for many codon sets, the third position is entirely redundant; all four GGN codons, for example, encode glycine. It thus seems that the genetic code *did* freeze. That is, sometime after the first two codon positions became important, but before all the third positions became distinct, organisms had become complex enough that any further changes (save the rare, limited exceptions described above) were prohibitive. The freeze occurred after only 20 amino acids had become encoded, presumably because this was the balance point between the selective pressures that pushed toward greater complexity (more diversity equals better ability to solve problems) and the increasing chance that any additional changes would prove fatal.

The near-universal spread of the standard code suggests that it was already in place in the time of LUCA. This, in turn, means that a full set of tRNAs—and the catalysts required to create them and charge them with the appropriate amino acids—were also in place. For example, all three domains of life process nascent tRNA into its final, functional length using the ribozyme ribonuclease P (RNase P), implying that we all inherited this catalyst from LUCA. Genomic comparisons likewise indicate that LUCA encoded between 40 and 45 tRNAs and 18 of the 20 tRNA synthetases in use by life today.*

Because LUCA was a bottleneck in evolution through which all life passed, it is often difficult to use evolutionary comparisons to peer any farther back. The complexity of the protein synthesis apparatus, however, allows researchers to study the evolution of the code beyond that time limit. In humans, for example, there are around 50 different tRNAs (not quite as many as there are codons since some can recognize several codons differing only in the third position, a phenomenon called "wobble") all sharing a common L-shaped structure and a variety of specific peculiarities such as the occurrence at specific points in the molecule of rare nucleobases—that is, bases other than the four common ones (e.g., a base called pseudouracil, which differs slightly from uracil). The high degree of sequence similarity among the different tRNAs suggests that they originally evolved from a single, progenitor "adaptor molecule" invented by evolution long before LUCA, which ultimately duplicated and diverged to create the 40 to 45 tRNAs that LUCA appears to have employed.

The tRNA synthetases tell a similar tale about the pre-LUCA evolution of translation. Researchers have found that the 20 synthetases are evenly divided into two distinct classes, with the 10 members within each sharing a common three-dimensional structure as well as similarities in the chemical details of their catalytic sites. The two families, however, do not appear to be even remotely related to one another; their amino acid sequences and final folded structures are

*Archaea lack the synthetase enzymes that charge the glutamine and asparagine tRNAs with these amino acids, suggesting that LUCA likewise lacked these. Instead archaea (and, presumably, LUCA) charge these tRNAs with the related amino acids glutamate and aspartate, then chemically convert them to glutamine and asparagine, respectively.

utterly dissimilar. The thinking is that evolution invented a protein to replace one of the 20 tRNA-charging ribozymes, and that through gene duplication and subsequent evolution, related protein sequences evolved that replaced more and more of the earlier synthetase ribozymes. But before evolution had succeeded in replacing all the ribozyme synthetases via duplications of the first synthetase protein, it developed a second, unrelated synthetase protein. Through gene duplication and drift, this, too, began to replace other ribozyme synthetases until, today, we find that all 20 synthetases belong to one of two unrelated families.

Archival Storage

In principle, RNA can serve as genetic material. In fact, it still does for many viruses, including HIV and the coronavirus that causes COVID-19. But the poor chemical stability of RNA limits the size that can be achieved using RNA genomes to a few thousand bases. Any longer, and the genome is too likely to suffer a fatal self-cleavage reaction via a mechanism described below. Thus, to achieve larger genomes—perhaps to accommodate a growing number of protein-coding genes—evolution had to invent a new archival storage material.

At first glance, RNA and DNA look very much alike (fig. 6.2). What is so special about the missing oxygen atom that puts the *D* into DNA (the *D* stands for "deoxy," i.e., lacking an oxygen)? Textbooks of biochemistry tend to start from DNA and then mention in passing that RNA has a hydroxyl group (—OH) at position 2′. (In nucleotides, the prime distinguishes the numbering of atoms in the sugar—ribose or deoxyribose—from the numbering of those in the base.) But it seems that, in the history of life, RNA is the earlier version of the idea, and DNA the deluxe edition introduced later.

Considering the structure of the ribose and deoxyribose alone, removal of one of the five oxygen atoms doesn't look like such a big deal (see fig. 6.2). But if you look at the ribose inside a nucleic acid polymer, you see that the 2′ oxygen is the only one that does not serve an immediate function in the primary structure. In RNA, the oxygen

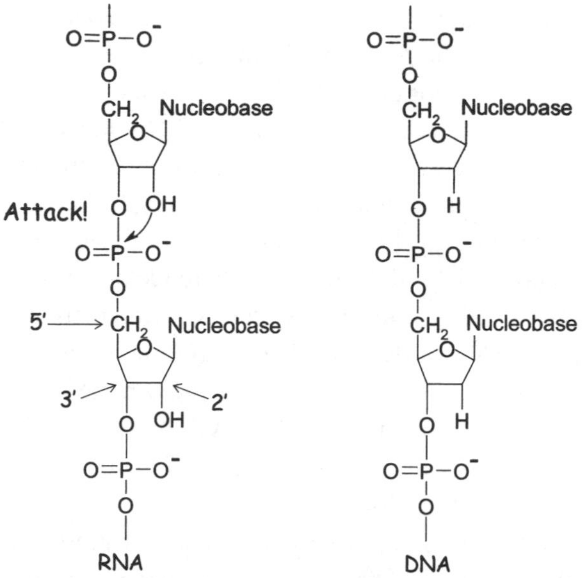

Figure 6.2 At first glance, RNA and DNA look very much alike: DNA is simply missing an oxygen atom at the 2′ position. (The carbon numbering system denotes atoms in the sugar with a "prime" in order to distinguish them from those in the nucleobase, which are denoted without primes.) This missing oxygen, however, makes a world of difference: the 2′ hydroxyl group in the RNA is well placed to attack the phosphate backbone, leading to a relatively rapid self-cleavage and putting a strong limit on the maximum size of an RNA-based genome.

in position 1′ is replaced by the nitrogen of the nucleobase, and the 4′ oxygen is holding the ring together. Oxygens 3′ and 5′ link to the phosphate groups leading to neighboring nucleotides. This leaves the oxygen of the 2′ hydroxyl group as the sole survivor, a chemically reactive group that might get into all kinds of mischief, including additional (branched) polymerization, hydrogen bonding, and—worst of all—autocatalytic hydrolysis of the phosphate-sugar bonds that hold the RNA together: the 2′ oxygen is, in fact, in the perfect geometry to attack the phosphate linkage in the RNA backbone, cleaving it. Because of this, RNA polymers are relatively unstable and are prone to breakage under even mild conditions over the course of days or weeks. DNA lacks the 2′ hydroxyl group and thus is enormously more stable.

So stable, in fact, that genome sequences have been obtained from the bones of hominins, the Neandertals and the Denisovans—although, sadly, not yet from dinosaurs—that became extinct more than 30,000 years ago. This enhanced stability renders DNA much better suited than RNA to the archiving of large amounts of genetic information. Moreover, as DNA is completely compatible with the RNA archives that presumably preceded it (DNA binds more strongly to a complementary strand made of RNA than to one made of DNA), it was presumably easy for this improved system of information storage to evolve from the RNA world. A vestige of DNA's takeover as the genetic material may be found in the way DNA is made in the cell: the synthesis of DNA is initiated using a short RNA "primer." The formation of RNA, in contrast, doesn't generally require a primer, and thus RNA synthesis can "bootstrap" itself.

As an additional bonus, DNA contains a new kind of nucleobase that facilitates the repair of damaged genetic material. Instead of uracil (U), DNA contains a methylated version of this base, known as thymine (T). The trouble with the nucleobases used in RNA is that cytosine sometimes spontaneously converts to uracil via a simple hydrolysis reaction that replaces an amino group on the nucleobase with oxygen. The additional methyl group in thymine allows the cell to distinguish this base from any uracil accidentally introduced into a DNA strand by the hydrolysis reaction. There is an entire tool kit of repair enzymes to fix this: they first cut off the uracil base, then open the damaged DNA strand, and finally restore the cytosine. Thus, the introduction of thymine and the associated set of quality controls is a valuable improvement over the RNA world in the fidelity of genetic inheritance, but it would be of much less value in "disposable" products, such as messenger RNA, which are therefore still made of RNA today.

Which Came First, Proteins or DNA?

With the advent of a more durable genetic material and the availability of highly efficient protein-based catalysts, the machinery that copied, repaired, and transcribed the cell's genome evolved to levels of complexity that were simply not possible in an RNA world. This complexity included not only enzymes that copy DNA to make either new DNA or RNA transcripts, but also enzymes that unwind DNA, repair DNA damaged induced by UV light or other sources of radiation, untangle DNA when it's all knotted up, cut DNA at specific sites and that regulate the length of the chromosome ends, and much, much more.

But this raises a question. Because of the obvious advantages that proteins and DNA possess, it is easy to rationalize why the protein-DNA world took over from the RNA world. But even if we accept that the RNA world begat the protein-DNA world, which of these new polymers came first? As yet, nobody knows whether the RNA world first recruited proteins or recruited proteins only after inventing DNA. A rather indirect argument in favor of DNA first is as follows: while we know LUCA used DNA, LUCA may not have contained the enzyme ribonucleotide reductase, which converts ribonucleotides to deoxyribonucleotides, because this enzyme is quite different in archaea, bacteria, and eukaryotes. Steven Benner has speculated that this is because LUCA's ribonucleotide reductase was still a ribozyme. That is, it was a ribozyme that catalyzed the reaction that made DNA, not a protein—supporting the argument that DNA came *before* proteins. Still, the ribonucleotide reductase reaction is a very difficult one to perform (it is one of the few biochemical reactions that relies on radicals), so there is some question as to whether it could have been performed by a ribozyme, and no such ribozyme has been created in the laboratory. Given this, it is probably best to say that the jury is still out on the question of which came first. Or, more accurately, second, after RNA.

Enzymes and Metabolic Networks

At this point in our story, we're talking about an organism with all the major molecular components of today's life in place, including RNA, proteins, and DNA. So, having reached this point, several natural questions now arise. How many proteins did this early organism have?

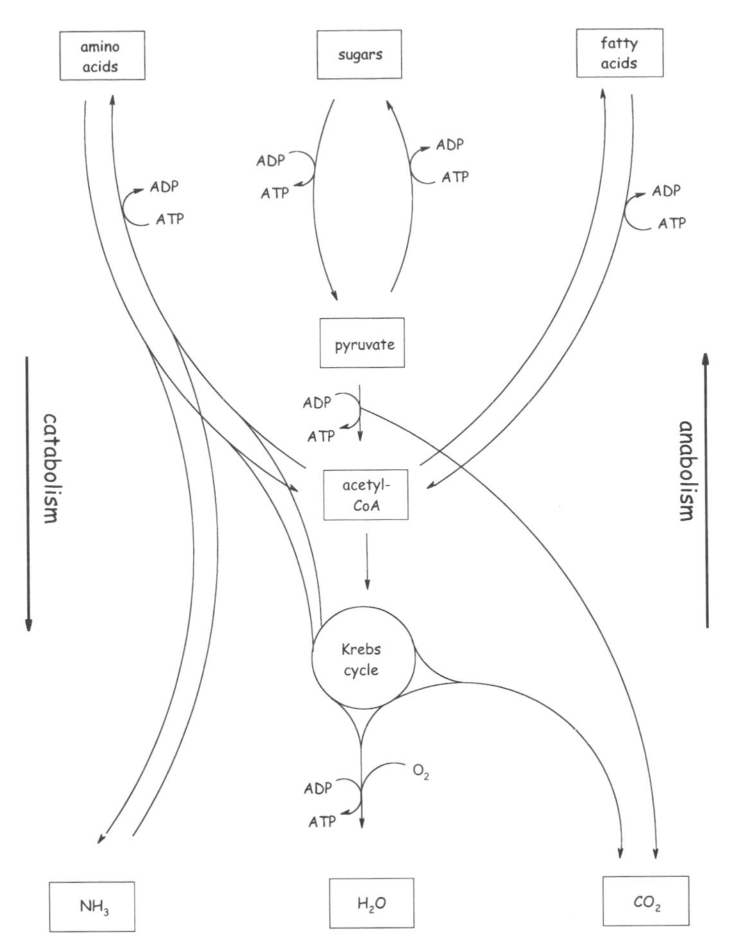

Figure 6.3 Even a minimalist Terrestrial genome encodes a very complex metabolic network. This includes *catabolic* pathways that break down large molecules, such as glucose, into smaller molecules, such as pyruvate, in order to generate energy; and *anabolic* pathways that use energy to synthesize amino acids, nucleobases, and the like from simpler precursors.

What were their likely tasks? And how did they link up into complex metabolic networks? In short, which of the metabolic pathways of today's cell (fig. 6.3) would already be listed in a biochemistry textbook had it been written by visiting aliens a few billion years ago?

Since the sequencing of entire bacterial genomes became feasible in 1995, genomics (the study of the full complement of genes that an organism carries) has increasingly provided researchers with new tools to address all these questions. Obviously, the availability of unprecedented amounts of gene sequences has aided molecular phylogeny—the construction of family trees for groups of related organisms, or even groups of related proteins, which we'll discuss in the next chapter. But on a higher organizational level, the comparison of complete genomes has made it possible to investigate the question of what makes up a minimal set of genes for a cellular organism.

The first two organisms to have their entire genomes sequenced, *Haemophilus influenzae* and *Mycoplasma genitalium*, were selected for these studies, in part, because their genomes are small. In fact, with just 480 genes, *M. genitalium* was assumed to contain precious little beyond the minimal set. Nonetheless, comparison of the genomes of these two organisms suggests that there is a shared set of around 260 genes that represent the minimum "essential" collection. Later studies focused on sequentially knocking out individual genes from the much larger genome (approximately 4,100 genes) of the bacterium *Bacillus subtilis*. These studies arrived at a list of just 271 essential genes. Still other studies have produced results ranging from 150 to 670 essential genes. But it is not so much the precise number that should interest us here. Instead, we should focus on the tasks that these genes fulfill, and their usefulness as a model for the simplest, and perhaps earliest, metabolisms.

And what do these simple, minimal genomes tell us? In the *B. subtilis* study, most of the 271 essential genes can be clearly grouped into the broad categories of information processing (DNA processing: 27 genes; RNA processing: 14; protein synthesis: 95), and the cell envelope (44), cell shape and division (10), and energetics (30). Not surprisingly, this suggests that DNA replication, protein synthesis, maintenance of the cell's physical integrity, and the metabolism required for energy production are all critical elements of the simplest complete metabolism. Still, the question remains as to which of these metabolic networks we inherited from LUCA and which, if any, were invented after her. To answer this question, we have to look at the bigger picture.

Steven Benner has likened the evolution of metabolism to a palimpsest, the word used in archaeology for a parchment that has been used more than once and from which traces of the imperfectly erased, earlier inscriptions can still be read. For example, as we described in chapter 5, ribonucleotide cofactors (e.g., ATP) can be interpreted as vestiges of the RNA world. Benner argues that all of metabolism can be viewed in this way, for, as we argued above, any metabolic pathway that is shared across all life-forms was probably inherited by them from LUCA. Using this approach to map out LUCA's biochemistry, Benner's group postulated that LUCA not only used DNA as her genetic material, but that she also employed a fairly modern-looking DNA polymerase, the enzyme that synthesizes DNA by stringing together nucleotides on a DNA template. As described above, LUCA also contained the transcriptional machinery by which RNAs are made from a DNA template and a full set of machinery for translating messenger RNA into the appropriate protein sequences.

It appears current life on Earth also inherited a fair bit of its metabolism from LUCA. That is, beyond transcription and translation, current life inherited from LUCA many of the myriad of biochemical pathways that it uses to convert food into organisms (and into heat). Although we humans, for example, can synthesize only 12 of the 20 amino acids found in proteins (due to our rich diets, we could afford to lose some of the pathways through mutation without taking too hard a selective hit), most organisms are not similarly handicapped. Looking across the tree of life, we find, in fact, not only that most organisms can synthesize all 20 amino acids, but also that the biosynthetic pathways by which most amino acids are formed are closely related. The only notable exceptions to this rule are the biosynthetic pathways for the three aromatic amino acids (phenylalanine, tyrosine, and tryptophan, each of which contains a benzene-like group in its side chain; see fig. 4.2), which are synthesized via very different pathways in eukaryotes and bacteria. So it seems that LUCA had developed an ability to synthesize most of the amino acids herself, if perhaps not the aromatic amino acids. By a similar argument, LUCA probably could synthesize the nucleobases; the biosynthetic routes by which both the purine and pyrimidine bases are manufactured in the cell are closely related across all life. Finally, LUCA also employed pretty much all the nucleotide-containing cofactors, including ATP, FAD, vitamin B12, S-adenosyl methionine, and coenzyme A, as might be expected given the putative connection of these cofactors to the earlier RNA world.

But what about the mechanisms by which the cell derives its energy? Humans obtain most of their energy via two metabolic pathways. The first is called glycolysis and involves the *nonoxidative* breakdown of glucose into the smaller molecule pyruvate. Such nonoxidative, energy-producing reactions are called fermentation.* The 10 enzymes involved in glycolysis share close relatives across the three domains of life, suggesting once again that LUCA contained this key metabolic pathway. In our cells, the pyruvate is then oxidized to carbon dioxide in a cyclic metabolic pathway called the Krebs (citric acid) cycle (which we've already encountered but will learn more about in chapter 7). The enzymes of our Krebs cycle, however, are not present in archaea, suggesting that LUCA did not contain this oxidative metabolic reaction.

Although LUCA's genome encoded the enzymes necessary to perform glycolysis, it is difficult to imagine that, by the time she was around, long after the origins of life itself, sugars like glucose were simply lying around like manna from heaven. Instead, she probably employed the enzymes of glycolysis in reverse, using them to synthesize sugars from simpler carbon compounds. Given this, though, what did she eat? That is, what was her source of these simpler carbon compounds? To answer this question, William Martin and coworkers at Heinrich-Heine-Universität in Dusseldorf, Germany, compared more than 6 million genes taken from 1,847 bacterial and 134 archaeal genomes to identify 355 genes that contributed to LUCA's genome. From this, they deduced that LUCA employed the Wood-Ljungdahl pathway (fig. 6.4), a metabolic route by which carbon dioxide is reduced into more useful compounds using hydrogen gas, which, in LUCA's case, would have been produced by the reaction of water with reduced iron compounds in the Earth's crust. Their work indicates that LUCA's

*Under anaerobic conditions (such as in a champagne bottle), yeast obtain their energy from the glycolytic pathway. As a last step in the pathway, they convert the pyruvate to ethanol and carbon dioxide. Cheers!

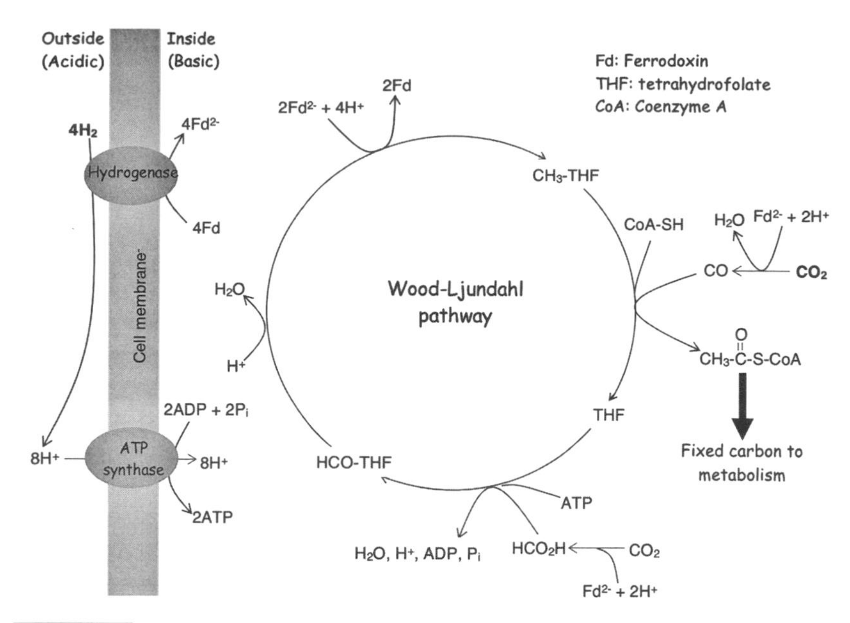

Figure 6.4 Comparative genomic studies indicate that LUCA encoded the Wood-Ljungdahl pathway (also known as the reductive acyl-CoA pathway), a metabolic route by which two molecules of carbon dioxide are reduced via the consumption of four hydrogen molecules to create one molecule of the central (see Figure 6.3) metabolite acetyl-CoA. The consumption of hydrogen takes place outside the cell, generating hydrogen ions there, and the reaction consumes hydrogen ions inside the cell. LUCA used the resulting hydrogen ion concentration gradient to produce metabolic energy in the form of ATP. Tetrahydrofolate and coenzyme A are small-molecule cofactors, and ferredoxin is a small, iron-and-sulfur-containing protein; these are used throughout biology to carry single-carbon units, acetyl groups, and electrons, respectively.

genome also encoded nitrogenase, the enzyme that uses molecular hydrogen to reduce nitrogen gas to the more useful ammonia, and glutamine synthase, the enzyme that converts ammonia into more useful (and less toxic) nitrogen-containing compounds. This ability to fix atmospheric nitrogen was lost in most organisms, including all eukaryotes; it presumably would have been of value to us, but oxygen is a potent inhibitor of this enzyme, rendering it difficult to keep active after the advent of the oxic atmosphere.

Membranes: Wrapping It All Up

The DNA-RNA-protein division of labor is, effectively, a universal feature of organisms living on our planet today. Another similarly near-universal feature is the barrier that separates the living from the nonliving world: the cell membrane. Life may have originated as a set of self-propagating chemical reactions on a solid surface, in liquid droplets, or in other kinds of media (as described in chapter 5), but it acquired the ability to grow and spread, independent of the medium, only when it succeeded in isolating itself from the environment by creating the cell membrane. In the RNA-world hypothesis, the membrane also serves the critical role of keeping the genetic material in physical association with the metabolic catalysts it encodes, without which, the metabolic networks of the cell would not provide a selective advantage for the organism (which is defined by its genes), and evolution would grind to a halt. In a sense, the membrane was the key step in creating cellular life as we know it today. (Note that viruses, many of which are also surrounded by a membrane, were a later development based on the DNA or RNA of already existing cellular organisms.)

But where did the membrane come from? The universality of the DNA-RNA-protein system strongly suggests that it predates the common ancestors of today's organisms. Similarly, all cells are surrounded and defined by at least one double-layer lipid membrane, supporting the argument that LUCA must have had a membrane too. The chemical composition of the lipids making up the membrane, however, differs fairly significantly among the three domains of life. Specifically, both bacterial and eukaryotic membranes (including our own) are made of diacylglycerides, which are composed of a glycerol molecule to which two fatty acid tails are linked via an ester bond to a backbone comprised of glycerol-3-phosphate (fig. 6.5). Archaea, however, cannot synthesize fatty acids, and instead build their membranes from a class of lipids called terpenoids, which they link to the backbone glycerol-1-phosphate via a chemically distinct ester bond. Given that LUCA appears to have been able to make terpenoids (all three of the

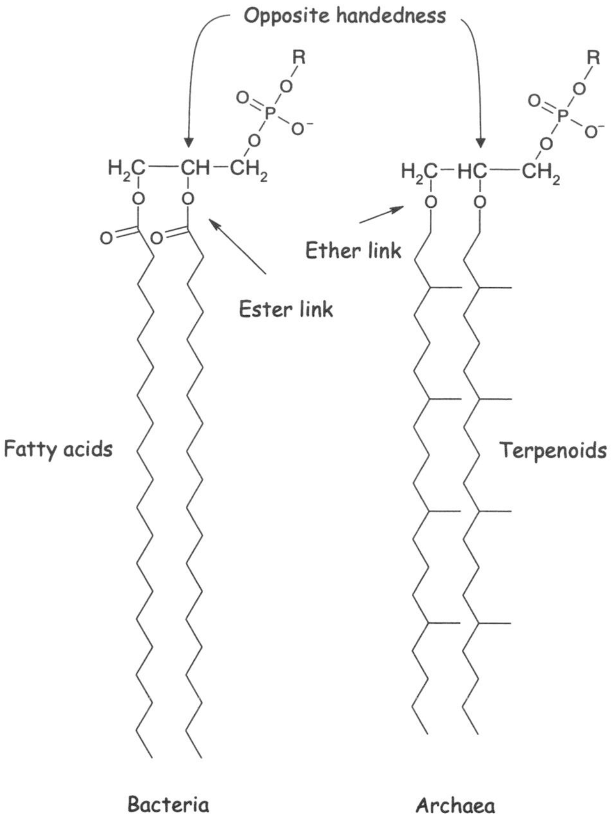

Figure 6.5 The cell membranes of most organisms, including bacteria, are composed largely of diacylglycerides, in which two water-hating fatty acid tails are linked by ester bonds to a glycerol, which in turn is connected to a water-loving "head group" comprised of modified phosphates (the "R" group denotes a family of molecules that vary in that position). The cell membranes of one large, microbial branch of the tree of life (the Archaea) are formed from branched hydrocarbons called terpenoids attached to glycerol by ether, rather than ester, linkages. A final, more subtle distinction is that the chirality (handedness) of the glycerol backbone is flipped between the two.

broadest branches of the tree of life can synthesize them), and given that archaea, and thus presumably LUCA, cannot synthesize fatty acids, the presumption is that LUCA employed terpenoids in her membranes and that these were later replaced with fatty acids in the bacterial and eukaryotic branches of the tree of life (see sidebar 6.2).

Sidebar 6.2

Mixing Membranes

Microorganisms in one of the main branches on the tree of life, the archaea (more about them in the next chapter) use ether-linked terpenoids as the lipids in their membranes. Species in the other two main branches (bacteria and eukaryotes) use ester-linked fatty acids. Archaea link their lipids to glycerol-1-phosphate. Bacteria and eukaryotes use glycerol-3-phosphate, a subtle distinction but one that flips the chirality (handedness) of the resulting product. From a biochemist's perspective, these are big changes, raising serious questions regarding how LUCA's offspring might have negotiated the problem of converting her terpenoid membranes into the fatty acid membranes we (and bacteria) employ. Highlighting how big this question is, a number of researchers had claimed that mixed membranes comprised of both components, a logical intermediate step, would be thermodynamically unstable. In 2018, however, experimentalists in the Netherlands succeeded in creating a variant strain of *E. coli* with a mixed membrane containing both types lipids, thus nicely addressing at least the last issue.

Improving on prior efforts, which had successfully "reverse-engineered" the synthesis of archaeal ether-linked lipids in the bacterium but only inserted them in the membrane at a level of less than 1%, the research groups of Arnold Driessen at the University of Groningen and John van der Oost at the University of Wageningen teamed up to grow *E. coli* that contained a large fraction of archaeal lipids in its membranes. To achieve this, they "boosted" the synthesis of terpenoids in the cell and then added the three genes required for archaeal ether lipid synthesis. After a bit of fine-tuning of the metabolic pathway, they succeeded in producing a cell in which all the phosphatidylglycerol lipids, the least complex of the membrane lipid structures (in which the "R" group in fig. 6.5 is simply a hydrogen), were replaced with archaeal terpenoids, amounting to 30% of the total membrane content. To the researchers' surprise, the bacteria even coupled the ether lipids to a glycerol-1-phosphate, as in archaea, in contrast to the glycerol-3-phosphate found with bacterial fatty acid chains. As bacteria were not known to be able to synthesize glycerol-1-phosphate, this was unexpected and warrants further research into the mechanisms involved.

The hybrid-membrane bacteria exhibited a notably long 16-hour delay ("lag phase") before they started to grow, but once over that hump, cells producing moderate levels of the novel lipids grew at rates comparable to those of their unmodified brethren. They even displayed a slightly increased robustness toward a range of stress conditions, including heat, freezing, and exposure to 1-butanol, an organic solvent. Thus, instead of being thermodynamically unstable, it appears mixed membranes could have helped an ancestor thrive even in some of the more inhospitable conditions assumed to have prevailed during the early evolution of life on Earth.

Caforio, Antonella, Melvin F. Siliakus, Marten Exterkate, Samta Jain, Varsha R. Jumde, Ruben L. H. Andringa, Serve W. M. Kengen, Adriaan J. Minnaard, Arnold J. M. Driessen, and John van der Oost. "Converting *Escherichia coli* into an Archaebacterium with a Hybrid Heterochiral Membrane." *Proceedings of the National Academy of Sciences USA* 115, no. 14 (2018): 3704–9.

Conclusions

From the first simple, self-replicating molecules to protein- and DNA-based organisms of breathtaking metabolic complexity, it's a fascinating story, the broadest sweep of which seems clear: the incessant push of selective pressure, fighting against the imperatives of chemical reactivity, guided the earliest, simplest life into the complex, robust life-forms we see around us.

As illustrated by the scientific biography of Francis Crick, the tale of how we came to understand the origins of cellular life is a similarly complex and fascinating story, one driven by episodes of startlingly original, outside-the-box thinking. And while we may be tempted to deride those wild ideas that turned out to be off the mark, at the frontiers of our knowledge many wild ideas turn out to be true. The origin and early evolution of life is one such frontier where this kind of unbridled creativity is still useful.

Further Reading

Directed Panspermia

Crick, F. H. C., and L. E. Orgel. "Directed Panspermia." *Icarus* 19, no. 3 (1973): 341–46.

RNA Evolution

Joyce, Gerald F. "The Antiquity of RNA-Based Evolution." *Nature* 418 (2002): 214–21.

Protein Biosynthesis

Ramakrishnan, Venki. *Gene Machine: The Race to Decipher the Secrets of the Ribosome.* London, UK: One World Publications, 2018.

LUCA's Metabolism

Benner, S. A., A. D. Ellington, and A. Tauer. "Modern Metabolism as a Palimpsest of the RNA World." *Proceedings of the National Academy of Sciences USA* 86, no. 18 (1989): 7054–58.

Weiss, Madeline C., Filipa L. Sousa, Natalia Mrnjavac, Sinje Neukirchen, Mayo Roettger, Shijulal Nelson-Sathi, and William F. Martin. "The Physiology and Habitat of the Last Universal Common Ancestor." *Nature Microbiology* 1, no. 9 (2016): 16116.

Chapter 7

A Concise History of Life on Earth

In 1969, Carl Woese (1928–2012), then a young professor at the University of Illinois, came up with an unusual way to investigate phylogeny, the study of the interrelatedness of organisms, and in doing so uncovered a rather surprising result. For the preceding century, biologists had categorized life into five kingdoms: four eukaryotic (cells with nuclei) kingdoms comprising animals, plants, fungi, and single-celled protists, and a single prokaryotic (cells lacking a nucleus) kingdom, the bacteria. This classification, which was based on gross cellular features, came about in part because it seemed a natural assumption from our perspective as big, lumbering eukaryotes that the major divisions of life on Earth would be weighted heavily toward us and our close cousins. Woese's ambition was to probe the relatedness of these kingdoms in more detail and to settle the question, once and for all, as to how bacteria fit into the bigger picture.

Because bacteria don't have gross structural features, such as hair or scales, that can be easily compared, Woese realized that he would have to study phylogeny at the molecular rather than the cellular or organismal level. Even as late as the late 1960s, though, biologists had depressingly few molecules with which to perform such comparative molecular biology; at the time, the atomic-resolution structures of only about 10 proteins were known,* along with the amino acid sequences of a few dozen more. Even simply establishing the sequence of a small protein took years of effort, and there were no viable methods for sequencing genes, much less entire genomes. How, then, could one compare organisms on a molecular level and work out their evo-

*In February 2020, the number was 128,000 and counting.

lutionary relationships beyond those that are obvious from outward appearance?

To start, Woese needed a molecule that is present in, and thus could be compared across, all living things. As all cellular organisms employ ribosomes to make their proteins, he chose the RNA strand of the small ribosomal subunit, known as the 16S RNA in bacteria. Reading the nucleotide sequence of the molecule was well beyond the technology of the day, so he shredded the molecule instead. More specifically, he digested the ribosomal RNA with an enzyme that cuts after guanosine (G), and only after guanosine. The resulting fragments, short enough to be sequenced with available methods, would all constitute "words" ending with G (and only containing that one G): AUG, CG, ACACACUUG, and so on.

After sorting the fragments by their length (using gel electrophoresis, a well-established method of separating molecules according to size), sequencing some of them, and thinking about the results, Woese found that the most useful words were those of six letters or more—shorter words were too common to provide clues as to who was related to whom. There are $3^5 = 243$ different six-letter words ending in G, and a typical 16S RNA contains around 25 of them. These short words thus provided good enough statistics to allow Woese to "fingerprint" organisms by their 16S RNA and to use these fingerprints to map out which organisms are most closely related.

Over many years, Woese and his graduate students compiled the "dictionaries" of 16S RNA words from different bacterial species and constructed family trees based on the degree of identity between the content of the dictionaries. To their surprise, a subset of the bacteria they studied—those that produce methane—turned out to be just as distantly related to other bacteria as they are to elephants, guinea pigs, or ourselves. Woese proposed that these microbes represented a new branch on the tree of life, equal in stature to the Eukarya and Bacteria, and he called this third branch the Archaebacteria (later shortened to Archaea, with an individual member of the group called an archaeon) because it seemed to represent an "ancient" form of life.

As mentioned in chapter 6, we now regard the domains of Archaea and Bacteria as the direct offspring of LUCA and the more complex eukaryotic cells as a later recombination of representatives of these two ancestral domains. But, back in the 1970s, the idea was heretical. After all, Woese's new-fangled "molecular fingerprinting" aside, archaea and bacteria look very similar from the perspective of us big, multicellular organisms, so, naturally, Woese's arguments were rejected more or less out of hand by many microbiologists. Only the small and initially marginal community of researchers exploring life under extreme physical conditions (see chapter 8), including the German biologists Holger Jannasch (1927–1998) and Karl-Otto Stetter, embraced the new order. It was not until the late 1990s, when the first archaeal genomes were sequenced, that Woese's claim that Archaea was a separate realm of its own and that life is better divided into two prokaryotic and one eukaryotic "domain" rather than four eukaryotic and one prokaryotic "kingdom" finally prevailed over the earlier, more traditional view. With literally thousands of gene sequences available, comparative molecular studies demonstrated quite compellingly that Woese was correct, and the archaea are, if anything, more distant from bacteria than they are from us. The wheels of science sometimes turn slowly, but turn they do.*

The Emergence of Life on Earth

On the one hand, the history of life on Earth is a bit of a parochial topic for wide-ranging types like us astrobiologists. Clearly, for example, the division of Terrestrial life into three domains would hardly be of relevance to the study of, say, life (if there is any) on Jupiter's moon Europa. On the other hand, the history of how the story unfolded here on Earth is the only example we have so far, so if we're careful not to be complacent and fall into thinking that how it occurred on

*To quote Carl Sagan: "In science it often happens that scientists say, 'You know that's a really good argument; my position is mistaken,' and then they actually change their minds and you never hear that old view from them again. They really do it. It doesn't happen as often as it should, because scientists are human and change is sometimes painful. But it happens every day. I cannot recall the last time something like that has happened in politics or religion." Carl Sagan, "The Burden of Skepticism," *Skeptical Inquirer* 12, no. 1 (1987).

Earth is the only way it could have occurred here or anywhere else, the topic seems worthy of serious consideration.

How and when did the history of life on Earth unfold? Although the Earth formed some 4.56 billion years ago and its crust solidified a few tens of millions of years later, after the formation of the Moon (see chapter 3), it was effectively uninhabitable for a long time after that. It was not until some 3.8 billion years ago that the late heavy bombardment and its planet-sterilizing impacts came to an end. The question, then, is, how quickly after the end of the late heavy bombardment did life arise on our planet?

The answer to this question is obscured, at least in part, by our lack of detailed knowledge of the Earth's early history. Even after the end of its turbulent formative years, the Earth remained (and remains) a highly dynamic planet, and few records of its early days have survived intact. Due to the incessant erosion brought on by the Earth's hydrological cycle and plate tectonics constantly recycling crust into the mantle, typical rocks on the surface of our planet are estimated to have a half-life of only a few hundred million years. The chances of the Earth's first rocks, those dating from the so-called Hadean eon, surviving this gauntlet and remaining unscathed from their formation up to the present approach zero. Despite significant effort on the part of geologists, only one or two rock outcrops have been reported to be much older than 4 billion years, with the oldest reported (the validity of this is debated) clocking in at 4.28 billion years. It is really only from the Archaean eon, which started 4 billion years ago, that any significant geological record has been preserved. (See table 7.1 for a time line of the geological intervals in relation to the history of life on Earth.)

The oldest known Earth rocks come from several locations in the far north of North America, with the oldest, well-accepted age being associated with the Acasta gneisses (metamorphosed igneous rocks, i.e., rocks that originated from molten material and were later altered) found near Great Slave Lake in northern Canada. Unfortunately, though, during the 4.03 billion years since these rocks solidified, they have been modified by heat and pressure to such an extent that they provide little information on what the Earth was like that early in its youth. More critical to our story are the somewhat younger, apparently supracrustal rocks of Isua and Akilia, West Greenland. The "supracrustal" tag denotes that these metamorphosed rocks were deposited as sediments or volcanic flows in shallow water when they were formed, some 3.7 to 3.8 billion years ago. If this mineralogical assignment is correct, these rocks confirm that liquid water (viewed, of course, as critical for the formation of life) existed on Earth at that time. But was there life in this water?

Sediments are formed from a steady rain of material—both organic and inorganic—that settles out of the water, so they provide an ideal place to look for signs of past life. Perhaps consistent with this, the Isua and Akilia rocks contain small globules of graphite, the pure carbon form used in pencil "lead." Is this evidence that life was flourishing more than 3.7 billion years ago? As it contains only carbon and none of the other chemical elements necessary for life, graphite is not usually associated with biology. On the other hand, this graphite might be. The reason is that the Isua and Akilia rocks are also significantly metamorphosed; had they originally contained life, the organic carbon would have been dehydrogenated to form graphite when the rocks became buried and "cooked" deep within the Earth. Based on this argument, the German geologist Manfred Schidlowski (1933–2012), and later the American geochemist Stephen Mojzsis, suggested that the carbon extracted from these rocks might have been derived from living things. Mojzsis, then a graduate student at the Scripps Institution of Oceanography working under Gustaf Arrhenius (grandson of Svante Arrhenius of panspermia fame; see chapter 5), characterized the ratio of the carbon isotopes ^{12}C and ^{13}C in these ancient materials by heating the rock and analyzing the carbon compounds that were driven off. What he found was carbon depleted in the heavier isotope as is observed today in the organic molecules produced by photosynthesis! (The more rapidly moving, lighter carbon isotope is preferentially "fixed" in the photosynthetic reaction.) Thus, Mojzsis suggested, not only had life arisen at the time the original rock was deposited—just a few hundred million years after the crust cooled and smack at the end of the late heavy bombardment—but it had evolved to such a high degree of complexity that photosynthesis was already a common form of metabolism.

Table 7.1
A chronology of life on Earth

Geological eon/era/period (starting age, years ago)	Time (years ago)	What was happening?
Hadean eon (4.56 billion)	4.56 billion	Accretion of the Earth
	4.53 billion	Formation of the Moon
	4.28 billion	Oldest claimed Earth rocks still in existence (not widely accepted)
	4.03 billion	Oldest Earth rocks still in existence (widely accepted)
Archaean eon (4 billion)	~3.8 billion	End of late heavy bombardment; oldest claimed evidence for life on Earth (not widely accepted)
	>3.77 billion	Oldest claimed microfossils (not widely accepted)
	~3.6 billion	Formation of first continents
	~3.4 to 3.5 billion	Reasonably accepted evidence for life (stromatolites, microfossils, isotopic signatures)
	3.2 billion	Fully accepted evidence for life (microfossils, stromatolites)
Proterozoic eon (2.5 billion)	~2.4 billion	Great oxygenation event
	1.65 to 2.2 billion	First recognizably eukaryotic fossils
	~1.1 to 1.7 billion	Divergence of major eukaryotic lineages
	>1 billion	Invention of sex
	1.05 billion	Oldest multicellular eukaryotic algae fossil
	560 million	Oldest widely accepted animal fossils

(continued)

But are these putative indications of life, and possibly even photosynthesis, on a firm footing? Within only a few years of the publication of Mojzsis's investigations, other researchers began to question the evidence on many grounds. First, the seemingly telling graphite occurs in veins of carbonate rock that were probably formed by the injection of hot fluids when the older host rocks were buried deep within the Earth, perhaps long after they were initially formed. Sec-

Geological eon/era/period (starting age, years ago)			Time (years ago)	What was happening?
Phanerozoic eon (541 million)	Paleozoic era (541 million)	Cambrian (541 million)		Atmospheric oxygen reaches about half of current level; Cambrian "explosion"
		Ordovician (485 million)		Cephalopods and jawless fishes rule the oceans; fossilized arthropod footprints
		Silurian (444 million)		Diversification of the bony fish; land plants; spiders and scorpions
		Devonian (420 million)		First land vertebrates (amphibians)
		Carboniferous (359 million)		Forests of tree ferns; first reptiles
		Permian (299 million)		Period ends with largest recorded mass extinction
	Mesozoic era (252 million)	Triassic (252 million)		Origin of dinosaurs (and mammals)
		Jurassic (201 million)		Beginning of the age of dinosaurs; first birds (which are a dinosaur lineage)
		Cretaceous (145 million)		Origin of flowering plants
	Cenozoic era (66 million)	Paleogene (66 million)	66 to 20 million	Mammals diversify after dinosaur extinction event
		Neogene (23 million)	20 to 3 million	Climate cools; grasslands expand
			~6 million	Human-chimp divergence
		Quaternary (2.58 million)	2.1 million	Rise of *Homo erectus*
			0 million	You're reading this

ond, the isotopically odd carbon was released at a temperature far too low to be from the graphite (which has the highest vapor point of any element) and thus could well be a more recent contaminant. Third, known abiological chemical processes can produce carbon similarly depleted in the element's lighter isotope. Finally, even the age of the relevant rocks has been questioned (see sidebar 7.1)—in fact, by members of Mojzsis's original research team—as has the original identifi-

Sidebar 7.1

The Dating Game

A small part of the debate on the ancient rocks of Greenland concerns their age. Are they really 3.8 billion years old? In chapter 3, we discussed radioisotope dating, but while this technique is straightforward and well established, applying and interpreting it sometimes are not. A big part of the problem is that this method dates the last crystallization of the rock. The formation of sedimentary rock—the kind that's likely to contain fossils, and the kind putatively observed on Akilia—does not involve melting and crystallization, and thus sedimentary rock cannot be dated directly. Sometimes, however, it is possible to define the "minimum age" of sedimentary rocks by dating igneous "intrusions," which are thick veins of minerals solidified from a molten state, cutting through the sedimentary rock. These igneous intrusions, which must have formed after the sedimentary rock to have cut through it, sometimes contain crystals of zirconium silicate ($ZrSiO_4$), called zircons, which can be accurately dated.

Scientists interested in dating rocks, grandly called geochronologists, often rely on zircons. This is because when zircons crystallize from magma, they contain uranium, which is similar in size to zirconium and thus fits into the crystal lattice, but they contain no lead, which is preferentially excluded from the crystal lattice. Why is this important? Because uranium decays into lead at a known rate (or rates, actually—there are two naturally occurring uranium isotopes). And given that all the lead trapped in the zircon crystal lattice originally must have come from uranium, the ratio of uranium to lead reflects the time since the zircon formed. More specifically, because uranium-238 decays into lead-206 with a half-life of 4.5 billion years, and uranium-235 decays into lead-207 with a half-life of 0.7 billion years, these uranium-lead "clocks" provide two independent measurements of a zircon's age. Thomas Krogh, who helped develop the uranium-lead zircon dating method at the Royal Ontario Museum in Toronto, Canada, says that even if the zircons are reheated, as happened at least once to the Greenland samples, they retain a "memory" of their first crystallization.

Obviously, though, dating igneous zircons can only set the minimum age of a sedimentary rock and, even then, only if the relationship between the igneous intrusion and the sedimentary substrate is well understood. But because the Greenland rocks were severely deformed during their nearly 4 billion years on Earth, their sequence—the information about which layers formed first and which came later—has become jumbled. And Stephen Moorbath, a geologist at Oxford University, contends that the sedimentary rocks were most likely deposited about 3.65 to 3.70 billion years ago. This more recent dating would explain the absence of the element iridium—rare on Earth but common in asteroids—and any other signs of the late heavy bombardment that would be expected if the rocks were more than 3.8 billion years old.

On a separate note, geochronologists working on Australian sediments have found a single, small zircon crystal that apparently withstood the erosion that created the original sediments and thus predates them. This micron-sized crystal has been dated at 4.4 billion years and, though small, is the oldest known Terrestrial "rock."

Moorbath, Stephen. "Palaeobiology: Dating Earliest Life." *Nature* 434, no. 7030 (2005): 155.

cation of the Isua and Akilia rocks as sedimentary (sediments are a great place to collect fossils; other rocks are not). It's probably safe to say that, at best, the jury is very much still out regarding the evidence for life in these 3.8-billion-year-old rocks.

Similar, if still debatable, evidence for life is claimed in studies of the Nuvvuagittuq supracrustal belt in northeastern Canada, a geological formation consisting of layers of basalt that, as described above, are debatably dated to around 4.28 billion years, and sediments dated to at least 3.77 billion years. The group of Dominic Papineau at University College London, UK, and others have analyzed the putative sediments, which, with the basalt, they believe formed the floor of an ancient ocean, and found chemistry that they argue is consistent with the rocks having formed in a hydrothermal vent environment. Looking closely, the researchers identified tubes and filaments of the iron oxide mineral hematite that exhibit the characteristic branching seen for filamentous, iron-oxidizing bacteria found near hydrothermal vents today, suggesting that these might be microfossils. The researchers also identified isotopically light carbon (as Mojzsis reported for the Greenland rocks) in both carbonates and in graphite. Anticipating the usual criticism that abiotic processes might be responsible for these structures, the authors rather bluntly argue that, "Collectively, our observations cannot be explained by a single or combined abiogenic pathway, and therefore we reject the null hypothesis."* Their confidence notwithstanding, the paleontological community remains far from consensus regarding the merits of their claims.

If the evidence for life at 3.8 billion years ago is contentious, how much more recently do we have to go before the evidence becomes firmer? Perhaps not much. In 1993, William Schopf, a professor of paleobiology at the University of California, Los Angeles, described 3.46-billion-year-old specimens from Western Australia (near a fiercely hot mining center, ironically named North Pole) that seemed to contain microscopic, tar-colored fossils of bacteria. The tiny organisms were encased in chert, an extremely fine-grained rock composed of

*Matthew S. Dodd et al., "Evidence for Early Life in Earth's Oldest Hydrothermal Vent Precipitates," *Nature* 543 (2017): 60–64.

silica that can preserve the smallest of details. Schopf sorted the bacteria into 11 taxa, or distinct groupings, based on the shapes of the fossils and claimed that, in terms of their shape, seven appeared to be early relatives of cyanobacteria. Raman spectroscopy of the samples, which crudely identifies molecular components, indicated that the tar-like substance defining the fossils was kerogen, a complex mixture of hydrocarbons that is typically produced when biological material is subjected to heat and pressure beneath the Earth's surface. Taken together, Schopf claimed, this provided incontrovertible evidence that complex ecosystems, likely to consist of multiple species of photosynthetic cyanobacteria, existed as early as about 300 million years after the end of the late heavy bombardment.

Following up on Schopf's claims, Donald Canfield and coworkers at Odense University in Denmark studied sulfur isotopic fractionation in the same rocks and found possible biogenic signatures in the sulfur-containing mineral barite. If their identification proves correct, it would both confirm the existence of life at 3.46 billion years ago and identify one of its key metabolic reactions: namely, the use of the electrons in sulfate (SO_4^{-2}) to oxidize hydrogen or hydrocarbons to produce sulfide (S^{-2}) and water or carbon dioxide.

Schopf's claims, however, have also found critics. One criticism, raised by Martin Brasier (1947–2014) of Oxford University, takes aim at the ambiguous shapes of the putative fossils; of the thousands of inclusions in the rocks, only a tiny fraction of them look like cyanobacteria or, indeed, any contemporary bacteria. Of course, although they fossilize better than other bacteria, cyanobacteria do not fossilize well, especially if they degrade between death and fossilization. A second criticism, also raised by Brasier, is that Schopf misunderstood the geology of the supposed microfossils, which were preserved not in marine sediments, which would have collected fossils, but rather in a hydrothermal vent or even in volcanic glass, in which fossils are much less likely to form. Once again, it seems that the jury is out on whether these are real microfossils, and if so, whether they are cyanobacteria, although the case in favor of life's remnants in these rocks seems significantly better established than the case for earlier life described above.

If even the evidence for life at 3.46 billion years is contested, when does the evidence for life on Earth become incontrovertible? That's not such an easy question to answer. As we move forward in the geological record, we see only slowly mounting evidence, without ever finding any single, incontrovertible "smoking gun." For example, 3.43-billion-year-old sedimentary layers in Strelley Pool, Western Australia, a few tens of kilometers away from the North Pole site, contain abundant, stromatolite-like formations (we discuss stromatolites—fossilized bacterial colonies—in more detail below) that look very similar to stromatolites frequently found in more recent geological settings and are thought to have formed in shallow marine environments. Given their abundance and preservation, there appears to be a reasonable degree of consensus among paleontologists that these strata are convincing evidence of life. They lack, however, any identifiable microfossils, which would help clinch the argument. In contrast, as we move still further forward in the rock record, we see more and more of what looks like microfossils, although this perhaps is not so much because the putative organisms had become more plentiful but rather because the rocks themselves becomes more plentiful. There are fine-grained, 3.40-billion-year-old cherts from South Africa, for example, that preserve many bacteria-sized spheres, some of which seem to have been caught in the process of dividing. And in early 2000, the Australian geologist Birger Rasmussen, now at Curtin University, Bentley, Western Australia, reported convincingly lifelike microfilaments in 3.2-billion-year-old Australian hydrothermal deposits.

The First Complex Ecosystems

Cyanobacteria seem to have invented the first organized supracellular structures and, with them, the oldest widely accepted fossils: the stromatolites we mentioned above (fig. 7.1). Stromatolites, dome-shaped or conical sediments of very finely layered sedimentary rock that can reach heights of decimeters to meters, are a dominant feature of many Precambrian sedimentary rocks. Today, however, living examples are rare and are limited to select niches, such as waters that

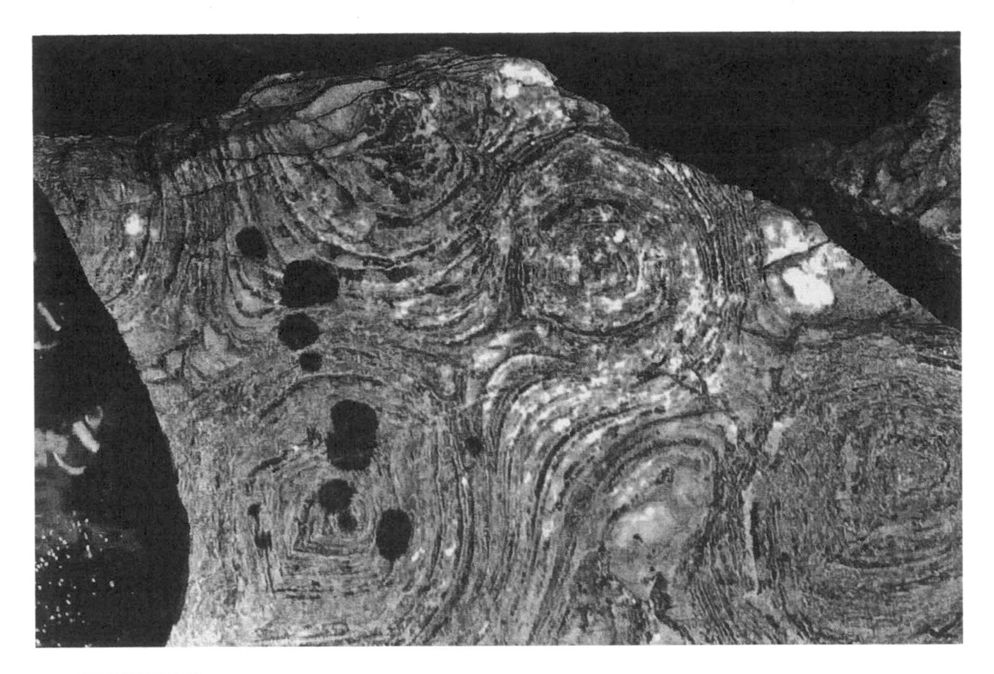

Figure 7.1 The edge of an author's shoe (left) on top of some billion-year-old stromatolite fossils in Glacier National Park in America's Rocky Mountains.

are too saline to allow animals to eat or otherwise disrupt them. With the rise of animals, it turns out that growing as a delicate and defenseless (and perhaps tasty) mat of bacterial cells is not as successful a lifestyle as it once was.

Hamelin's Pool, in Shark Bay on the coast of Western Australia, is just such a niche. In this hypersaline bay, which is twice as salty as seawater, stromatolites thrive, providing an opportunity to see in detail how their ancient brethren might have formed. Here, complex communities of microorganisms, usually cyanobacteria along with other bacteria and some microbial eukaryotes, spread out in coherent mats across the surface of sediments or rocks. The cells produce a thick, mucus-like material that glues them to one another and affixes them to the substrate. The mucus also traps fine sediments carried in the waves and currents. As this layer of sediment accumulates, the cells grow or migrate upward in order to continue photosynthesis.

Cells remaining behind are cut off from the light and die. Other organisms consume the organic material from the dying cyanobacteria, in turn liberating carbon dioxide. The carbon dioxide reacts with water to form carbonic acid, which binds calcium and precipitates out as a layer of limestone.

The oldest putative stromatolite fossils are those found in the 3.43-billion-year-old rocks of Strelley Pool, a site that, ironically, is just 800 km northwest of the stromatolites living today in Hamlin's Pool. And while the biological interpretation of these formations remains at least a bit contentious, it is hard to see how nonbiological processes, such as sedimentation or precipitation, could create such steeply sloping, conical layers. Thus, as we noted above, there is a fair degree of consensus regarding the biological origins of the Strelley Pool stromatolites despite their lack any clear microfossils. In contrast, the stromatolite fossils that become increasingly common over the early part of the Proterozoic eon, from about 2.5 to 1.6 billion years ago, often contain fairly convincing microfossils. These dominate the fossil record until some 600 million years ago, just before the end of the Proterozoic and the start of the Cambrian explosion, during which, as we describe below, animals proliferated and presumably began to graze on the stromatolites or, at the very least, disturb them by crawling over them and disrupting their fragile organization.

Photosynthesis Changes the World

Sunlight interacts with living organisms in a wide variety of ways. On the simplest level, the sunlight that hits the dayside of our planet delivers an energy flow of more than 170,000 TW (a terawatt equals a trillion watts), corresponding to the electricity output of 200 million nuclear power stations. Even though nearly one-third of this energy is reflected straight back into space, the other two-thirds stays with us and keeps us warm. It is the biggest contribution to our planetary energy balance by more than three orders of magnitude (followed by geological heating, human activity, and tidal friction). All life-forms, including the earliest and most primitive ones, must have benefited from this heat supply as it raised the surface temperature of our planet into the range that allowed water to remain liquid and life to originate and evolve.

At the most sophisticated level, light reflected from objects around us allows us to perceive our environment with our eyes. Although it is a relatively recent development compared with the evolutionary timescale dominated by bacteria, it appears that evolution is reasonably efficient at discovering and rediscovering vision. Enough so that the ability has arisen independently many times over the course of animal evolution. Octopus eyes, for example, look remarkably like our own despite the fact that the last common ancestor we shared with our excessively armed friends is thought to have been sightless.

Between these two uses of light—one very general, the other highly specific—evolution developed a third, equally important way of making use of light, thereby starting a revolution that arguably changed the nature of life on Earth more than any other single event since its origins. At some point, a group of bacteria, probably most closely related to today's cyanobacteria, came up with the two-step photosynthetic method that uses the energy of light to pull electrons from water, creating oxygen in the process.

The selective pressures in favor of photosynthesis are clear: quite simply, as LUCA and her offspring consumed the various reduced materials that were available in the environment, eventually a new source of energy had to be found. Less clear is how the modern photosynthetic apparatus might have arisen in the first place. As any student who has tried to memorize it will remember, the standard photosynthetic apparatus is incredibly complex. It must have evolved in a stepwise fashion from simpler mechanisms that already harvested sunlight without necessarily producing oxygen. Fortunately, some of the simpler (and presumably older) versions of bacterial photosynthesis are still around today and can help us understand how life came up with this new technology.

The simplest and best understood light-harvesting system is that of the extremely halophilic (salt-loving) archaea from the genus *Halobacterium*. The membranes of these cells contain characteristically colored patches known as the purple membrane. Its main component is a remarkably robust protein, bacteriorhodopsin, which uses light to pump protons out of the cell. With this activity, it creates a proton gradient across the membrane, which is a primitive means of storing the energy in an electrochemical form. Another membrane protein, called the F-ATPase, uses this proton gradient to drive the synthesis of the energy currency ATP (adenosine triphosphate). This mechanism constitutes a relatively inefficient, "hand-to-mouth" use of solar energy as it creates chemical fuel but no permanent chemical bonds, so it doesn't count as full-fledged photosynthesis.

In contrast, five major groups of bacteria have learned how to use the energy of sunlight to obtain the reducing potential (again, high-energy electrons) needed to synthesize carbon-containing molecules. Four of these do so without producing oxygen in the process. These are the Chlorobiaceae (green sulfur bacteria), Thiorhodaceae (purple sulfur bacteria), Chloroflexaceae (green nonsulfur bacteria), and Athiorhodaceae (purple nonsulfur bacteria, including *Rhodopseudomonas viridis*, which provided the first ever atomic-resolution structure of a photosynthetic reaction center). Like green algae and the higher (eukaryotic) plants, these non-oxygen-producing photosynthetic bacteria use light to drive a redox reaction that reduces carbon dioxide (carbon at oxidation state +4) to carbohydrates (carbon at oxidation state 0).* In green algae and the higher plants, the ultimate source of the electrons is water (oxygen at oxidation state −2), which is oxidized to molecular oxygen (O_2 at oxidation state 0). By contrast, the sulfur bacteria use sulfides as the electron source (leaving elemental sulfur as waste), and the nonsulfur photosynthetic bacteria employ hydrogen and small, reduced carbon compounds such as isopropanol. These materials are so easily oxidized that the energy of a single photon is sufficient to extract electrons from them, and thus only a single copy of the photosynthetic machinery (called a "photosystem") is required to exploit them.

Although sulfides and hydrogen are easy to oxidize and were probably abundant in the relatively reduced environments available on the

*The numbers here denote the number of electrons lost (positive oxidation states) or gained (negative oxidation states).

young Earth, they were ultimately in limited supply, and, when they ran out, photosynthetic organisms found themselves in need of a new source of electrons with which to reduce carbon dioxide. A promising, if rather ambitious, solution to this problem was to use water as the reductant since its relatively unfavorable oxidation potential is balanced by its extremely favorable abundance. But there was a hurdle to overcome: oxygen, the second most electronegative element, has a very high affinity for electrons, and thus water holds its electrons so tightly that the energy in a single photon of visible light cannot wrest them free. Cyanobacteria—the fifth, final, and most successful group of photosynthetic microbes—came up with the solution to this by linking two photosystems in series, each of which can trap one photon and convert its energy into chemical energy (fig. 7.2). This allows the energy in two photons to be summed in order to achieve the oxidation of water to oxygen and the efficient exploitation of the liberated electrons and protons.

The resulting light-driven "two-stroke engine" is one of the most complicated molecular machines we know. Essentially, the light energy captured by the chlorophyll molecule of photosystem II (PSII) lifts an electron to a higher energy level. In a slower reaction, the re-

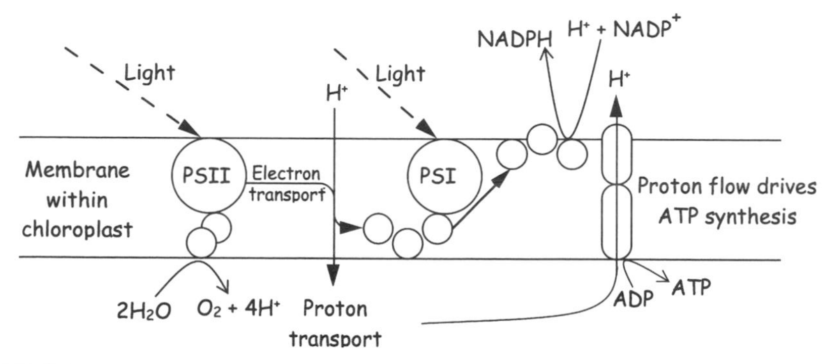

Figure 7.2 A simplified overview of the photosynthetic apparatus. The water-splitting reaction (bottom left) provides the electrons, which eventually serve to reduce $NADP^+$ to NADPH (top right). This process produces an excess of protons on one side of the membrane and consumes them on the other. In parallel, electron transport drives the movement of still more protons across a membrane. The resulting concentration gradient later drives ATP synthesis. (PSI and PSII are photosystems I and II, respectively.)

sulting "hole" is filled with an electron pulled from water (ultimately, after four electrons have been sequentially removed from two water molecules, releasing O_2). A pair of high-energy electrons flows from PSII down a cascade of reactions with the overall effect that the electrons are ultimately transferred to photosystem I (PSI), with the net production of one molecule of ATP. And while some of each electron's newfound energy is given up to form the ATP, it still arrives at PSI in a more energetic state than it had in the water molecule. In the second light reaction, the active center of PSI absorbs a photon and lifts this now higher-energy electron to a still higher energy state. The electron, combined with a proton derived from the split water molecule, can then reduce the redox carrier $NADP^+$ to form NADPH, the ribonucleotide that stores reduction potential in all cells.

In the next step of photosynthesis, NADPH moves over to the "dark reactions" (called this because they do not directly require the input of light), where its reduction potential is used in the Calvin cycle to "fix" carbon dioxide into reduced compounds that the cell can use. This biochemical cycle is named after its discoverer, Melvin Calvin of the University of California, Berkeley, whose early prebiotic chemistry studies we discussed briefly at the beginning of chapter 4. Calvin and his team elucidated the dark reactions of photosynthesis using radioactive ^{14}C to trace the path of carbon from carbon dioxide to the final sugar in what was one of the first instances of using a radioactive tracer in biochemistry. His work in this area, for which he won the 1961 Nobel Prize in Chemistry, was aided by the fact that ^{14}C was first synthesized at the Berkeley synchrotron just a few years earlier.

The Calvin cycle fixes carbon dioxide by covalently attaching it to the sugar ribulose-1,5-bisphosphate, splitting the sugar into two 3-phosphoglycerate molecules. The 3-phosphoglycerate is, in turn, reduced to glyceraldehyde-3-phosphate, from which the fixed carbon enters a complex metabolic shuffle among 10 sugar intermediates in all. Ultimately, for each six carbon dioxide molecules that enter the pathway, two glyceraldehyde-3-phosphates are split off (to serve as the starting material for the synthesis of other sugars, such as the hexoses, as well as amino acids, nucleobases, and all the other carbon-containing molecules the organism needs), and one molecule of ribulose-1,5-bisphosphate is regenerated. The latter allows the cycle to start anew. Overall, the formation of one six-carbon hexose, such as glucose, requires 12 molecules of NADPH, which in turn were synthesized from 24 electrons (generated using 48 photons) from the light reactions.

Comparison of cyanobacterial photosystems with those of the more primitive bacteria mentioned above shows that those of the green bacteria resemble PSI, while those of the purple bacteria resemble PSII. Studies of the similarities and differences among the various photosynthetic systems suggest an evolutionary history. Photosystems I and II arose in different branches of the bacterial family tree. The ancestors of modern cyanobacteria first had just one of these systems and only later acquired the second by horizontal gene transfer (cyanobacteria are notoriously efficient at pirating genes). After acquiring the second system, they managed to couple the two in a manner that enabled them to use water as a reductant.

The "Great Oxygenation Event"

The invention of photosynthesis and the subsequent advent of our oxic atmosphere had wide-ranging consequences for the anaerobic bacteria that until then ruled the biosphere. Indeed, the advent of an oxygen atmosphere—the "great oxygenation event"—has been called the greatest environmental catastrophe in the history of our planet, a catastrophe that killed off many branches of the tree of life and pruned back many others to a few anaerobic niches (anoxic mud, for example). The change was sufficiently drawn out, though, that there was time for some life to adapt to oxic conditions and, indeed, to benefit from the new opportunities it created.

The geological record provides a history of free oxygen on Earth, although one that remains the subject of significant debate among geologists.* Most experts believe that oxygen levels were quite low before about 2.2 billion years ago—at the time, the Earth's atmosphere

*Geological debates being *very* slow.

had presumably evolved away from its more reduced starting point but had become neutral rather than oxidized. Specifically, uraninite and pyrite (uranium and iron minerals, respectively) present in ancient "fossilized" river deltas provide the key evidence. These minerals are soluble in oxygen-rich water and would not have survived the trip downriver if the oxygen concentration were even a tiny fraction of today's value. Similarly, James Farquhar of the University of Maryland has noted that the isotopic ratios in sulfates found in rocks provide a clue as to the free oxygen content of the Earth's early atmosphere: photochemical reactions can shuffle the isotopic composition of atmospheric sulfur dioxide in a characteristic manner, but free oxygen would destroy the pattern before the sulfur could make it down to the surface to become locked into rocks. Based on this, Farquhar argues that the free oxygen concentration in the atmosphere could not have risen above a paltry one part per million before some 2.4 billion years ago.

A final line of evidence was the formation of banded iron formations, a process that started more than 3 billion years ago and lasted for at least a billion years. These enormous formations consist of alternating layers of red ferric oxide (rust) and the white silicate mineral chert and serve, incidentally, as the dominant commercial source of iron today. They form when highly soluble ferrous iron (Fe^{+2}) is oxidized to ferric iron (Fe^{+3}), which in turn forms an insoluble precipitate—as anyone who's tried to wash rust off a car well knows. Current thinking is that banded iron formations represent the global transport of iron in the ancient ocean to sites where oxygen was being produced (by photosynthesis, or photolysis of water in the atmosphere, or perhaps both), oxidizing the iron and causing it to fall out of solution. After about a billion years or so, the ocean's iron became depleted and the formation of banded iron formations stopped. No doubt linked to this depletion, oxygen levels seem to have started a rapid rise some 2.4 billion years ago, up to a few percent of total atmospheric pressure. Perhaps in concert with plants' colonization of land, another large increase in oxygen began about 500 million years ago, leading, over a few hundred million years to oxygen levels of 20% to 30% (as compared to the current value of 21%).

The Evolution of Aerobic Metabolism

In learning how to turn sunlight, carbon dioxide, and water into carbohydrates and oxygen, the ancestors of cyanobacteria created new ecological niches that were probably quickly filled by organisms running the reverse reaction: burning carbohydrates to produce energy. That is, oxygen-producing photosynthesis paved the way for oxygen-consuming metabolism, including our own. This, in turn, provided the metabolic basis for the evolution of multicellular organisms. Without the oxygen revolution, metabolism would be limited to relatively low-energy fermentation reactions and life would probably have remained limited to simple, single-celled organisms.

Photosynthesis and oxidative metabolism both involve circular biochemical pathways in which a small organic molecule acts as a matrix to which carbon is added, only later to be removed in some other guise. In photosynthesis, the foundation molecule is the five-carbon sugar ribulose, to which carbon dioxide is added to form two compounds with three carbons each, which ultimately feed the production of a six-carbon sugar (fructose) and restoration of the ribulose carrier. Oxidative metabolism uses the four-carbon compound oxaloacetate as a matrix, which reacts with two reduced carbons in the form of an activated acetic acid to form citric acid, which gives the cycle one of its several names. It is also known as the TCA (tricarboxylic acid) cycle and the Krebs cycle, after Hans Krebs (1900–1981), who discovered it in 1937 following up on his earlier discovery of the urea cycle. At the time, some two decades before Calvin's elucidation of the cyclic metabolic pathways of photosynthesis, such circular metabolic pathways were a revolutionary concept. Enough so that Krebs's original publication on his now eponymous cycle, which was to earn him the 1953 Nobel Prize in Physiology or Medicine, was rejected outright when he submitted it to the journal *Nature*.

In the Krebs cycle, a six-carbon citric acid is oxidized in a series of 10 steps producing two carbon dioxide molecules and, ultimately, another molecule of the four-carbon oxaloacetate to continue the cycle. In eukaryotes, this occurs in the mitochondria, which are "fed" acetyl-

CoA, representing an activated form of the two-carbon molecule acetic acid attached to an RNA "handle," by the pathways degrading sugars (glycolysis) and fatty acids in the cytoplasm. The cycle uses the reduced electrons it extracts from citric acid to synthesize NADH (reduced nicotine adenine dinucleotide) and $FADH_2$ (reduced flavin adenine dinucleotide), which are the "reduction currency" of the cell, delivering reducing power to any reaction that needs it. Both are also ribonucleotides and thus, as we've noted, may also represent vestiges of the RNA world. The cycle also delivers electrons (plus two protons) to oxygen to produce water in a process called oxidative phosphorylation, which uses the resultant energy to pump protons across the mitochondrial membrane to produce a concentration gradient. The energy released when protons flow back across the membrane is then harnessed by the F-ATPase to produce ATP, a potential vestige of the RNA world that is the energy currency of the cell: reactions that produce metabolically useful energy almost always produce it in the form of ATP, and metabolic processes that consume energy almost always use the energy stored in ATP.

What is the advantage of the Krebs cycle to the individual species that adopt an aerobic lifestyle? Let's look at the numbers of high-energy molecules to be gained from the degradation of one molecule of glucose in each process. The anaerobic metabolism using glycolysis alone—that is, splitting glucose into two molecules of pyruvate—produces only two ATP. In contrast, the Krebs cycle followed by the electron transport chain squeezes 24 molecules of ATP out of each and every molecule of glucose. In combination with glycolysis and the decarboxylation of pyruvate to acetyl-CoA, the aerobic metabolism of glucose produces 36 ATP per sugar molecule, achieving an eighteenfold increase in energy yield over anaerobic glycolysis alone.

Because of the tremendous energy it can tap into, it is not surprising that the aerobic lifestyle evolved and spread soon after the atmosphere became oxic. Like photosynthesis, the Krebs cycle was invented by bacteria, some of which later joined forces with other cells in an odd form of intracellular symbiosis. Moreover, with the advent of this fusion, a much higher level of metabolism became possible and with that came the possibility of *multicellular organisms*. But first, a more complex class of single-celled organisms had to evolve, a topic we'll return to in a moment.

When Did LUCA Live?

The paleontological record is not the only record we have of the history of life on Earth. By identifying metabolic pathways that are held in common across all life, we have defined the likely metabolic "tool kit" of LUCA, the last common ancestor of all life on Earth (see chapter 6). Inspection of what LUCA's metabolism did and did not contain allows us to hazard a guess as to when she lived. For example, a number of LUCA's metabolic pathways involved iron-containing enzymes that, on careful reflection, might seem like unfortunate choices to a contemporary biochemist. The problem is that in today's oxic environment, iron is quickly oxidized to the ferric state, which, as described above, is *extremely* insoluble. Because of this, iron is the limiting nutrient in the ocean, and marine microorganisms have had to invent an impressive arsenal of "chemical warfare agents" (called siderophores) with which to steal iron from the grasp of other bacteria. If iron is so hard to get that its use represents a selective disadvantage, though, why did LUCA use it? The thought is that LUCA lived when soluble iron was plentiful, before rampant photosynthesis produced our now iron-limited, oxic environment. In contrast, the most oxidized form of copper (Cu^{+2}) is much more soluble than the more reduced forms, so while many more recently invented branches of metabolism employ copper-containing enzymes, LUCA seems to have avoided that element. From these and similar arguments (e.g., LUCA expressed the nitrogen-fixing enzyme nitrogenase, which is poisoned by even small traces of oxygen), it seems clear that LUCA predated the formation of the oxic atmosphere, about 2.2 billion years ago. But there is a big gap between that date and the date of the earliest widely accepted evidence for life on Earth (about 3.5 billion years ago). Thus, although we have bounded the problem, we do not know when in this more than 1-billion-year span LUCA lived.

Eukaryotes: Bigger and Better

The invention of oxygen-producing photosynthesis paved the way for the evolution of multicellular life, a niche that was taken over in its entirety by eukaryotes. Look at cells through a microscope and you'll be able to tell whether they are eukaryotes or not. Typically, our cells and those of most other eukaryotes are 10 times longer, and thus 1,000 times greater in volume, than the simpler prokaryotes. Peering down a light microscope, this makes the difference between seeing a cell with internal structures and just seeing a dot (fig. 7.3).

The defining difference that gives eukaryotes their name (*eukaryote* means "true nucleus") is that their genetic material is isolated from the rest of the cell in the nucleus, which is surrounded by a membrane. The DNA in eukaryotes is typically organized in several long, linear units called chromosomes, whose coiling and packaging is usually controlled by histones, a class of protein that does not exist in bacte-

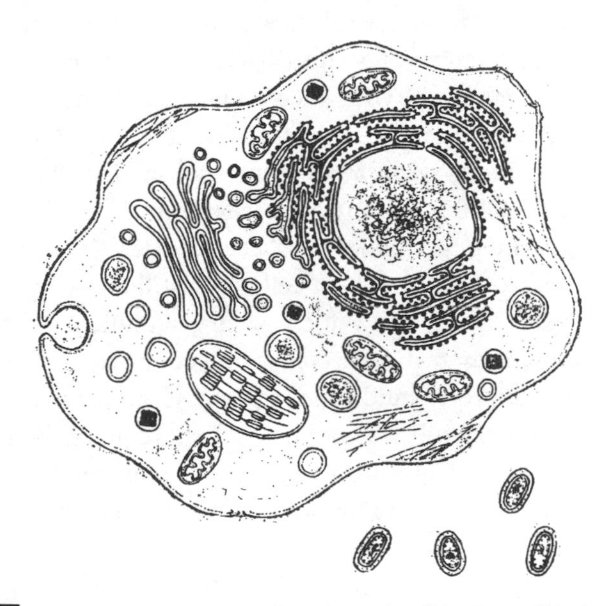

Figure 7.3 A schematic cross section of a eukaryotic cell (a plant cell) is shown in comparison with several bacterial cells. Note that eukaryotic cells tend to be an order of magnitude larger in each dimension, so their volume exceeds that of a bacterium by several thousandfold.

ria (but does occur in some archaea). The synthesis of messenger RNA (transcription) also takes place in the nucleus, while protein synthesis (translation) is carried out by ribosomes in the cytoplasm. This separation creates additional logistical problems and appears rather troublesome at first glance. Given this, what is the evolutionary advantage that enticed eukaryotes to keep their DNA wrapped up? The clue may lie in the additional editing of messenger RNA that is made possible by this separation. In bacteria, the front end of a messenger RNA can go into the translation machinery while the rear end is still being transcribed. Eukaryotes, in contrast, synthesize messenger RNA in the cell's nucleus before transporting it to the cytoplasm, where it is used to guide the synthesis of proteins. This separation in space and time allows the cell to introduce additional mechanisms to control gene expression, such as RNA editing, which in turn sets the stage for vastly more complex organisms.

The nucleus is not the only membrane-bound compartment (organelle) within a eukaryotic cell. The mitochondria, chloroplasts, lysosomes, and endoplasmic reticulum are also membrane bound and further distinguish eukaryotes from bacteria and archaea. Each of these compartments brings with it a complete set of important functions. Several of the organelles, most notably the mitochondria and the chloroplasts (in charge of aerobic metabolism and photosynthesis, respectively), are thought to have arisen from formerly independent bacterial symbionts that, over the generations, became better and better integrated into the host cell and lost their independence, along with most of their genes. Which, when you think about it, brings us to the question: where did the eukaryotes come from?

Comparison of protein and gene sequences has enabled researchers to trace back the family tree of life much more precisely than was possible using outward appearance (phenotype) alone. At first, they studied enzyme sequences but found that their efforts were limited by the effects of convergent evolution (remember: similar environmental requirements can lead to similar adaptations, at both the organismal and protein levels), which can produce similarities that do not imply relatedness. As described at the beginning of this chapter, Carl Woese focused his sights on ribosomal RNAs and came to the conclusion that, whereas living things had previously been divided only by the presence (eukaryotes) or absence (prokaryotes) of a nucleus, the oldest and deepest division between species separates the tree of life into three main branches. In the decades since, researchers have switched to the analysis of genes that, unlike even ribosomal RNA, contain variability that is under almost no selective pressure at all—such as codons where the third base is redundant and thus changing it doesn't change the amino acid sequence of the encoded protein (see chapter 6). The ever-increasing numbers of protein, gene, and genome sequences that have become available provide more and more convincing evidence in favor of this tripartite tree of life.

The three domains of life are fundamentally different in many ways. In some aspects, there are resemblances between two of them that exclude the third (such as the facts that only bacteria and eukaryotes synthesize fatty acids, only archaea and eukaryotes wrap their DNA around histones, and only bacteria and archaea lack nuclei), but there is no longer a case for a grouping into two domains. The very last doubts about that were removed by the first complete genome sequence of an archaeon (*Methanococcus jannaschii*), which was published in 1996 and illustrated that archaea are pretty much as distinct from the *E. coli* living in our guts as they are from us.

Even the most modern analytical methods, however, have failed to answer two important questions: first, how do the three largest branches on the tree of life relate to each other; and second, where is the root? Comparisons based on different genes or groups of genes yield very different answers to these questions. From these contradictions, it seems increasingly likely that "vertical" descent of species from earlier species along the direct lines of a family tree does not account for the whole story of life on Earth (fig. 7.4). Exchange of genes between separate species living at the same time must have played an important role.* This kind of horizontal gene transfer can still be observed among microbes, for example, when researchers study the

*Gene exchange also weakens our argument that any traits shared across all three domains of life must have been inherited from LUCA; some of the shared traits could have arisen in one branch, after LUCA, and been horizontally transferred to the others. Still, most authorities seem to agree in broad detail with the description of LUCA we've given here.

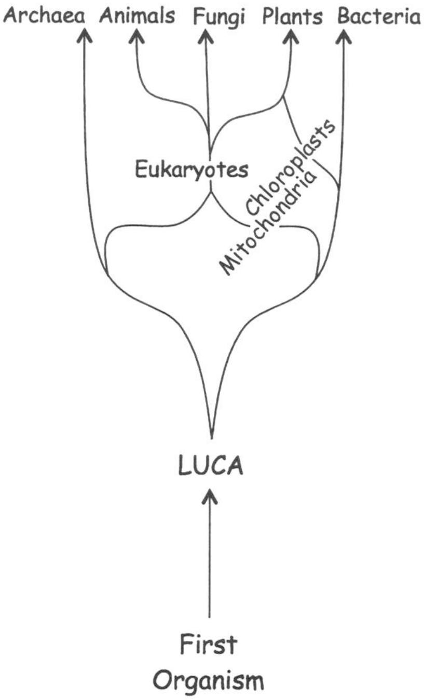

Figure 7.4 The traditional view of a "family tree" of life has been replaced by a more complex representation that takes into account horizontal gene transfer and the likely merger of members of the archaeal and bacterial branches to produce eukaryotes, with the bacterial member of this fusion becoming our mitochondria. A second, later fusion with a photosynthetic bacterium produced the higher plants.

spread of genes that confer resistance to antibiotics or other crucial survival skills. Given the frequency of such horizontal gene transfer across the history of life, the attempt to draw simple family trees relating all living species to a smaller set of ancestors and ultimately to a common root was destined to fail.

But all is not lost. In 2004, Maria Rivera and James Lake at the University of California, Los Angeles, delivered a different description for the crucial early phase of evolution when they applied a new set of algorithms to its modeling. They compared the genomes of 10 organisms representing all three domains of life using an algorithm ("conditioned reconstruction") designed to cope with both horizontal and vertical gene transfer without discrimination. In this method, one of the genomes (the "conditioning genome") is picked as a standard that does not enter the resulting tree as it only serves as a reference point for the others. Thus, there is a very simple inbuilt control: one can construct trees based on different choices of conditioning genome and then overlay them. Rivera and Lake first tested this method on the parts of the prokaryotic family tree that were already well described, then applied it to the question of how the deepest branches—the three domains of life—relate to each other. Sifting through the results with the highest statistical significance parameters, they realized that all of them were permutations of a single pattern that can best be described as a ring (see fig. 7.4).

Biologically, this finding implies that Eukarya arose from both Bacteria and Archaea, possibly through the fusion of two early cells. This interpretation is consistent with earlier results of genome comparisons revealing that eukaryotic genes in charge of information processing (ribosomes, translation factors, enzymes of DNA processing) are more closely related to their archaeal counterparts than to the bacterial ones, while the reverse is true for genes in charge of metabolism. With several independent studies now confirming a "mixed" origin of eukaryotes, it seems almost certain that they arose from some kind of marriage between the two older, more "primitive" domains.

Genomic analysis suggests that the Alphaproteobacteria are the extant bacteria most closely related to the bacterial symbiont whose vestiges remain in us today in the form of mitochondria. Perhaps not coincidentally, this bacterial class includes the Rickettsiales, a group of pathogens that are obligate intracellular parasites, that is, parasites that live inside the cells of their eukaryotic hosts and are responsible for such diseases as typhus and spotted fever.

In contrast to the bacterial ancestor of the mitochondria, the archaeal partner in the marriage was harder to find. In 2015, the group of Christa Schleper at the University of Vienna, Austria, identified a new archaeal phylum, called Lokiarchaeota, via genomic sequencing because the microorganism could not, at the time, be cultivated in the laboratory (see sidebar 7.2). Identified in mud samples collected near the Loki hydrothermal fields at the mid-Atlantic spreading center, the archaea feature a number of genes hitherto considered specific for eukaryotes, including those for cytoskeletal components such as actin. Further analyses of environmental DNA revealed other related lineages in the phylum, all of which are now named after Nordic deities (Loki being a god in Norse mythology) and thus collectively

Sidebar 7.2

Metagenomics

Historically, microbiologists defined new bacterial species by cultivating ("culturing") them in the laboratory. This, however, is a painstaking procedure as it requires that one figure out what conditions and nutrients are required in order to encourage a given, often highly specialized species to thrive. Indeed, scientists often failed; it had long been clear that the majority of the bacteria in, say, an environmental soil sample were recalcitrant to growth in the laboratory. Given this, our understanding of microbial diversity had long been rather shallow and left out many species thriving in extreme conditions, some of which we will meet in chapter 8. Indeed, it is estimated that nearly three-quarters of all bacterial phyla lack *any* species that can be grown in the laboratory.

The situation began to change, however, in the early twenty-first century, when the price of sequencing and the cost of the computer power necessary to process "big data" had dropped so low that it became possible to simply scoop up an environmental sample, extract all the DNA from all the organisms in it, and sequence the whole mess simultaneously, counting on computers to reassemble the pieces into individual genomes. This approach, now called metagenomics, circumvents the need to isolate and culture individual species in the laboratory, vastly improving our understanding of microbial relationships both in terms of their evolution and in the context of how they work together in microbial ecosystems.

called the Asgard archaea. Genomic comparisons suggest that, among these, the phylum Heimdallarchaeota is the closest known archaeal relative to the eukaryotes.

Still, all these microorganisms were characterized on the basis of genomics. It was only in 2020 that the first species of the Asgard archaea was reported as having been successfully cultivated in the lab. Hiroyuki Imachi and his colleagues at the X-star institute in Yokosuka, Japan, had originally started cultivation attempts for unrelated reasons, even before Lokiarchaeota were identified. Their work eventually led to the characterization of *Prometheoarchaeum* as the first cultivated member of the Lokiarchaeota. The cells turned out to be very small but are equipped with spectacular membrane extrusions that, intriguingly, allow them to perform nutrient exchange with two associated, hydrogen- and methane-consuming bacterial species in a manner reminiscent of how eukaryotic cells "feed" reducing potential to their internal mitochondria.

If endosymbiotic fusion with one bacterial cell is good, wouldn't fusion with two be better? The plants seem to represent just such thinking; sometime after the origins of the eukaryotes, a second endosymbiotic event occurred (see fig. 7.4), this time bringing in a cyanobacterium which then "devolved" (transferred many of its genes to the nuclear genome) into today's chloroplasts, the photosynthetic organelles of the eukaryotic plants. In 2017, genomic analysis was provided suggesting that, among known bacteria, the closest living relative to the chloroplast ancestor is *Gloeomargarita lithophora*, a cyanobacterial species found in Lake Alchichica, an alkaline crater lake in the high desert of Puebla, Mexico.

And when did our eukaryotic branch first sprout off the tree of life? That's not so easy to say since the nucleus, which is, by definition, the thing that separates eukaryotes from prokaryotes, doesn't readily fossilize, and thus paleontologists base their claims on the sizes and shapes of microfossils, which are far from unambiguous indicators. Using this metric, for example, eukaryotic cell–sized microfossils have been claimed from as far back as 3.2 billion years. The oldest microfossils that appear to be widely accepted as eukaryotic, however, are large (>100 μm), spiny, organic-walled microfossils dating to 1.65 billion years ago. However, we're on firmer footing if, instead of trying to identify the first eukaryote (which depends on what, precisely, it means to be a eukaryote), we focus on dating the last eukaryotic common ancestor (LECA), the most recent common ancestor of all living eukaryotes. Since almost all eukaryotes are aerobic, and those few that are not show clear signs of once having had (but then lost) mitochondria, it's fairly clear that LECA lived more recently than 2.2 billion years ago, the time after which atmospheric oxygen climbed to an appreciable level. Consistent with this, molecular phylogeny studies generally suggest LECA lived between 1.1 and 1.7 billion years ago. Finally, efforts to trace the paleontological record using molecular markers unique to eukaryotes, such as the degradation products of cholesterol and its relatives, suggest that at least the sterol-synthesizing eukaryotes (which includes all current eukaryotes) did not arise, or at least did not become common, until more recently than 1.1 billion years ago. This said, by 650 million years ago, algae-derived sterols become common in marine sediments. Enough so that this is thought to have marked a transition from a biosphere that was predominantly driven by prokaryotes to one dominated by eukaryotes.

Multicellular Life and the Cambrian Explosion

The rise of the multicellular plants and animals that we tend to think of as "biology" occurred rather late in the history of our planet. Within a few hundred million years of Earth becoming inhabitable at the end of the late heavy bombardment, single-celled life seems to have been abundant. In contrast, the "next step," the creation of multicellular life, took several billion years. In part, this was because multicellular life likely requires an energetic metabolism and thus had to await the advent of an oxic atmosphere. But multicellularity also requires a good deal of complexity at the cellular level, which might likewise have been slow in coming.

Green algae were among the pioneers of higher organization in colonies but didn't immediately make the transition to developing a body plan. Not until several hundred million years later did things start to really take off, when the earliest known multicellular eukaryotic, the red alga *Bangiomorpha pubescens*, showed up in the fossil record some 1.05 billion years ago. The first clear evidence of animals crops up some 400 million years later, first as enigmatic forms that left behind a variety of tubular and frond-shaped fossils. These simple forms, with just two layers of cells, later evolved into forms with three layers, then three layers with a cavity, and, finally, onward to the human body plan of three layers of cells surrounding an internal cavity with both a mouth and an anus. This represented a significant advance; less-developed animals, such as the flatworms, have a mouth but lack an anus, and thus anything they can't digest has to be vomited back up.

Why did bacteria fail to master this move? Cyanobacteria, for example, are amazingly sophisticated, having mastered photosynthesis, nitrogen fixation, symbiosis with fungi to form lichens, and even development of circadian clocks. So why did some obscure eukaryote steal the limelight from them, increasing the complexity of living beings and inventing the higher plants and animals? There are clearly many contributing factors, no doubt including some that we still don't understand. But a few of the features that enabled the jump are clear: organization of space, organization of the genome, and sex.

The most striking difference that distinguishes eukaryotic cells from bacteria is not just the volume of their interior space, which is typically a thousandfold larger, but the way this space is organized into compartments of well-defined function. Bringing together the information processing of the nucleus, the aerobic metabolism of mitochondria, possibly the photosynthesis of chloroplasts, and other functions in separate entities, even a formally single-celled eukaryote is effectively a "multicellular" organism.

More importantly, eukaryotes abandoned the circular DNA that bacteria still use today and invented linear DNA strands wrapped around histones and capped with telomeres, resulting in the familiar chromosomes. Even the humble, single-celled eukaryote baker's yeast has 16 separate linear chromosomes. There is a limit to how much DNA can be stored and processed in a single ring without ending up in a lethal tangle, but the organization into chromosomes offered not only more storage space but a natural way of expanding the space, namely by adding new chromosomes. Thus, the number of chromosomes varies widely among different eukaryotic species.

But, most importantly, their style of genome organization enabled eukaryotes to embark on a completely new way of fostering genetic diversity while ensuring the genetic stability that keeps the species together. This wonderful tool of evolution, invented around 1 billion years ago, is known as meiosis on the cellular level, but on the organismal level it is called sex. Biologists have argued about its usefulness. In comparison with an asexual reproduction mechanism, where every individual can have offspring without the need of a partner, the maintenance of a second gender contributing only a minor part to the reproductive process (e.g., the human male) is a complete waste of energy, one might argue. However, the success of sexual reproduction throughout the animal kingdom and in much of the plant world shows that the benefits to the species more than compensate for this energy loss.

So, equipped with these advantages, some protists (the generic name for single-celled eukaryotes) finally got their act together and became the founders of zoology, sometime before 600 million years ago. We know very little about their first efforts as they don't show up in the fossil record very clearly, and the molecular analyses carried out so far have not resulted in a convincing reconstruction of the first animal. But these events laid the groundwork for something big because, only 50 million years later, the fossils document a sheer explosion of animal diversity.

Now isn't that ironic? You wait some 3 billion years for animals to come along, and then they all arrive at once. Or so it may seem. Some 650 million years ago, several billion years after the first traces of life on Earth, there were still no traces of animals in the fossil record, and yet, 200 million years later, they were everywhere. In fact, more than a hundred different orders of animals have already been found in the fossil record within 200 million years of the first animal fossils, almost as many orders as exist today.* During the 100 million years of the Cambrian (541 to 485 million years ago) and Ordovician periods (485 to 444 million years ago) alone, evolution invented essentially every single body plan in use today and many more bizarre body forms that are not. This unrivaled burst of evolutionary inventiveness, known as the Cambrian explosion, has mystified biologists from the time of Darwin to the present. Today there are two fundamentally different schools of thought on this issue, each with its own tool kit of possible explanations and interpretations.

The "late arrival" school, represented by Stephen Jay Gould (1941–2002) from Harvard University, maintained that what we see in the fossil record is essentially true and that the diversification did indeed happen unusually fast, creating as many as 50 new orders in just 10 million years. One plausible explanation sees the explosion as an arms race triggered by the use of biomineralization by animals. The controlled deposition of minerals from body tissues enabled animals not only to develop skeletons (which allowed them to diversify into more complex

*An *order* is a phylogenetic group that is smaller than a *class* and larger than a *genus*. For example, Mammalia (mammals) is a class that contains the orders Primates (e.g., ourselves), Rodentia (e.g., mice), and Proboscidea (the elephants, of course). Likewise, the order Primates includes such genera as *Homo* (us), *Pan* (the chimpanzees), and *Gorilla* (the lowland and highland gorillas).

shapes and larger sizes) but also to grow claws, rasps, and teeth with which they could prey on other animals. Predatory lifestyles opened up additional ecological niches and triggered defensive measures in the animals threatened by them. And biomineralization allowed animals to build protective shields, such as mollusk shells. Similarly, the development of eyes may have intensified the arms race between species.

To us, who can only see these animals when they are fossilized, the impression of an "explosion" is made even more dramatic by the fact that mineralized tissues such as bones, teeth, and shells have a much better chance of being preserved than the soft tissues of the animals that lived before the onset of biomineralization. This argument leads us to the second school of thought, the "early arrival" model (favored by Darwin), which claims that animal diversity existed for hundreds of millions of years before the Cambrian period but didn't show up in the fossil record because either the animals were too soft to fossilize properly or the conditions were unfavorable for their preservation. Consistent with this, "molecular clock" studies that date divergences by counting the number of mutations suggest that some animal diversity existed before the explosion. Nevertheless, most researchers would not allow more than 150 million years for this hidden period of animal evolution. Several authorities, for example, place one of the deepest divisions in the family tree of multicellular life, the one between protostomes (including mollusks, insects, crustaceans) and deuterostomes (including vertebrates but also the echinoderms, such as starfish*), at 670 million years ago, some 130 million years before the start of the Cambrian.

Conquering Land

The Cambrian explosion happened in water. But in addition to providing enough energy to support the formation of multicellular life, the oxygen revolution also ultimately set the stage for the coloniza-

*This is about the anus and the mouth again. In protostomes, the first opening to form during early embryonic development becomes the mouth, but in deuterostomes it develops into the anus; the mouth comes second.

tion of dry land: the only reason living organisms can thrive above the water line without suffering catastrophic DNA damage is the ozone content of the stratosphere. Although the phenomenon is often referred to as the "ozone layer" and is even measured in terms of the thickness it would have if there were such a thing, the ozone is in fact rather dilute and the compound as such is highly unstable. But its fleeting presence in the stratosphere (which extends between the heights of 16 and 50 km) is sufficient to absorb the damaging, far-UV component of sunlight.

The earliest non-marine eukaryotes appear to have been multicellular algae, which first came ashore about a billion years ago. These were followed by the higher plants, with the oldest known land-plant fossils dating to 420 million years ago. Fungi may have come along for the ride since phylogenomic studies of plants and the fungi that live symbiotically in their roots suggest that the two helped each other colonize this new world. For example, Susana Magallón, from the National University of Mexico in Mexico City, reported in 2018 that the phylogenies of both sides of the symbiosis go back 700 million years, likely before either group had moved onto dry land. Between them they made the bare surface of the continents habitable for the animals, with the first known land animals arising some 405 million years ago, and the first terrestrial animals in our line, the amphibians, climbing onto the land only 370 million years ago.

Conclusions

It has been a few chapters since we last said it, so it probably bears repeating here: biology is a parochial science. Given that all life on Earth arose from a biochemically complex common ancestor, it's not so easy to figure out which aspects of our biochemistry and cell biology reflect adaptations to the fundamental issues related to life on a terrestrial planet and which are merely historical artifacts of evolutionary chance. But then again, we biologists have to play the cards we are dealt. From that perspective, detailed studies of the evolution of life on Earth are probably the best approach we have toward understanding how life is defined, constrained, and encouraged by the physical reality of growing up on a small, rocky planet.

What does this ultra-compressed history of life teach us in the context of astrobiology? That, at least on this planet, it took a good fraction of the age of the Universe itself for intelligent life to arise (see sidebar 7.3). The evidence that life existed by 3.43 billion years ago is strong and is pretty much rock solid by 3.2 billion years ago. It took another 1 or 2 billion years, however, for evolution to invent eukaryotes, and the first multicellular life didn't arise until several hundred million years later still. In the end, it was not until at least a billion years after the first multicellular organism, and likely more than 3.5 billion years after the start of life itself, that an obscure, mammalian branch on the eukaryotic domain of the tree of life evolved a species that could spend endless years arguing in the scientific literature before ultimately deciding that Woese was right: the Archaea truly are distinct from the Bacteria.

Further Reading

Earliest Life / Evolution of Life / Origins of Eukaryotes

Knoll, Andrew. *Life on a Young Planet: The First Three Billion Years of Evolution on Earth.* Princeton, NJ: Princeton University Press, 2015.

Evidence for Life at 3.8 Billion Years

Dalton, Rex. "Fresh Study Questions Oldest Traces of Life in Akilia Rock." *Nature* 429, no. 6993 (2004): 688.

Dodd, Matthew S., Dominic Papineau, Tor Grenne, John F. Slack, Martin Rittner, Franco Pirajno, Jonathan O'Neil, and Crispin T. S. Little. "Evidence for Early Life in Earth's Oldest Hydrothermal Vent Precipitates." *Nature* 543, no. 7643 (2017): 60–64.

Fedo, Christopher M., and Martin J. Whitehouse. "Metasomatic Origin of Quartz-Pyroxene Rock, Akilia, Greenland, and Implications for Earth's Earliest Life." *Science* 296, no. 5572 (2002): 1448–52.

Sano, Yuji, Kentaro Terada, Yoshio Takahashi, and Allen P. Nutman. "Origin of Life from Apatite Dating?" *Nature* 400 (1999): 127.

Sidebar 7.3

Weighing the Probabilities

The evidence—albeit weak—that life may have taken only tens of millions of years to both arise and proliferate after the end of the late heavy bombardment is often taken to imply that the formation of life is a very likely event. But this logic is as flawed as any statistics based on only one sample. The logic is further undermined if rapidity is an imperative for the formation of life, and it may well be. Although Jeffrey Bada has shown that Miller-Urey chemistry can, under some circumstances, proceed even in a relatively oxidizing atmosphere, it is clearly much more efficient in a reducing atmosphere. And due to the loss of hydrogen via photolysis, the Earth's atmosphere started to oxidize from the day it was formed. This oxidation was sufficiently rapid that the earliest rocks recording evidence of the Earth's surface suggest that it was fairly oxidized by the time they formed. Thus, quite possibly, life on Earth captured a very narrow window of opportunity between the end of sterilizing impacts and oxidation of the primordial atmosphere.

In contrast to the potentially rapid origins of life on Earth, the development of complex cells, the step from complex cells to complex organisms, and the step from complex organisms to intelligent life proceeded at a much more leisurely pace. But does the fact that it took 4 billion plus years for intelligence to arise on Earth imply that this would require so long everywhere? Or are we just slow? The answer to this question is at least a qualified "it takes time." For example, intelligence no doubt requires multicellularity (or, probably harder, the equivalent complexity in a unicellular organism). This level of complexity, in turn, probably requires the very active metabolism provided by oxygen. And, as first pointed out in the late 1960s by the eminent biogeologist Preston Cloud (1912–1991), formation of an oxic environment probably requires billions of years. The reason is that for the first billion or so years, all the oxygen produced by photosynthesis (and by abiological photolysis) will be consumed by the oxidation of rocks. Only after enough reducing material has been removed from the system (carbon in the form of oil and coal, and any hydrogen lost to space) can a *net* flux of oxygen be achieved. On Earth this did not occur until our planet was middle-aged. When it finally did occur, it took less than 200 million years for multicellular organisms to evolve into every current phylum (and many that are long extinct) and another 600 million years for us humans to arrive on the scene.

What, then, were the chances that multicellular life would evolve on the Earth, culminating (or so we like to think) in intelligence? This is, of course, impossible to determine. But that doesn't mean we can't have some fun speculating.

A potentially important input to this speculation is that intelligence didn't show up on our planet until at least 2.7 billion and possibly as long as 3.8 billion years after the origins of life itself. These numbers are rather surprisingly high, even when compared with the 13.8-billion-year age of the Universe, much less the approximately 5-billion-year span over which the Earth will remain inhabitable (we've only got about 2 billion years before the Sun's increasing brightness boils the oceans), suggesting that we are a lucky break. If it really requires billions of years to evolve from the first, simplest organisms to something smart enough to read a book about astrobiology, what are the chances that something won't come along and kill life off before it gets there? Given the probability of sterilizing impacts and the certainty of moving habitable zones (see chapters 2 and 3), it is not at all a trivial thing for a planet to remain habitable for a quarter of the age of the Universe.

Evidence for Life at 3.46 Billion Years

Brasier, Martin D., Owen R. Green, Andrew P. Jephcoat, Annette K. Kleppe, Martin J. Van Kranendonk, John F. Lindsay, Andres Steele, and Nathalie V. Grassineau. "Questioning the Evidence for Earth's Oldest Fossils." *Nature* 416 (2002): 76–81.

Schopf, J. William. "Microfossils of the Early Archaean Apex Chert: New Evidence of the Antiquity of Life." *Science* 260, no. 5108 (1993): 640–46.

Evidence for Life at 3.43 Billion Years

Allwood, Abigail C., Malcolm R. Walter, Balz S. Kamber, Craig P. Marshall, and Ian W. Burch. "Stromatolite Reef from the Early Archaean Era of Australia." *Nature* 441 (2006): 714–18.

Origins of Photosynthesis

Gross, Michael. *Light and Life.* Oxford, UK: Oxford University Press, 2003.

Archaeal Ancestors of First Eukaryotes

Imachi, Hiroyuki, Masaru K. Nobu, Nozomi Nakahara, Yuki Morono, Miyuki Ogawara, Yoshihiro Takaki, Yoshinori Takano, et al. "Isolation of an Archaeon at the Prokaryote-Eukaryote Interface." *Nature* 577 (2020): 519–25.

Evolution and Phylogenetics of Eukaryotes

Gross, Michael. "The Success Story of Plants and Fungi." *Current Biology* 29, no. 6 (2019): R183–85.

Gross, Michael. "The Genome Sequence of Everything." *Current Biology* 28, no. 13 (2018): R719–21.

Chapter 8

Life on the Edge

In the spring of 1977, the geologists Jack Corliss of Oregon State University and John Edmond (1944–2001) of the Massachusetts Institute of Technology boarded the research submarine *Alvin* for humankind's first direct look at a mid-ocean ridge. They were following up on observations made two years earlier in the Atlantic that these ridges—a globe-girdling chain of mountains beneath the sea—seemed to consist of freshly solidified basalt, suggesting that an active source of lava was nearby. The researchers were on a hunt for "spreading centers," where new crust is formed. The existence of such centers was predicted by plate tectonics, a theory that was first proposed by the German geophysicist and meteorologist Alfred Wegener (1880–1930) back in 1912 but, in the mid-1970s, was only recently beginning to achieve widespread acceptance.*

When *Alvin* reached the slope of the ridge some 2,500 m beneath the surface of the Pacific Ocean, the geologists noticed that the water temperature was five degrees higher than the normal 2°C of the ocean's depths. At the time, marine geologists had already theorized that the entire volume of the oceans somehow flows through hot volcanic rocks every 8 million years or so—only this could account for the dissolved solids in seawater, which are drastically different from the residue you'd get by evaporating river water to dryness—but no one had yet identified the hydrothermal features that might account for the cycling. This hint of hot springs on the ocean floor was thus a sensa-

*More precisely, Wegener had proposed continental drift based on the close fit of eastern South America with the western coast of Africa and coincident mineral formations on either side of the Atlantic, but his theory did not provide details as to why or how the continents might be moving. The idea of plate tectonics as we know it was developed in the 1950s.

tional discovery, and the researchers excitedly took samples so that they could later determine the chemical composition of this unexpectedly warm water. Still excited, they piloted *Alvin* up to the top of the ridge, where a much bigger sensation was waiting for them. Where they had expected to find a stark "desert" of bare, lifeless basalt freshly erupted from the spreading center atop the ridge, they instead found an oasis 100 m across that had warm water seeping through every little crack of the seafloor and was richly populated with clams, crabs, sea anemones, and large pink fish. As Edmond later recalled in *Scientific American*, they spent the five remaining hours of their dive in frantic excitement. They measured temperature, conductivity, pH, and oxygen content of the seawater, took photographs, and collected specimens of as many of the animal species as they could catch.

Holger Jannasch (1927–1998), a German marine biologist working at the Woods Hole Oceanographic Institution, was one of the first to hear the news. He recalls that he "got a call from the chief scientist, who said he had discovered big clams and tube worms, and I simply didn't believe it. He was a geologist, after all."*

The Art of Living Dangerously

Living organisms tend to be sensitive to drastic changes in their environments. Heat and cold, pressure, drought, salinity, and acids and bases all disrupt the crucial interactions that keep biomolecules folded and functional and quickly put an end to the fragile state of chemical disequilibrium we call life. Therefore, scientists have tended to assume that there are strict boundaries to the biosphere that are imposed by Terrestrial life's requirement for a rather narrow and specific range of physical conditions (see sidebar 8.1).

Discoveries in the past few decades, however, have shown that life isn't always as sensitive as we might have imagined and that the limits of life on Earth are far from well defined. Historical notions of what is, or is not, a hostile environment have turned out to be erroneous.

*Quoted in *TIME*, August 14, 1995.

Sidebar 8.1

Stress Responses

Adaptation on an evolutionary timescale is one way of responding to extreme environmental conditions. But given that not all habitats are equally stable, some organisms can temporarily find themselves in "hot water" on a much shorter timescale. Thus, the ability of organisms (both mesophiles, those of us who live under "normal" conditions, and extremophiles) to survive short-term deviations from the conditions they've evolved to live under is a fundamental factor in defining the range of conditions in which life can survive.

Here on Earth, the most important and universal form of this response to environmental stress is the production of specific stress proteins, of which the heat shock proteins are among the best studied. This is a large family of proteins that cells produce in response to heat shock. The heat shock response had been known and characterized as a phenomenon of gene regulation decades before researchers understood the primary functions of these proteins. Finally, in the late 1980s, it was established that several of the main components of the heat shock response act as molecular chaperones—that is, they protect other, newly synthesized or partially unfolded proteins from intermolecular interactions that would favor aggregation (which leads to loss of function) over folding (or refolding).

The "classic" chaperones DnaK, DnaJ, GroEL, and GroES, which in *E. coli* are combined in a complex, efficient protein-processing pathway, were studied intensely in the 1990s. Over time, a range of additional functionalities were discovered in other heat shock proteins. For example, the major protein component of the vertebrate eye lens, alpha crystallin, has a chaperone function and is related to the small heat shock proteins (it is presumably in the lens to prevent the aggregation of lens proteins, which if left unchecked would lead to cataracts). Several other heat shock proteins take part in the destruction of discarded proteins, and at least two (Hsp31 and DegP) can switch between the functions of chaperone and destroyer. In cooperation with other, more conventional chaperones, Hsp104 can even rescue proteins from aggregates (the biochemical equivalent of "unboiling an egg"). Hsp90 specifically chaperones transcription factors (proteins that regulate gene expression), folding them even if they are mutated and thus helping to silence the effects of mutations. But when an acute heat shock requires Hsp90 to be used elsewhere in an emergency response, it stops chaperoning transcription factors. This allows the "expression," as it were, of previously silent mutations in the transcription factors, which in turn can result in a spectacular increase in developmental variation. This unexpected finding from the group of Susan Lindquist (1949–2016) at the Whitehead Institute in Cambridge, Massachusetts, suggests that Hsp90 has an important role linking environmental stress to the generation of new biological functions.

Responses to other kinds of stress typically involve a combination of heat shock proteins and proteins more specific to the given type of stress. Among the better-studied examples of non-heat-related stress proteins are the cold shock proteins. The prototype, CspB from *Bacillus subtilis*, is known to serve as an RNA chaperone in that it keeps messenger RNA from folding into loops that might inhibit its translation. Only limited information is available on proteins specific to other kinds of stress, but a family of universal stress protein genes, first described in *E. coli*, has emerged as an important factor in complex cellular responses to a wide range of different stress conditions.

Stress proteins are universally present in all organisms that we know about and are thus presumed to be ancient in evolutionary terms (the ability to survive temporary environmental change provides a significant selective advantage). Moreover, their involvement with key regulatory proteins suggests that stress proteins played an important role on those occasions when stressful environmental change made relatively rapid adaptation necessary. So they might provide additional clues to the question of how life on Earth managed to adapt to some surprisingly unstable habitats.

Methods routinely used for sterilization, including boiling, freezing, and γ-ray treatments, turn out to be deadly for most—but not all—microbes. For every extreme physical condition investigated, extremophilic organisms have shown up that not only tolerate those conditions but often require them for their survival. With these discoveries, the expanse of the known biosphere has grown and the putative boundaries of life have expanded. The decades following that *Alvin* dive, in particular, have witnessed substantial shifts in what scientists consider the limits of habitable environmental conditions.

While there are certain hard physical limits to the existence of DNA-based cellular life, these limits are far from the normality of common garden-variety organisms such as *E. coli* and *Homo sapiens* (a normality admittedly defined by anthropocentric thinking!). Here we take a very brief look at some of the extreme conditions faced by organisms living on our planet, considered under the fundamental astrobiological question of what these findings tell us about the prospect of finding life elsewhere in the Solar System and beyond.

Some Like It Hot

Microbial activity at temperatures above the "normal" range of 20°C to 40°C was reported in the nineteenth century, but the upper limit of life's known temperature scale has risen rapidly over the past five de-

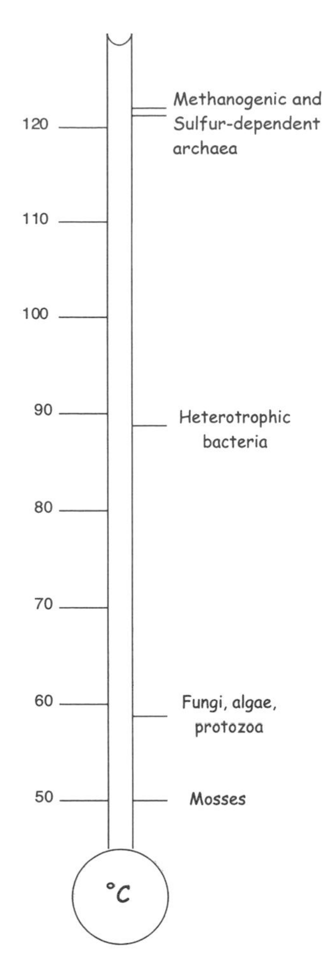

Figure 8.1 The temperature scale of the best heat-adapted groups of organisms.

cades (fig. 8.1). An important foundation was laid in the 1960s when Thomas Brock started culturing heat-resistant bacteria isolated from the hot springs of Yellowstone National Park. His discoveries included *Thermus aquaticus*, which was to become the first and most common source of heat-stable DNA polymerase for the polymerase chain reaction (PCR) two decades later (see sidebar 8.2). The temperature records set by Brock's organisms, however, were not destined to last, as even more hostile habitats remained to be discovered.

Two years after geologists using the research submarine *Alvin* discovered warm springs on the seafloor that hosted a surprisingly rich

Sidebar 8.2

Commercial Interest in Extremophiles

Because conditions that count as extreme for biologists are often fairly standard in industrial settings, astrobiologists aren't the only people interested in extremophiles. Many processes in the chemical industry, for example, are routinely run at high temperatures and pressures. The food industry uses all kinds of extreme conditions to exclude food pathogens. Not surprisingly, the discovery of extremophiles has thus spawned interest from industry.

A branch of industry that extremophiles have helped since long before science even knew of their existence is mining. The traditional "bioleaching" methods used in copper and gold mining, for instance, depend on naturally occurring acidophilic, and often thermophilic, microbial strains. Dump or heap bioleaching to extract copper from low-grade ores happens in places like Bingham Canyon, Utah, Unites States; Chuquicamata, Chile; and Vlaikov Vrah, Bulgaria, where extended irrigation is used to stimulate microbes already resident on the mineral ores. These organisms oxidize the iron and sulfur present in the insoluble chalcopyrite ore ($CuFeS_2$), liberating soluble Cu^{+2}, which is then collected from the effluent at the bottom of the dump.

Scientifically guided bioleaching only developed in the second half of the twentieth century, after the microbes involved had been characterized. *Acidithiobacillus ferrooxidans*, first identified in 1947, is the most studied and best understood. This rod-shaped bacterium is often the dominant species found in moderate-temperature bioleaching heaps (25°C to 35°C). Under aerobic conditions, it derives energy from the oxidation of iron ($Fe^{+2} \rightarrow Fe^{+3}$) and reduced sulfur compounds, but when oxygen is in short supply, it can reverse the first reaction and use oxidized iron as its electron sink. Not surprisingly, given where it lives, the bacterium is remarkably resistant to toxic metal ions, including arsenic and uranium. It is also frugal in its nutrient requirement. It can fix nitrogen and derive all its needs from a well-aerated heap of crushed pyrite in acidified water. Other iron- and sulfur-oxidizing species were discovered from 1965 onward, some of which were later recognized as archaea.

Our perpetual quest to break our dependency on fossil fuels has also focused attention on the extremophiles. Thermophiles such as *Caldicellulosiruptor saccharolyticus* and *Thermotoga elfii*, for example, have been used in biotechnological efforts to produce hydrogen from plant material, and *Thermoanaerobacterium saccharolyticum* has been used for the production of ethanol at high temperatures.

Enzymes from thermophilic organisms have also proven of significant interest in the chemical and pharmaceutical industries, where they are used in processes that combine industry-friendly high temperatures with the specificity and efficiency of enzymatic catalysis. The single most successful of these is the polymerase chain reaction (PCR), which revolutionized molecular biology in the 1980s. The original PCR protocol, for which Kary Mullis (1944–2019) received the 1993 Nobel Prize in Chemistry, involved the polymerase of *Thermus aquaticus*, now widely known as Taq polymerase, which remains stable at 90°C, the temperature needed to fully dissociate a DNA double helix so that it can be copied. Thus, repeated temperature cycling can be used to amplify even a single copy of a DNA sequence into a macroscopic amount. In the 1990s,

(*continued*)

Sidebar 8.2 *(continued)*

rapidly growing competition in the PCR field produced a range of enzymes derived from other thermophiles, in part to get around the original PCR patents.

This goes to show that immense benefits can come from research into the apparently "offbeat" areas of science such as survival under extreme conditions. Further discoveries will no doubt prove useful for research, medical applications, and indeed everyday life. One day, our knowledge of the extremophiles' survival strategies might even help us brave the extreme conditions on other planets.

ecosystem, another *Alvin* expedition found the first hydrothermal vents. Here water erupts from chimney-like deposits rising several meters above the seafloor, emerging at temperatures as high as 350°C. (Under the elevated hydrostatic pressure at such depths, the boiling point of water is raised considerably; it reaches 340°C at a depth of 2 km.) The drastically increased solubility of certain minerals in the hydrothermal fluid leads to instant precipitation when the mineral-rich fluid mixes with cold seawater, producing both the characteristic chimney walls and the "smoke plume" that earned these vents the nickname "black smokers" (fig. 8.2). Around these springs, complete ecosystems with complex food webs flourish without daylight or any carbon source more exotic than carbon dioxide.

Detailed investigation of the ecology of hydrothermal vent communities revealed that at the base of their food chain are single-celled chemotrophs, which are organisms that live by taking in abiological nutrients as raw materials to build their cells. These single-celled organisms are, in turn, either bound in a close, mutual symbiosis with, or are eaten by, higher eukaryotes such as clams and the now well-known tube worms that can grow to lengths of more than a meter. These macroscopic, multicellular organisms, however, live in the much cooler water centimeters to meters away from the vents and thus are not themselves particularly thermophilic.

Adaptation to survival at high temperatures is closely linked to the chemical needs of these organisms, specifically to the sulfides they use as their source of reducing potential. These sulfides are provided by the seafloor springs and hydrothermal vents but tend to be insoluble at the 2°C temperature of the deep ocean. The organisms are there-

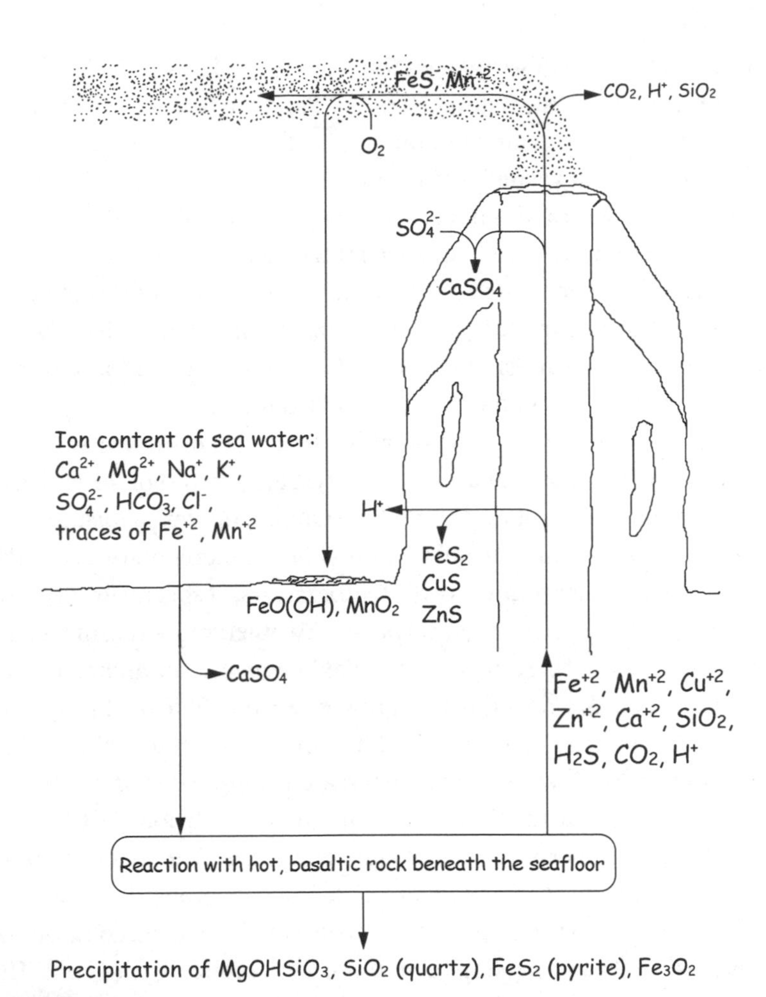

Figure 8.2 A schematic cross section of a hydrothermal vent, or black smoker, showing the reaction paths of the most important minerals. Even before black smokers were discovered, geologists had predicted that such reactions must be occurring at the ocean floor, to account for the unusual salt content of seawater. It is estimated that the entire water volume of our oceans runs through black smokers once every 8 million years.

fore able to harvest more sulfide the closer they live to the hot spring, and this produces a strong selective pressure for the evolution of extreme thermophilic properties.

Other hot environments that have yielded new, extremely thermophilic microbes include solfatare fields, which are volcanic soils permeated by hot vapors that are found in such places as Iceland, Sicily, and America's Yellowstone National Park. Surface waters in solfatare fields are often near the boiling point and are rather caustic as well, with pH values ranging from an extremely acidic 0.5 to a moderately basic 9.0. Per their name, they also usually contain a diverse mix of sulfur compounds, with deeper layers being anaerobic and thus rich in sulfides. Under such reducing conditions, the sulfides of heavy metals tend to precipitate out of solution, coloring the soils a gooey black. Some of the most extreme thermophilic and acidophilic microbes have been isolated from these environments.

A number of thermophilic organisms have been studied in detail, and several of them are of particular interest to us astrobiologists. These include two of the most thermophilic bacteria, *Aquifex pyrophilus* and *Thermotoga maritima*. The 80°C optimal growth temperature and the 89°C and 96°C maximum growth temperatures, respectively, of these microbes, however, have been surpassed by other, more recently characterized species. These include *Pyrolobus fumarii*, an archaeon isolated from a black smoker, with a growth range of 90°C to 113°C, and "strain 121," which grows at 121°C (a temperature routinely used in autoclaves to "sterilize" medical or research equipment) and can survive several hours at 130°C but does not grow at all below 85°C. Strain 121 has since been classified as an archaeon of the Crenarchaeota phylum and baptized *Geogemma barossii*. It is an anaerobe that seems to make its living by extracting electrons from simple reduced compounds in thermal vents and passing them on to Fe^{+3}, reducing it to Fe^{+2}. The current record holder, though, appears to be *Methanopyrus kandleri*. In 2008, Koki Horikoshi (1932–2016) and his group showed that, as long as the pressure is elevated to the 400 atm of its natural, black-smoker habitat, this carbon-dioxide-and-hydrogen-eating archaeon continues to reproduce at temperatures as high as 122°C.

Life under conditions of extreme heat, especially when coupled with low pH, requires that an organism's proteins remain folded and active in these seemingly more challenging environments. Most eukaryotic proteins, for example, unfold at modest temperatures or modest deviations from neutral pH, as seen when you boil an egg or when a Peruvian chef makes *ceviche* by soaking raw seafood in lime juice. The molecular basis for the adaptation to high temperatures has been studied in more detail than any other extreme condition, but no straightforward and universally applicable rules have yet been uncovered that define whether a protein will remain folded and active. One interesting observation has come to light, though: enzymes derived from thermophilic organisms are just as stable at high temperatures as our enzymes are at our body temperature. For example, testing each enzyme at the optimal growth temperature of its source organism yields similar values for key parameters, including thermodynamic stability. This constant stability is achieved through a large number of subtle molecular interactions that combine to produce a rather complex overall picture of thermostabilization.

A final twist regarding the thermophiles is that *A. pyrophilus* and *T. maritima* seem to be among the most "ancient" of all prokaryotes. That is, they may be the living organisms most closely related to the putative branch organism from which the domains of Archaea and Bacteria diverged. That these thermophiles might be LUCA's closest living relative has widely been taken to suggest that life arose under high-temperature conditions, perhaps at the deep-sea vents. Of course, we have to remember that LUCA was quite advanced and that a great many organisms had evolved and disappeared again long before she came on the scene. An alternative scenario is that life on Earth arose someplace far removed from the vents and evolved into organisms that filled many niches, including thermophiles living in the vents, and then the non-vent organisms became extinct. How might this have happened? A meteor or comet impact large enough to boil most of the volume of the Earth's oceans—an impactor a few dozen kilometers across would do the trick—would kill off everything but the deep-ocean thermophiles.

Cold Adaptation

While hot springs and hydrothermal vents are very important both for biology and geology, most of the water on our planet is rather cold. The bulk of the deep sea is at a constant temperature of 2°C, and in the vicinity of the polar icecaps, liquid seawater may even be cooled to below 0°C as the typical salt content of seawater (3.4%) lowers the freezing point to −1.8°C. When seawater freezes, the crystallization of pure water increases the salinity of the remaining liquid to as high as 15%. Under these conditions, the freezing point may be depressed to as low as −12°C. Numerous species, collectively known as psychrophiles, are now known to thrive under these harsh conditions.

The challenges associated with cold adaptation are very different from those associated with a thermophilic lifestyle. While high temperatures speed up reactions, including those that convert the biological macromolecules into smaller and much more stable molecules such as carbon dioxide and water, low temperatures slow reactions down and thus bring biochemical systems to a halt but do not destroy them. The real danger sets in when water is allowed to freeze since its significant expansion during freezing exerts shear forces that can easily cause mechanical damage to cells and their components.*

Two broad solutions to the problem of ice-induced damage have evolved on Earth. The first is the production of antifreeze compounds to prevent the formation of ice. Different species have come up with different ways to avoid freezing. Some organisms simply produce large quantities of osmolytes, such as glycerol, which suppress the freezing point of their tissues by several degrees. Fish of the polar waters have several different kinds of antifreeze proteins (AFPs), ranging from the simple type I AFP, which consists only of a single α helix, through to complex glycoproteins. These proteins typically recognize and bind to structural features of nascent or growing ice crystals, thus blocking their further growth. In contrast, some species of frogs and turtles have found an alternative solution by going in the opposite direction. They have ice nucleation proteins, which facilitate freezing of the animal's body liquids. By triggering ice formation in many different places at once, these proteins ensure that crystals cannot grow large enough to cause mechanical damage. Essentially, the

*If this kind of problem is excluded, many kinds of microorganisms can be frozen and thawed without harm. The crucial point is that a sufficiently low temperature has to be reached without any formation of ice crystals, which in the laboratory is generally achieved by flash freezing in liquid nitrogen at −200°C.

freezing process is similar to what happens when a small sample of biological material is thrown into liquid nitrogen: the water freezes instantly in all places at the same time, which minimizes the damage to cell structures.

Ice-induced damage, however, is not the only problem that cold-adapted organisms face. Many of the molecular interactions that define our cells are affected by cold, and thus many aspects of biochemistry must be modified for an organism to thrive at low temperatures. These adaptations include the modification of membrane lipids (to avoid stiffening, just as oils often solidify in the refrigerator) and the production of proteins that stop RNA molecules from binding to themselves to form complex structures that could inhibit translation.

A particularly successful strategy of adaptation to the coldest climates on Earth involves the collaboration between two species. Lichens, which colonize much of the rock surfaces in Antarctica, are famously resistant to the dry, cold conditions of their habitat. Although many lichens look like plants, they represent a symbiotic life form made up of a fungus and a photosynthetic organism. The ability to form lichens is spread widely across many families of fungi. In fact, it is estimated that one in five fungal species can perform the trick. Their symbiotic, photosynthetic partners can be algae (e.g., with the fungus *Prasiola crispa*) or cyanobacteria (e.g., with the fungus *Nostoc commune*), representing the domains of Eukarya and Bacteria, respectively. The symbionts make a very hardy team. In the wet and relatively mild climates found on the coast of Antarctica, the fungus *P. crispa* survives on its own by harvesting chemical energy from its environment, but living as a lichen provides an alternative source of energy that allows the fungus (and its algal partner) to thrive even on dry rock faces despite the more limited nutrients and drastic temperature swings typical of this environment. Cold-resistant lichens, forming characteristically colorful, plantlike structures, thrive at very high polar latitudes—even, for example, some 2,200 m up Mount Czegka, which, at a latitude of 86°20′ south, is just 400 km from the pole. This said, the southernmost exposed rock on our planet, some 110 km closer to the pole, on Mount Howe, appears to be completely lacking in multicellular life.

Apart from molecular adaptations and symbiosis, organisms can also exploit their local or even microscopic environment to obtain protection against freezing. Algae, for example, have been shown to survive in tiny liquid pockets encapsulated in the sea ice that surrounds Antarctica during the winter. As the freezing of a closed volume of seawater brings with it an increase in salt concentration, the freezing point is suppressed in these environments. This prevents the organisms from freezing, but as a consequence they have had to adapt to high salinity as well as low temperatures and low light. Other species may "hibernate" most of the year in the sea ice and only spring to life when their habitat is defrosted for a brief period during the southern summer. Similarly, several species of algae have been found to grow a few millimeters under the surface of translucent rocks in Antarctica's "dry valleys," the harsh, cold, and extremely dry areas that are widely regarded as the closest thing to Mars on Earth. The overlying rock acts like a greenhouse, significantly extending the growth season of these endolithic (from the Greek *endo*, meaning inside, and *lithos*, meaning "rock") organisms and protecting them from drying out.

A final "ice niche" is located far below the surface of the ice cap in permanently liquid lakes that may have existed for millions of years. Kept warm by geothermal energy (helped by the insulating properties of a kilometers-thick ice cap), these lakes are thought to have been isolated from the rest of the biosphere for the entirety of their existence. Although they were only discovered in the 1990s, during ice-penetrating radar studies, these subglacial lakes are vast. The largest one, Lake Vostok, is located at 77° south beneath the eponymous Russian research station and is thought to contain as much water as North America's Lake Ontario.

The motivation for exploring Lake Vostok and the other subglacial lakes is clear. The Antarctic lakes, and the methods developed to explore them, are of interest as potential models for the saltwater oceans suspected to be hiding under the ice crusts of several moons in the outer Solar System, which we'll discuss in detail in chapter 10. Survival in these cold, subglacial lakes may also be relevant for understanding life's history on our own planet, which is believed to have undergone several "snowball Earth" episodes as recently as 600 million years ago when the carbon dioxide cycle (see chapter 3) that regulates the Earth's temperature may (or may not—this is still controversial!) have gone haywire.

Motivated by the astrobiological relevance of the lake, researchers gradually advanced drills toward its surface, proceeding with extreme caution so as not to damage the biotope by contaminating it with microbes from the outside world—or by releasing the high pressure, thereby diminishing the freezing point depression and causing the lake to solidify. In 1998, and then again in 2010, drills were halted just 100 m above the water. Ice cores consisting of refrozen lake water were shown to contain evidence of cellular life. Only in February 2012 did research teams punch a hole all the way through the nearly 4 km of ice to the lake water. They allowed the pressurized lake water to rise up by a few meters and then freeze in place, returning a year later to retrieve the frozen lake water for analyses. Severe contamination of the samples with both kerosene and microbes has cast a shadow over the validity of any results obtained, although the researchers have claimed to have identified at least one non-contaminant species. A metagenomic study (see, again, sidebar 7.2) of the ice naturally accreted from Lake Vostok published in 2013 yielded taxonomic classifications for more than 1,600 gene sequences, of which 6% were assigned to eukaryotes and the rest to bacteria.

By contrast, researchers from the Whillans Ice Stream Subglacial Access Research Drilling (WISSARD) project used a specially developed hot-water drill, designed to achieve contamination-free access, to retrieve lake water and sediment samples from Lake Whillans, which forms part of a dynamic drainage system in West Antarctica underneath more than 800 m of ice. In 2014, the team reported the first comprehensive description of a subglacial ecosystem. In the lake water samples they characterized, the researchers found around 130,000 cells per milliliter, which is comparable to cell densities in the deep sea. Analyzing ribosomal RNA genes, they discovered a large number of microbial species in both the water and sediment samples, representing both bacteria and archaea, but not eukaryotes.

Detailed chemical analyses combined with the taxonomic relations of the new species to known microbes suggest that the ecosystem in the lake is able to thrive on a chemoautotrophic basis, that is, deriving energy and nutrients from minerals. Experiments measuring nutrient incorporation rates by the cells retrieved from the lake led the researchers to conclude that the chemoautotrophic production is sufficient to serve as the basis of a complex food web also including heterotrophs, organisms who, like us humans, derive their nutritional needs from consuming complex carbon compounds. Heterotrophs were also found in the lake sediment, presumably feeding on organic material there. A key aspect is the internal nitrogen cycling in the system, which appears to include nitrifying species in the water column that convert abundant ammonium ions into more oxidized nitrogen compounds, along with other species in both the sediment and water column that reduce nitrogen compounds to ammonia.

Drought and Salinity

As we have seen, many extremophiles can survive and even thrive at temperatures above 100°C or below 0°C as long as water remains liquid (because of high pressure or salinity or both). Generally, the availability of liquid water is widely seen as a key requirement for life. Deserts and salt lakes illustrate the struggle for survival in environments where water is either absent or unavailable for chemical reasons.

All known organisms need water as a solvent for their biochemical reactions. But some have evolved ways to survive long periods of drought in a passive state and then carry on living when the water returns. The bacterium *Deinococcus radiodurans*, for instance, attracted scientists' attention with its extraordinary resistance to ionizing radiation. First discovered in a can of corned beef that had been "sterilized" with γ-rays, it was also isolated from cooling baths of nuclear reactors. Research showed that the microbe hosts a highly efficient DNA repair mechanism that enables it to survive radiation levels a thousand times higher than those that would kill a human. As it is clear that radioactivity was not the selective factor that produced this mechanism (nuclear reactors being a relatively new habitat), the question as to why a bacterium would have developed this trait was at first a mystery. The consensus now, however, is that the DNA repair mechanism of *D. radiodurans* evolved as a means of surviving DNA damage induced by severe drought since both irradiated and dried cells suffer similar types of DNA strand breakage due to the presence of active chemical species such as oxygen radicals.

Another impressive defense against the threat of drying is the one employed by the tardigrades (more colloquially known as water bears), a group of microscopically small animals that are found on all continents. Tardigrades are typically found in water droplets suspended in moss and lichens. Remarkably, these organisms have two separate, highly original emergency routines. One is for the case of flooding and associated oxygen shortage and involves inflating to a balloon-like state that allows the tardigrade to float at the surface of the water. The second emergency plan is for the opposite case. When their habitat dries out, tardigrades shrink into a spore-like granule known as the tun state (*tun* is an archaic word for "barrel," which is what it looks like). In the tun state, the tardigrade replaces most of its water with the sugar trehalose. When trehalose solutions evaporate, the remaining sugar forms an amorphous glass rather than sharp, damaging crystals, and thus trehalose is an ideal medium for dehydrating proteins and DNA without damaging them.

Once the tardigrade converts into a tun, it is one of the toughest animals on Earth. Tuns found in moss samples from museums have been revived after more than a century of drought. Among other improbable stress treatments, researchers have suspended tuns in perfluorocarbon solvents and subjected them to hydrostatic pressures of up to 6,000 atm, and more than 80% of the animals survived. Tuns can similarly survive temperatures as high as 150°C and as low as 0.2°C above absolute zero. If anybody wanted to send animals to Mars (and who doesn't?), perhaps tardigrades would be the most likely to survive the trip without much in the way of life support. Indeed, in September 2007, researchers from the University of Kristianstad in Sweden actually sent tardigrades into Earth orbit on the *Foton-M3* mission. Twelve days and 189 orbits later, the spacecraft landed safely back on Earth with its cargo safe and sound; the water bears had withstood the vacuum and cold in Earth orbit without breaking a sweat. The hard UV radiation that pervades the Solar System outside the Earth's ozone layer (which protects us) proved their only real threat: whereas most of the animals in a group protected by UV filters survived, only a few survived in a second group lacking this protection.

In a bittersweet follow-up in April 2019, the Israeli spacecraft *Beresheet* attempted—but failed—to make a soft landing on the Moon. Along with a "Lunar library," a 30-million-page archive of human knowledge on a metal disc, it carried more than a thousand tardigrades, sent on the trip by the Arch Mission Foundation, which aims to find a backup habitat for Terrestrial life elsewhere in the Universe. The organization reckons that both the library and the tardigrades may well have survived the crash landing, albeit the latter presumably not for long.

The harshest classical desert on Earth is northern Chile's Atacama. With the 6,000 m high Andes on the east, which prevent moisture flow from the Amazon, and the Pacific's cold Humboldt current and the 2,500 m high Chilean Costal Range limiting moisture flow from the west, the heart of the Atacama is considered to be the most Mars-like environment found on our planet. On average, the core of his hyperarid desert receives 1 or 2 mm of rain *per decade*, and as we can confirm from personal observation, one does not see even a single dead blade of grass over hundreds of kilometers (fig. 8.3). The area is so lifeless, in fact, that NASA uses it as a challenging case study for the development of new life detection technologies, which, by the way, often fail here. (Foreshadowing chapter 10, there are so few bacteria in the soil, and the organics in the soil are so sparse and so refractory, that, had they landed in the Atacama, it is thought that the life detection technologies sent to Mars with the *Viking* missions would not have found life.) Despite this, in 2006, astrobiologist Chris McKay reported that, even in the driest parts of the Atacama, colonies of cyanobacteria live just a few millimeters below the surface of halite evaporites (rock salt), occupying spaces between the salt crystals. It seems the hygroscopic nature of salt ensures that the crystals absorb some atmospheric water even at the low humidity found in this harsh desert.

Figure 8.3 The heart of the Atacama Desert in northern Chile is considered to be the most Mars-like environment found on our planet. And yet, even in this hyperarid desert, where only a few millimeters of rain falls a few times a decade and one does not see even a single dead blade of grass over many hundreds of kilometers, there is life: endolithic colonies of cyanobacteria living millimeters below the surface of halite (rock salt), occupying spaces between the salt crystals and living off atmospheric water vapor drawn in by the salt crystals. (Photo credit: Jade Dutcher)

Living in condensed dew that's wicked between salt crystals presents another problem: while water may be present, it is not chemically available. Think of the so-called Dead Sea, which despite its name is very much alive. Evaporation lakes like the Dead Sea are saturated with salt, so all the water is essentially taken up with the task of solvating salt ions. Living organisms have to compete with these ions for water, lest the osmotic gradient suck them dry. Several species of algae, along with bacteria and archaea, have adapted to such high-salt environments. While the algae tend to be halotolerant, meaning that they can live with the salt but may be better off without it, many of the archaea found in salt lakes are obligate halophiles—that is, they grow better, or only, when high salt concentrations are present. In response to the salt, all these adapted organisms maintain very high concentrations of other solutes in the cell to keep their insides in osmotic balance with the outside world. While salt-adapted algae and bacteria tend to use small organic molecules for this purpose, highly halophilic archaea fight fire with fire by keeping extremely high concentrations of potassium chloride within their cells. This literally takes away the pressure from the membrane, but it shifts the stress and the requirement for adaptation onto the molecular machinery of the cell. All the proteins in a halophile have to be optimally folded and functional at saturated salt conditions, much the same way that the proteins of hyperthermophiles must function near 100°C. Researchers have studied the amino acid sequences, structures, and functional characteristics of halophilic proteins in comparison with thermophilic and mesophilic ("normal") proteins in order to gain some insights into the evolutionary strategies used to adapt proteins to stress conditions, but the picture remains far from complete.

Extremes of pH

Another part of the "normality" that we rarely question is that the pH of most liquids in a biological context is fairly neutral, typically ranging from 7.0 to 7.4 (fig. 8.4). Where acids come into play, such as in the saliva and stomach fluids, they are meant to destroy biological material and any surviving foodborne organisms. However, there are habitats (such as the Yellowstone solfatare fields described above) where both fungi and archaea thrive at pH values around zero, which is even more acidic than pure stomach fluids. Although extremely acidophilic organisms have been known for decades, the nature of their specific adaptation is far less studied than that of thermophiles.

In contrast, the base-loving alkaliphiles have been studied in some detail, particularly by researchers interested in bioenergetics. The issue is that most aerobic organisms generate energy in the form of ATP via reactions that actively pump protons out of the cell or out of the mitochondria (rendering the pH inside more basic than the pH outside)

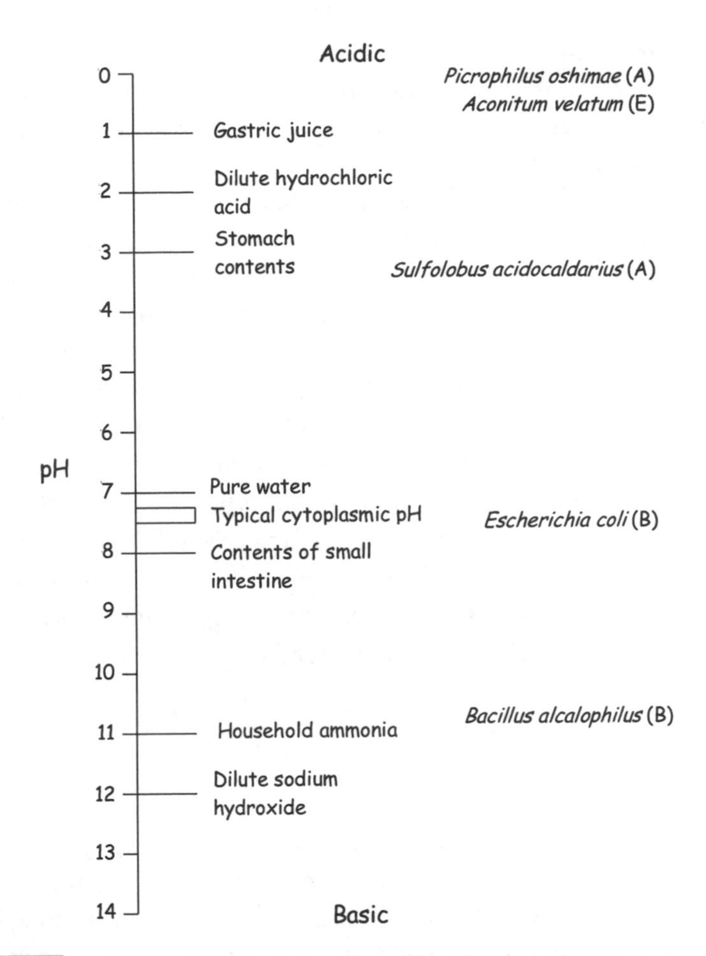

Figure 8.4 The pH scale is shown with examples of acidic and basic fluids and some pH extremophiles (A, archaeon; B, bacterium; E, eukaryote).

and then generate ATP utilizing a millstream-like flow of hydrogen ions back down this gradient and through the enzyme F-ATPase.* This seemingly odd arrangement allows energy-producing reactions whose energy is not quite enough to make one ATP, or not quite enough to make two ATP, to donate their energy to a single energy

*The millstream analogy is better than you might imagine. In 1997, Masasuke Yoshida and his research team filmed the F-ATPase rotating under the influence of this proton flow as a critical step in the enzyme's synthesis of ATP.

"reservoir" from which the F-ATPase can then extract exactly one ATP's worth of energy at a time. This ensures the efficient utilization of the available energy and allows many diverse reactions to all contribute to making ATP.

The mystery with the alkaliphiles is that, if the "normal" pH gradient is reversed, the F-ATPase reverses, consuming ATP and pumping protons out of the cell. That is, if a normal cell found itself in a high pH environment, it might end up wasting its energy trying to acidify the outside world. Adaptation to high pH thus likely involves fairly drastic reorganization of metabolic electrochemistry and, perhaps for this reason, is found in only a few branches of the tree of life. The standard model organism for alkaliphile studies is *Bacillus alcalophilus*, an obligate alkaliphile that can only live at pH values between 9 and 12. To do so, it actively pumps sodium ions out of the cell and then uses the resulting charge imbalance to drive protons back inside, thus maintaining the pH of its cytoplasm below 9.5. How it generates its ATP with the pH gradient "backward," though, remains a mystery.

Going Deep

The effects of pressure on organisms are not so widely appreciated. Changes in atmospheric pressure may be indicative of a change in the weather, but otherwise they don't normally affect organisms that, like us, live on dry land. When we climb a mountain, the decline in the partial pressure of oxygen affects us quite strongly, but the overall change of atmospheric pressure does not have any notable effects.

The situation changes drastically when we start diving into the ocean's depths, where the hydrostatic pressure increases by about an atmosphere for every 10 m of descent (fig. 8.5). The increasing pressure and associated changes in the solubility of gases and toxicity of oxygen limit the reaches of recreational scuba divers to around 40 m and divers with optimized gas mixtures, to around 300 m. Beyond that, humans need submarines to explore life under hydrostatic pressure, which is one of the reasons we know a lot less about barophiles (pressure-adapted organisms) than about thermophiles.

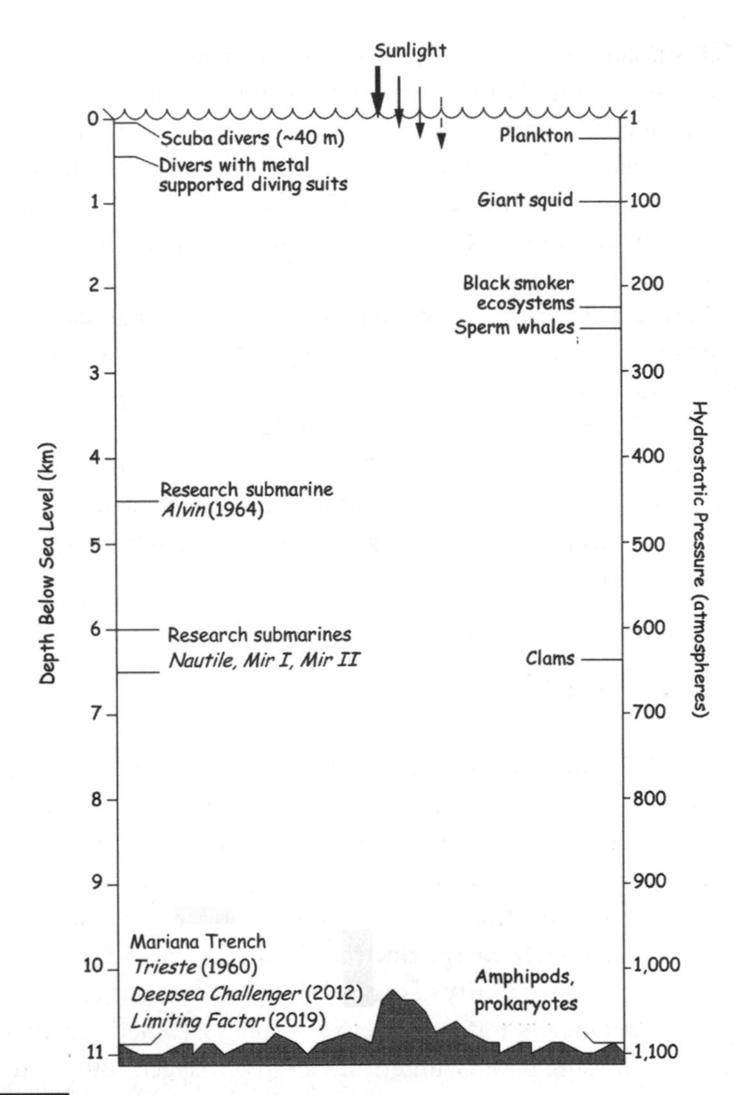

Figure 8.5 The depths below sea level that can be reached by divers, undersea vessels, and some pressure-adapted organisms are shown with the associated hydrostatic pressures.

There is no doubt that barophilic adaptation exists and is widespread in the oceans. Research vessels have consistently found life on the seafloor over the whole depth range of the oceans down to the 11 km maximal depth of the Mariana Trench, where the pressure reaches 1,086 atm. Most non-pressure-adapted microorganisms, however, stop growing at pressures above 500 atm, and mammals like us, with lungs and an entire metabolism depending critically on the equilibrium of gases (oxygen and carbon dioxide) transported through the blood, are more pressure sensitive still. Among mammals, the sperm whale, with its 2,440 m (230 atm) diving range, is a very lonely record holder.

The difficulty in obtaining samples from deep-sea habitats and the challenges of conducting biochemical experiments under high pressure conditions in the laboratory have conspired to make this research field one of the less comprehensively studied. For example, not until 2000 did researchers identify the first genes unequivocally involved in adaptation to high hydrostatic pressure. Most of the mechanisms that allow life to thrive even at the greatest depths of the oceans remain to be discovered.

Intraterrestrial life

Almost all life on Earth depends, ultimately, on the Sun for its energy. Plants utilize solar energy directly, obviously, and we do so indirectly by eating plants (or things that ate plants) and breathing oxygen. And while the "fixed" carbon and oxygen produced by plants are not the only sources of reduction potential (high-energy electrons) and oxidative potential (high-energy electron sinks) that are employed in the biosphere, the vast majority of organisms use at least one of these two (fig. 8.6). We humans, for example, oxidize photosynthetically derived fixed carbon using photosynthetically derived oxygen. But there are other organisms that, instead of using oxygen as their electron sink, use minerals such oxidized iron (Fe^{+3}) or manganese (Mn^{+4}). Not surprisingly, these organisms are found in environments lacking free oxygen, such as the deep mud at the bottom of lakes. But they are still heterotrophs, however, since they feed off the organic material that rains down from the water column, and thus their metabolism is, like ours, ultimately coupled to sunlight. The hydrothermal vent communities described in the opening of this chapter are likewise coupled to sunlight; while they use the reduction potential available from sulfides (such as H_2S) extracted by the hot vent water from the crust, the reason this provides energy is that a mixture of sulfides and oxygen is out of equilibrium relative to sulfate (SO_4^{-2}) and water. The chemical disequilibrium on which the entire vent ecology is founded thus relies on the availability of photosynthetically produced molecular oxygen, without which the vent water would presumably be too close to chemical equilibrium to provide enough energy for anything but the simplest of living communities.

If life depended on light, either directly through photosynthesis or indirectly via reactions such as those presented in fig. 8.6, it would

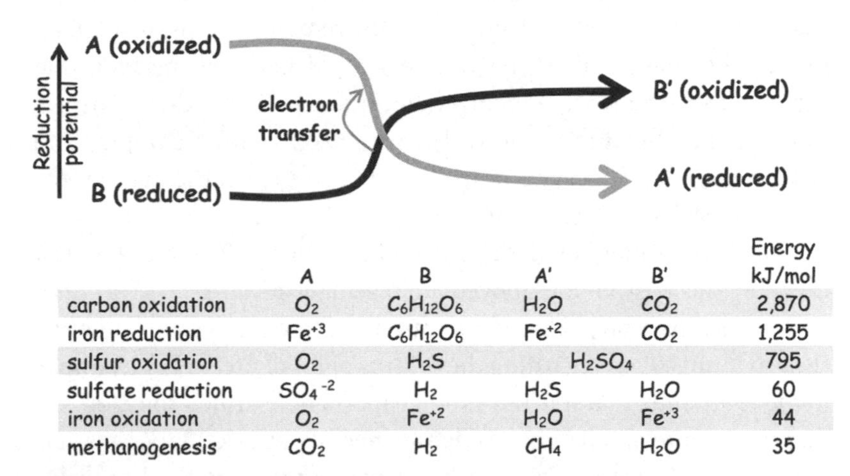

	A	B	A'	B'	Energy kJ/mol
carbon oxidation	O_2	$C_6H_{12}O_6$	H_2O	CO_2	2,870
iron reduction	Fe^{+3}	$C_6H_{12}O_6$	Fe^{+2}	CO_2	1,255
sulfur oxidation	O_2	H_2S	H_2SO_4		795
sulfate reduction	SO_4^{-2}	H_2	H_2S	H_2O	60
iron oxidation	O_2	Fe^{+2}	H_2O	Fe^{+3}	44
methanogenesis	CO_2	H_2	CH_4	H_2O	35

Figure 8.6 The redox interchange of life. Energy metabolism in all life on Earth involves something getting oxidized and something else getting reduced in exchange. Heterotrophs, for example, oxidize the "organics" (fixed carbon) obtained ultimately from plant material using either molecular oxygen (as we do) or oxidized minerals (as many microorganisms do in anoxic lake bed and seafloor muds). Deep-sea vent archaea, in contrast, harvest their reduction potential from abiological sources, such as sulfides or other reduced minerals leaching from the crust. These organisms nevertheless use photosynthetically produced oxygen as their oxidizer, which links their metabolism, like ours, to the energy of Sun. Only a few subsurface life-forms, including the subsurface methanogens and sulfate-reducing microorganisms, use geology rather than photosynthesis to provide their sources of both reductant and oxidizer, and thus live completely uncoupled from sunlight. This independence is achieved, however, at some cost: these reactions provide relatively meager amounts of energy.

be limited to the surfaces of planets. Given, as we'll see in chapter 9, that the surfaces of most planets are too hot or too cold to support liquid water, this would be a major constraint on potential habitats. In the late twentieth century, however, our view of habitability was greatly broadened with the discovery that there are biotopes completely cut off from the energy of the Sun. They reside in the hot, permeable rocks deep beneath our feet.

The idea of abundant microbial life deep within the Earth's crust was first promoted by the Cornell University astrophysicist Thomas Gold (1920–2004), who, in 1992, published a paper in the *Proceedings of the National Academy of Sciences* entitled "The Deep, Hot Biosphere." Gold's ideas regarding life below the Earth's surface were an extension of his fairly heretical theories about the origins of fossil fuels. The traditional view is that coal and oil are the remains of long-dead plant matter and marine algae, respectively. Gold believed, instead, that these hydrocarbons were incorporated into the Earth during accretion and that they provide a ready carbon source for a complex and rich underground ecosystem that could even exceed the mass of the surface biosphere we see around us.

And is there any evidence for Gold's claims? Over the last few decades, the answer has moved increasingly toward "yes," at least with regard to a biosphere eking out a living deep within the crust (less so for his ideas regarding the origins of fossil fuels). For example, in the 1990 study that spurred Gold's paper, a hole drilled some 6.7 km into the granite of Siijan in Sweden turned up a thick, black, strongly odiferous fluid from which two different species of thermophilic, iron-reducing bacteria were later cultured.* About the same time, complex ecosystems were being reported within rocks freshly harvested from 3 km down in South African gold mines. But the idea of microbes living in what appeared to be solid rock far beneath the surface didn't sit well with the microbiology community, most of whom seemed to believe that the purported subsurface microbes

*An important caveat is that, back then, prior to metagenomics, microbiologists could identify new bacterial species only if the organisms could be cultured (grown) in the laboratory. As many extremophiles do not survive under laboratory conditions, there may well have been more than two species in this mud.

were simply normal surface organisms that had contaminated the relevant samples after they were harvested. In 1993, however, Todd Stevens and James McKinley of the Pacific Northwest Laboratory probed for ecosystems deep below eastern Washington state and turned the tide of scientific consensus.

Stevens and McKinley's identification of subsurface organisms relied on two approaches. First, they went to exceptional lengths to collect their samples aseptically from a thick bed of basalt several hundred meters down. On retrieving their samples, they isolated methanogens, microbes that use the (quite limited) energy available from the conversion of carbon dioxide and hydrogen to methane and water (see fig. 8.6). In many, perhaps better-known, habitats (such as swamp mud and cow stomachs), methanogens consume hydrogen that is produced biologically, but the water samples from the deep drill contained more hydrogen than could be explained by any biological mechanism. Investigating further, the researchers found that the hydrogen had been produced by the reaction of reduced iron from the basalt with hot, oxygen-free groundwater. Putting these observations together, the researchers theorized that their subsurface ecosystem was fueled by abiologically produced hydrogen and carbon dioxide, thus rendering it completely independent of solar energy in any of its many guises. Second, to clinch their hypothesis, Stevens and McKinley demonstrated that their bacteria could survive for more than a year when sealed in a flask with nothing more than hot water, carbon dioxide, and basalt to munch on. Indeed, the microbes not only survived but actually multiplied under these seemingly extreme culture conditions, suggesting that they were well adapted to their diet of reduced rocks, water, and carbon dioxide. With this second line of evidence in hand, the case for subsurface ecosystems was clinched and today is no longer in doubt.

Following on the work of Stevens and McKinley, in 2008, Dylan Chivian of the Lawrence Berkeley National Laboratory in Berkeley, California, circled back to those South African gold mines, where he discovered a subsurface biotope in fractured rock some 2.8 km deep. Metagenomic sequencing of DNA extracted from water collected from the fracture revealed that at least 99.9% of the microorganisms in the sample belonged to a single bacterial species, which he named *Desulforudis audaxviator*, from a Latin quote in Jules Verne's *Journey to the Center of the Earth*: "Descende, audax viator, et terrestre centrum attinges" (Descend, bold traveler, and you will attain the center of the Earth).

The lack of species other than *D. audaxviator* in the deep rock of South Africa is odd; the vast majority of living things on Earth today rely on other living things for at least some of their nutrients. Plants, for example, can fix carbon from carbon dioxide, but they generally receive much of their nitrogen as nitrates produced by soil bacteria. The genome of the gold mine microbe, however, suggests that this alkaline-resistant (to pH 9.3) thermophile (to 60°C) harvests its carbon and nitrogen from inorganic materials leached from the rock and uses hydrogen produced by the radiolysis (radiation-induced decomposition) of water as its electron source and mineral-derived sulfate as its electron sink. Chivian and colleagues couldn't exclude the presence of other species, which might, for instance, cling to the rocks so tightly that they weren't retrieved during the sampling process. But they concluded that the species they discovered may well constitute an ecosystem all by itself, living off only the inorganic materials that geology provides for it. And it seems to be good at it; in 2019, researchers reported isolating the same species from 2 km down in western Siberia, suggesting that this deep crustal organism has somehow spread around the globe. Indeed, the South African and Siberian strains differed by only 800 nucleotides across their entire genomes, suggesting that the microbe can migrate from environment to environment relatively rapidly or that its extremely slow replication rate (brought on by its exceedingly slow metabolism) resulted in the observed high genetic similarity despite the 200 million years that have passed since the Pangea supercontinent broke up and Eurasia separated from Africa.

Building on the work of Stevens, McKinley, Chivian, and others, in 2009, a group of several dozen scientists from three continents banded together to create a virtual institute called the Deep Carbon Observatory, the goal of which was to systematically characterize the distribution of carbon within the Earth's crust. This interdisciplinary program, which ran until 2019, addressed four key areas: the ecology of subsurface life (the Deep Life directorate), the reservoirs and fluxes of carbon within the planet, their implications for energy, and the extreme chemistry and physics of carbon materials in a geological context.

The Deep Life directorate, chaired by Isabelle Daniel from the Université Claude Bernard in Lyon, France, set out to find, metagenomically identify, and, if possible, culture microbial samples from deep subsurface habitats around the world. Drilling 2.5 km into the seafloor and sampling microbes from continental mines and boreholes more than 5 km deep, for example, the researchers used the samples from hundreds of sites around the globe to construct models of the ecosystems that lie deep within our planet's crust. From these studies, the directorate estimates that, at about 2 billion cubic kilometers, the volume of habitable rock beneath our feet is almost twice the volume of all the oceans and that the total carbon mass of the biosphere inhabiting this zone weighs in at 15 to 23 billion tons, which is around 300 times the mass of all humans alive. Among the billions of tons of life in the deep biosphere, the researchers identified representatives of all three domains of life, but, not surprisingly, found that bacteria and archaea were predominant.

Another surprise that the effort uncovered is that the microbial biodiversity found below ground is comparable to that found on the surface. This despite the fact that, at first glance, one might assume that the environmental diversity is lower below ground than above. The microbial diversity is likewise a surprise due to the slow reproduction of deep biosphere organisms, which should slow their evolutionary drift and subsequent speciation. Although few of these deep biosphere microorganisms have been cultured in the laboratory, and thus we do not know their replication rates, measurements of the rate with which they increase their mass suggest that their doubling times are often measured in decades if not centuries. Life is not easy when nutrients are so limited.

And how deep does the deep, hot biosphere go? As of yet, we do not know: even at the bottom of the deepest drillings performed by the Deep Carbon Observatory, some 5 km down, researchers still found life, and thus the true depth of the biosphere remains unplumbed.

Conclusions

In some sense, the Terrestrial extremophiles simultaneously tell us both a great deal and very little about the ability of life to arise and thrive on other planets. The ability of life to survive in such unlikely, resource-poor environments as solid rock kilometers below the surface does suggest that the range of niches suitable for life may be much greater than many scientists previously suspected. Similarly, tardigrades' ability to survive desiccation down to a water content of 3% suggests that oceans and rivers may not be required even for advanced, multicellular organisms. It is quite plausible that the rock-eating archaea that live within the Earth's crust could survive within the crust of Mars as well, and, although probably less likely, perhaps the subglacial microbes of Lake Vostok could survive in the ocean hidden beneath the ice shield of Jupiter's moon Europa (which we'll learn about in the next chapter). But even if such life could survive on Mars or Europa, could it have arisen there in the first place (see sidebar 8.3)? Or was there something special about the early Earth that

Sidebar 8.3

Weighing the Probabilities

The discovery that life could exist in seemingly solid rock, deep beneath the Earth's subsurface, opened the door to a great deal of speculation about how life might exist in similar habitats throughout the galaxy. This was particularly true given that the subsurface ecosystem found in Washington State, as described in the text, survives without recourse to solar energy in any of its many guises; every other ecosystem on Earth either requires light for photosynthesis or uses oxygen, a by-product of photosynthesis. Since Mars, for example, may be geologically active enough to support liquid water within its crust (more on this in the next chapter), the discovery of rock-eating organisms on Earth seems to suggest that similar locales could be inhabited on Mars. But is it that simple?

Probably not. The problem is that, even if a planet contains environments that could support life today, it is still very much a question as to whether life could have arisen there originally. As we saw in chapter 5, very special conditions might be required for the spark of life to arise from inanimate matter. In short, the enormous range of bizarre environments that life can inhabit on Earth probably reflects life's extraordinary ability to diversify to fill any available niche, rather than the ease with which life can arise in the first place.

rendered it much more suitable than elsewhere in the Solar System for the formation of life? Perhaps the answers to these questions will be found as humanity reaches out to explore the cosmos—which is what we explore in the remaining chapters.

Further Reading

Discovery of Seafloor Biotopes

von Damm, Karen, and John M. Edmond. "Hot Springs on the Ocean Floor." *Scientific American* 248, no. 4 (1983): 70–85.

Extremophiles

Gross, Michael. *Life on the Edge*. Cambridge, MA: Perseus, 2001.

Strain 121

Kashefi, Kazem, and Derek R. Lovley. "Extending the Upper Temperature Limit for Life." *Science* 301, no. 5635 (2003): 934.

Life under Extreme Aridity

Wierzchos, Jacek, Carmen Ascaso, and Christopher P. McKay. "Endolithic Cyanobacteria in Halite Rocks from the Hyperarid Core of the Atacama Desert." *Astrobiology* 6, no. 3 (2006): 415–22.

Subsurface Ecosystems

Chivian, Dylan, Eoin L. Brodie, Eric J. Alm, David E. Culley, Paramvir S. Dehal, Todd Z. DeSantis, Thomas M. Gihring, et al. "Environmental Genomics Reveals a Single-Species Ecosystem Deep within Earth." *Science* 322, no. 5899 (2008): 275–78.

Gold, T. "The Deep, Hot Biosphere." *Proceedings of the National Academy of Sciences USA* 89, no. 13 (1992): 6045–49.

Magnabosco, C., L.-H. Lin, H. Dong, M. Bomberg, W. Ghiorse, H. Stan-Lotter, K. Pedersen, T. L. Kieft, E. van Heerden, and T. C. Onstott. "The Biomass and Biodiversity of the Continental Subsurface." *Nature Geoscience* 11 (2018): 707–17.

Stevens, Todd O., and James P. McKinley. "Lithoautotrophic Microbial Ecosystems in Deep Basalt Aquifers." *Science* 270, no. 5235 (1995): 450–55.

The Deep Carbon Observatory

https://deepcarbon.net/

Chapter 9

Habitable Worlds in the Solar System and Beyond

On the summer night of July 16, 1969, Neil Armstrong (1930–2012), Buzz Aldrin, and Michael Collins blasted off from Kennedy Space Center in Florida for humanity's first visit to the surface of another celestial body. Three (Earth) days later, the spacecraft dropped into Lunar orbit, from whence Armstrong and Aldrin split off from Collins and took the Lunar module *Eagle* down to the surface. On July 20, uttering his historic—if slightly misspoken—words, "That's one small step for [a] man, one giant leap for mankind," Armstrong stepped off the *Eagle*'s ladder and onto the face of our Moon. Aldrin joined him shortly thereafter, and after a scant two and a half hours of exploring the dusty Lunar surface on foot, the duo returned to the Lunar module. Some 18 hours later, after an awkward, noisy "night's" sleep, the two launched the *Eagle* back into space to rendezvous with Collins. Firing up the command module's single rocket, the reunited astronauts spent three more days on the return home, ending the historic *Apollo 11* mission by splashing down in the Pacific on July 24.

On their return, the three astronauts were immediately ushered into a fully self-contained and hermetically sealed isolation chamber (a converted recreational vehicle!), physically cut off from the rest of the world. Heroes they may have been, but they would remain behind tightly sealed doors and windows as if they, themselves, were some exotic Lunar samples. The reason? A fear that "Lunar bugs" that might have infected them could escape to wreak havoc on our planet.

After three days, the trailer arrived in Houston, Texas, where the three astronauts joined three others (a doctor, a NASA public affairs officer, and a film technician who was accidentally exposed to some Lunar dust while handling a film canister) in more spacious quarantine facilities at the Lunar Receiving Laboratory. While the doctor kept close watch on the health of the astronauts, scientists in other, equally well-isolated laboratories incubated Lunar dust in nutrient broths and injected it into mice to see whether any microorganisms could be cultured. After two and a half weeks, neither the astronauts nor any of the mice seemed the worse for their exposure, and the quarantine was lifted. The astronauts were released to a world tour, months on the banquet circuit, and changed lives.

In December 1969, some six months after *Apollo 11* achieved President Kennedy's goal "of landing a man on the moon and returning him safely to the Earth," the *Apollo 12* mission set off for the Moon's prosaically—if inaccurately—named Oceanus Procellarum (Ocean of Storms). One of the many goals of this mission was to demonstrate that the *Apollo* technology could achieve a precision landing (*Apollo 11* had missed its planned landing site by several kilometers) so that future missions could explore more interesting, but more difficult, terrain. The all-Navy crew pulled it off and landed within 200 m of their target, the no-longer-functioning, robotic *Surveyor 3* spacecraft that had landed on the Moon just two and a half years earlier. The astronauts, Pete Conrad (1930–1999) and Alan Bean (1932–2018), spent almost eight hours exploring the Lunar surface, during which they collected 34 kg of rocks and soil.* They also snipped off several pieces of *Surveyor 3* for return to Earth so that engineers back in Houston could see how the various materials had fared after so many months under the harsh Lunar conditions. And what happened when the *Apollo 12* astronauts returned? They too were ushered into quarantine. Once again, though, no Lunar life was found, save, perhaps, for some Terrestrial bacterial spores that may (or may not; there is some debate as to whether they were picked up after the sample was re-

*Conrad, ever the joker, snuck an automatic timer for his camera with him to the Moon, hidden in a rock sample bag, so that he could perch the camera on *Surveyor 3* and snap a picture of himself and Bean together. His hope was that, after the film was developed, people would eventually, with shock, ask the question: "If both astronauts are *in* the picture, *who took the picture*?" Alas, though, when the time came, Conrad could not find the timer in the sample bag where he'd hidden it.

turned to Earth) have survived in hibernation, buried under some insulation deep within *Surveyor*'s camera.

The lack of Lunar life was, in reality, to be expected. The argument against life on the Moon was and remains, basically, that if you wanted to build a really good sterilizer, you'd make something like the Moon's surface: no atmosphere, no water, extremes of heat well past the boiling point, intense radiation, and intense UV light. Still, even with the insights provided by the intervening decades, the quarantine precautions seem prudent. After all, what do we really know about the range of conditions that life in its broadest scope might find suitable? How does the broad range of environmental conditions to which life has adapted on Earth correspond to the conditions that exist elsewhere in the Solar System and beyond (again, given the caveat that the set of environmental niches that life can evolve to fill may be much larger than the set of conditions under which life can arise in the first place)? Our question is, in a nutshell: are there habitats—that is, habitable environments—beyond Earth? Here we explore this issue in detail for the Solar System and, in necessarily less detail, for the rest of the cosmos.

Potential Abodes of Life Elsewhere in the Solar System

Let's start the search in the immediate neighborhood. Although the first astronauts to return after a Moon landing were placed under quarantine for fear of an infection with Lunar life forms, it is now clear that our satellite never had the rich mix of volatile elements that are almost certainly required for life. If there are any microbes on the Moon, it is safe to assume that they traveled there with the astronauts and that they've been dead for several decades now.

Moving inward from our home planet we have Venus, the planet most like Earth in terms of mass and composition, then tiny, rocky Mercury. Being so small and so close to the Sun, Mercury is hard to observe and thus was never much in mind during early speculations on life beyond the Earth. Just as well, since we now know that Mercury is a hot, dead world devoid of any appreciable atmosphere (see fig. 3.9). Venus, in contrast, is our closest neighbor in space and thus relatively easy to scrutinize through a telescope. Yet our sister planet presents a featureless, cloud-covered visage when viewed in visible light, an observation that encouraged endless speculation: the obvious interpretation being, of course, that clouds mean abundant water! Basing his hypothesis solely on Venus's thick cloud cover and relative nearness to the Sun, Svante Arrhenius (you'll remember him as the Nobel laureate and panspermia promoter) speculated in 1918 that "everything on Venus is dripping wet" and that "the vegetative processes are greatly accelerated by the high temperature; therefore, the lifetime of organisms is probably short."* Sadly, though, spectroscopic observations conducted in 1922, just a few years after Arrhenius's optimistic statement, failed to detect any water vapor in the Venusian atmosphere. As we now know, the clouds of Venus are composed not of water but of sulfuric acid droplets. The problem, as we explored in chapter 3, is that due to photolysis and the escape of hydrogen to space, Venus lost its water long ago, and the runaway greenhouse effect that ensued raised its surface temperature well above the melting point of lead. Precisely how long this took is still a matter of speculation. But the consequences for the possibility of life on our nearest planetary neighbor are clear: we don't know of any chemistry that is of sufficient complexity to support life that would be able to withstand Venus's hotter-than-an-oven conditions (see fig. 3.10). And while there have been suggestions of cooler, more favorable conditions in the Venusian clouds, we should probably conclude that the overall probability of life on Venus is very low. Indeed, if there are any other denizens of the Solar System, it seems likely that they live out beyond the Earth's orbit.

Mars: From Canals to Rovers

From early on it was clear that, after Earth, Mars is the most promising potential habit in our Solar System, a designation that it continues to hold today. It is, for example, the second closest planet to the Earth. Even the modest telescopes available in the late eighteenth century

*From his book: Svante Arrhenius, *The Destinies of the Stars* (New York: G. P. Putnam and Sons, 1918).

were sufficient to let astronomers define the length of its day (just 37 minutes longer than ours), identify clouds and polar ice caps, and discover that Mars's axial tilt is similar to Earth's, giving it Earth-like seasons. More tantalizing still, by the late nineteenth century astronomers had noted that with these seasons came a "wave of darkening" occurring on a global scale. Similar variations are seen on Earth in, for example, New England or Scandinavia in the autumn. Could the Martian color changes also represent seasonal variations in plant coverage?

Attempts to actually detect life on the Red Planet, rather than simply speculate about it on philosophical and theological grounds, date back to at least 1877. In that year, the Italian astronomer Giovanni Schiaparelli (1835–1910) turned his telescope on Mars and thought he spied extended networks of trenches, which he termed *canali*, the rather innocuous Italian word for "channels." And although there is no indication that Schiaparelli meant to imply that these structures were the handiwork of intelligent Martians, others—in particular, the independently wealthy and exceptionally self-promoting American astronomer Percival Lowell (1855–1916)—eagerly translated the Italian *canali* into the English *canals*. That is, artificial structures. In 1894, Lowell built himself one of the finest observatories of the era in the mountains of northern Arizona and spent decades visually mapping out the features on Mars and concocting a detailed hypothesis about a dying, intelligent race that had heroically built the canals to carry water from the poles as their planet fell into drought. This theory, he felt, explained much about the appearance of Mars. Indeed, in 1916, Lowell summed up his more than two decades of work on the subject by boasting, "Since the theory of intelligent life on the planet was first enunciated 21 years ago, every new fact discovered has been found to be accordant with it. Not a single thing has been detected which it does not explain. This is really a remarkable record for a theory. It has, of course, met the fate of any new idea, which has both the fortune and the misfortune to be ahead of the times and has risen above it. New facts have but buttressed the old, while every year adds to the number of those who have seen the evidence for themselves."*

*From his article: Percival Lowell, "Our Solar System," *Popular Astronomy* 24 (1916): 427.

Given what the scientists of the day were saying, it's not surprising that Mars became a popular location of extraterrestrial life-forms in the imagination of science fiction writers in the first half of the twentieth century. The invaders trying to colonize Earth in H. G. Wells's classic *War of the Worlds* (1898) were only a vanguard of the many different civilizations placed on Mars by writers. In 1908, nearly a decade before the Russian revolution would bring forth another kind of red star, the Russian author Alexander Bogdanov (1873–1928) published the novel *Red Star*, in which the Red Planet is used as a canvas to describe a socialist utopia. In 1917, Edgar Rice Burroughs (who later created Tarzan) published his very first novel, *A Princess of Mars*, in which a Civil War–era prospector is mysteriously transported to Mars, where many a daring deed ensues. Even as late as 1938, an American radio broadcast of Wells's *War of the Worlds* was so believable that it drove much of the country to panic. The possibility—nay, the certainty—of life on Mars was part and parcel of nineteenth- and early twentieth-century thinking about our next neighbor from the Sun.

Sadly, with the advent of improved telescopes and, perhaps not coincidentally, photographic methods that replaced visual observations, the case for intelligent life on Mars dropped away. Lowell's canals have never been captured on film (fig. 9.1): in this enlightened—if less romantic—age, we know that they were the creation of an overactive imagination, poor "seeing," and no doubt a sincere belief that we are not alone. Still, the same telescopic observations that killed the canals supported the observation of widespread seasonal color changes mentioned above, suggesting that, even if it lacked civilizations, Mars at least harbored simple plant life. But cracks were apparent in this hypothesis, too, as far back as 1909. In that year, William Campbell (1862–1938), who'd had a number of impressive "fights" with Lowell in the literature, led an expedition to the top of California's 4,421 m Mount Whitney to collect spectra of the Martian atmosphere.* From this vantage point, situated above half of the Earth's atmospheric mass

*Two notes for our Californian readers: (1) If you've ever climbed Mount Whitney, the stone structure still there at the top was erected for Campbell's expedition and paid for by the Smithsonian. (2) Campbell went on to become the tenth President of the University of California, which is why there are buildings named "Campbell Hall" on some UC campuses.

(Whitney was the highest point in the then 46 US states, and remains the highest point in the United States outside Alaska), he first collected a spectrum of sunlight reflected by the Moon, which he knew to lack an atmosphere, as a means of correcting for any absorbance caused by the water in our atmosphere. Then turning his instruments toward Mars, he found that its water "absorbance bands" were the same despite the fact that the light had passed through the Martian atmosphere twice: once on the way in from the Sun and once on the way back out to the observers on the mountain. From this it was clear that the water content of the Martian atmosphere was far less than even that in the thin, dry atmosphere above his perch, casting serious doubt on Lowell's vision of a planet covered not only with life, but with a globe-spanning civilization.

Mars is our second-closest neighbor in space, though, and has weather and seasons like the Earth's, so surely it might be an abode for life? Although spectroscopic studies following on Campbell's pioneering work put the carbon dioxide content of the Martian atmosphere at about 1% of the total atmospheric pressure on the surface of Earth, most of the Earth's atmosphere is nitrogen, which is invisible in the infrared and thus not easily detected. Might Mars have a thick—if largely invisible from Earth—nitrogen-rich atmosphere that contains only a trace of carbon dioxide? Even early users of telescopes had noted clouds, suggesting weather, and the annual waxing and waning of the polar ice caps, hinting at Earth-like conditions. And, of course, there is that seasonal wave of darkening. Even as late as 1958, the American astronomer William Sinton (1925–2004) reported the spectroscopic detection of organic material on Mars, quite possibly as chlorophyll. Sadly, however, all this proved untrue. Sinton's spectroscopic "bands," for example, were soon to be argued away as experimental artifacts caused by similar features in the spectrum of deuterated water (water with one hydrogen atom replaced by a deuterium) in the Earth's atmosphere. And that seasonal wave of darkening? Improved telescopic observations throughout the 1950s proved that much, if not all, of the proposed color change was due to seasonal changes in the distribution of dust associated with massive, sometimes globe-

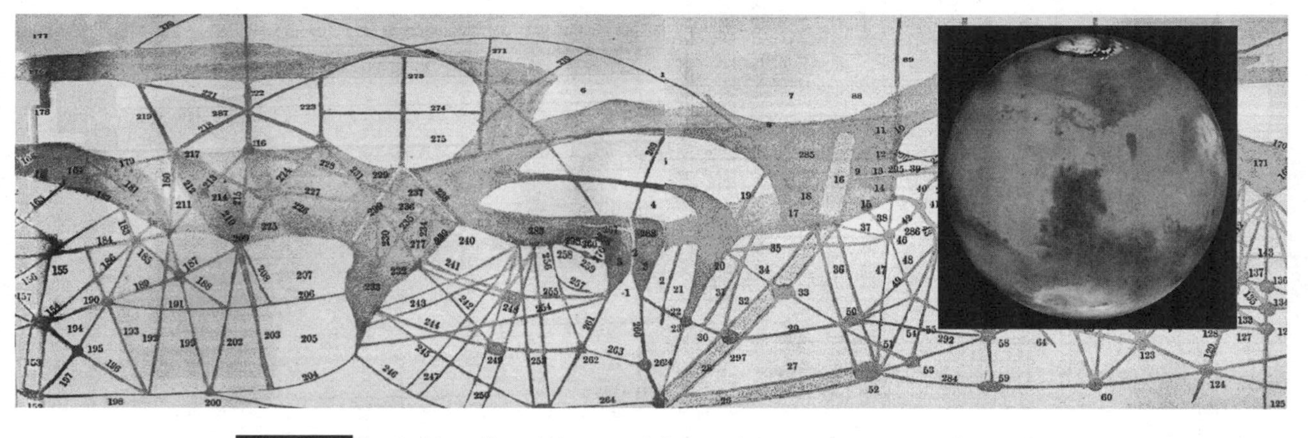

Figure 9.1 Percival Lowell used his state-of-the-art observatory to map out the canals of Mars in exquisite, hand-drawn detail. As shown in the inset, though, Lowell's linear canals sadly did not survive the advent of astrophotography. (Courtesy of NASA/STScI)

Table 9.1
Past successful robotic missions to Mars

Mission	Flyby dates or dates active in orbit or on surface	Notes
Mariner 4 (flyby)	July 15, 1965	First Mars flyby; 22 images of southern hemisphere reveal cold, thin atmosphere
Mariner 6 (flyby) *Mariner 7* (flyby)	July 31, 1969 August 5, 1969	Confirm thin atmosphere, low temperatures, and crater-saturated southern landscape
Mariner 9 (orbiter)	November 13, 1971 to October 27, 1972	Medium-resolution map of entire planet indicates northern hemisphere more geologically active, shows signs of fluvial (fluid-carved) features
Viking 1 (orbiter) *Viking 2* (orbiter)	June 19, 1976 to August 17, 1980 August 7, 1976 to July 25, 1978	Higher-resolution images confirm abundance of fluvial features
Viking 1 (lander) *Viking 2* (lander)	July 20, 1976 to November 13, 1982 September 3, 1976 to April 11, 1980	First successful Mars landers; carried cameras, meteorological instruments, and suite of life-detection technologies
Mars Pathfinder (lander and rover)	July 4, 1997 to September 27, 1997	Technology demonstrator
Mars Global Surveyor (orbiter)	September 11, 1997 to November 5, 2006	Technologically advanced orbiter carrying an altimeter (to measure topography), an infrared spectrometer, and cameras with 1.5 m resolution
Mars Odyssey (orbiter)	October 24, 2001 to present	Carrying an imaging infrared spectrometer and a γ-ray and neutron spectrometer
Mars Express (orbiter)	December 25, 2003 to present	Carrying a 2 m resolution camera, a high-resolution infrared spectrometer, and a subsurface-sounding radar to search for groundwater

(continued)

spanning dust storms. There are neither green fields nor golden autumnal forests on the Red Planet.

The final nail in the coffin of Mars's romantic, life-filled image came with the first close-up investigations of the planet. In 1962, after one American and several Soviet attempts had failed, the US spacecraft *Mariner 4* flew 9,800 km above the planet's southern hemisphere (table 9.1 provides a summary of humanity's successful robotic missions to Mars). Along the way, it snapped 22 black-and-white images, 200 × 200

Mission	Flyby dates or dates active in orbit or on surface	Notes
Spirit (rover)	January 4, 2004 to March 22, 2010	Explored Gusev Crater for almost five Earth-years before becoming stuck in soft sand and continuing operations at that fixed location
Opportunity (rover)	January 25, 2004 to June 18, 2018	Explored hematite deposits of Meridiani Planum
Mars Reconnaissance Orbiter	March 10, 2006 to present	Carrying a 0.3 m (!) resolution camera, a high-resolution imaging infrared spectrometer, and a shallow-subsurface radar
Phoenix (lander)	May 25, 2008 to November 2, 2008	Stationary lander; explored the soil chemistry at far northern latitudes
Curiosity (rover)	August 6, 2012 to present	A nuclear-powered, 900 kg rover with spectroscopic and chromatographic equipment for chemical analyses and sniffing out potential habitats; findings suggest conditions at Gale Crater were once wet; discovered most significant organic chemistry yet found on Mars
Mars Orbiter Mission	November 5, 2013 to present	Indian Space Agency's first interplanetary spacecraft; technology demonstrator
MAVEN (orbiter)	September 24, 2014 to present	Mars Atmosphere and Volatile Evolution mission's goal is to determine how the Martian atmosphere and hydrosphere, presumed to have once been substantial, were lost over time
ExoMars Trace Gas Orbiter	March 14, 2016 to present	This European Space Agency mission includes high-resolution spectrometers aimed at assaying the trace gas components of the Martian atmosphere
InSight (lander)	November 26, 2018 to present	Landed in the Elysium Planitia region; investigating Mars geophysics

pixels, each of which required nearly 10 hours to send home via the spacecraft's eight bits per second radio link. With this, centuries of speculation were supplanted by the first, close-up glimpses of the surface of Mars.

Sadly, far from the dynamic, life-filled world of lore, the pictures revealed an ancient, heavily cratered landscape reminiscent of the Moon (fig. 9.2, left). The Mars of *Mariner 4* appeared geologically dead, without even the dynamism to erode away the many craters dotting its surface. Equally troubling, as the spacecraft passed behind Mars (from the viewpoint of Earth), refraction of its radio signal through the Martian atmosphere provided measures of the temperature and total pressure of the Martian atmosphere. The temperature, it found, was a chilly −100°C, far colder than previous, Earth-based estimates. Worse, the total atmospheric pressure, less than 1% of Earth's, corresponded closely to the carbon dioxide pressure determined spectroscopically via telescopic observations—no thick nitrogen atmosphere for the Red Planet. Disappointingly, this pressure is so low that water *cannot* exist on the surface of Mars in the liquid state: under present Martian conditions, water sublimes directly from ice to vapor without passing through a stable, liquid phase. Two more flybys, by *Mariner 6* and *7* in 1969, returned another 70 close-up pictures of ancient, heavily cratered terrain across the southern hemisphere and further strengthened the case for a thin, cold, predominantly carbon dioxide atmosphere. So much for Lowell's heroic civilization.

But perhaps not all is lost. The generally pessimistic funk that Mars research fell into after the flyby missions lifted a little bit in 1971, when, after a launch failure left *Mariner 8* resting on the bottom of the Atlantic Ocean, *Mariner 9* became the first spacecraft to orbit another planet. Upon the approach, the Martian disk appeared strangely featureless; the planet was engulfed in a dust storm of global proportions. When the dust cleared a few weeks later, however, the outlook was brighter than expected. *Mariner 9* found evidence of a far more active planet than dreamed of after the earlier flybys; only the *southern* hemisphere of Mars is old and cratered (even now, a half century later, the origins of this massive, global asymmetry remain an open question). Much of the remainder of the planet shows abundant signs of past geological activity, including extinct, or at least dormant, volcanoes up to twice as high as Mount Everest and a canyon system that would dwarf the Earth's Grand Canyon.

Of particular interest to us, some of the exciting geology that *Mariner 9* turned up seems to have been created by a fluid, perhaps water (fig. 9.2, right). This evidence included what looked like vast sedimentary deposits in the polar regions and extensive networks of valleys that looked very much as if they were formed by rivers. But the orbiter

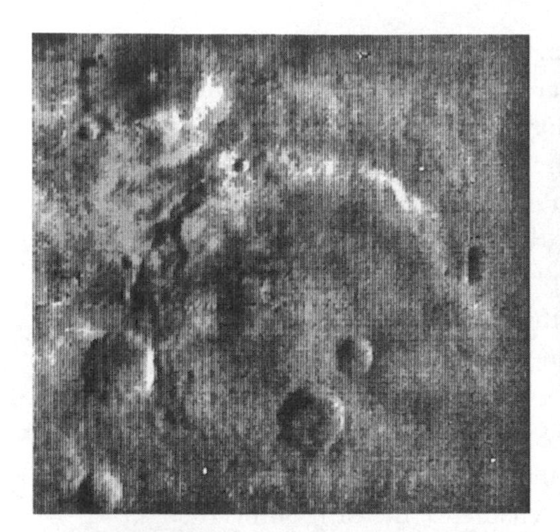

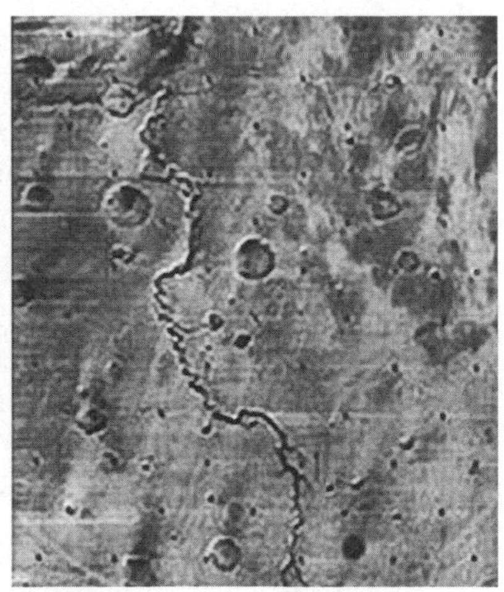

Figure 9.2 *Mariner 4*, the first successful Mars flyby spacecraft, returned pictures of a bleak, geologically dead landscape reminiscent of the Moon (left). The more global picture provided from orbit by *Mariner 9* is quite different (right). Mars's northern hemisphere is covered with the remnants of once active geology, including many seemingly water-carved features such as the putative river channel shown here. For scale, the crater in the image on the left is approximately 150 km across, and the image on the right covers approximately 400 km of the 600 km long Nirgal Vallis. (Photos courtesy of NASA/JPL)

also made detailed studies of the Martian weather and confirmed that, although the Martian tropics can top out at a downright pleasant 24°C during the very warmest summer days, the temperature on Mars is typically quite far below freezing, and the planet's atmospheric pressure is too low to allow water to exist as a liquid. The post–*Mariner 9* view was of a now frozen and exceptionally dry planet with only ancient relics of a presumably warmer, wetter past.

Given the *Mariner 9* views, optimism about Mars ran fairly high during the 1970s when follow-on studies from the *Viking 1* and *2* orbiters (we discuss their associated landers in the next chapter) carried much better cameras into Martian orbit. With the higher resolution provided by these cameras, scientists back on Earth identified a good many apparently fluvial (fluid-formed) features, including presumably water-carved, teardrop-shaped "islands" within massive outflow channels that themselves appeared to have been carved by intensive, if brief, floods, perhaps exceeding by a factor of 10,000 the flow rate of the mighty Mississippi River. As further evidence of these massive floods, the orbiters spied vast areas of chaotic terrains reminiscent of the scablands of the Pacific Northwest, which were generated at the end of the last ice age when an ice dam broke and a volume of water the size of Lake Superior flooded large parts of what are now Idaho and Washington states. There were even claims that the *Viking* orbiters had spied fossil remnants of the shorelines of some ancient Martian ocean. Based largely on these images, the warm and wet early period of Martian history seemed assured.

Unfortunately for those of us who would like to find Mars inhabited, the tide has once again largely turned against the warm, wet early Mars hypothesis. One of the most fundamental problems with this idea is that, as we discussed in chapter 3, the early Sun is thought to have been some 20% dimmer than it is today. And if sunlight cannot now heat Mars above the freezing point of water, how could it have done so billions of years ago? A thicker carbon dioxide atmosphere could have bumped the temperature up in the past, only to be lost later. Mars lacks a magnetic field, and thus the solar wind collides with the Martian atmosphere, providing a mechanism for such loss. The *MAVEN* spacecraft, in orbit since 2014, has measured the resulting erosion of the atmosphere and found it amounts to a ton a day, rising to 10 tons a day during solar storms. Even still, though, going through the numbers in detail, it has proven hard to come up with a physically plausible scenario by which Mars remained largely above freezing during the earliest days of the Solar System only to fall later into its present-day cold funk.

Further evidence against the warm, wet early Mars scenario was provided by the high-resolution infrared spectrometer on board another orbiter, *Mars Odyssey*, launched in 2001 and thus named after the film *2001, A Space Odyssey*. Mapping the mineralogy of the entire planet, the spectrometer failed to find any carbonates. Given that carbon dioxide forms these (e.g., limestone) whenever liquid water is present on Earth, their absence suggests that Mars has never hosted significant bodies of surface water. Following on *Odyssey*, the *Mars Reconnaissance Orbiter* arrived in orbit around the Red Planet in 2006 and continues to map the planet with the highest-resolution cameras ever sent into deep space (its surface resolution is 30 cm) and a high-resolution imaging spectrometer that should be wonderful for spotting the signatures of—to speculate a bit—hydrated minerals surrounding some ancient hydrothermal vent. Despite many years of searching, though, it has not turned up any clear signs of water-derived minerals other than some clays and a few small areas containing hydrated silica and hydrated sulfates. None of these, however, is thought to require long-term liquid water to form, much less oceans of it. It now seems generally accepted among planetary scientists that Mars never had a long, warm, wet spell.

If Mars never harbored long-term liquid water, then where did its fluvial erosion features come from? There's been no end of speculation about this. The suggested possibilities range from wind erosion, to temporary running water—either underneath thick, insulating ice caps or after a meteor strike vaporized some ice and cranked up the atmospheric density—to sapping (erosion from within) caused by the eruption of geothermally heated groundwater. Mars does, after all, have enormous and relatively fresh-looking volcanoes, so a liquid, subsurface aquifer is definitely within the realm of possibility, even if Mars never had a consistently warm and wet period in its history.

This vision of intermittent and brief bouts of liquid water on Mars has gotten some boosts in recent decades. In early 2004, abiding by its Mars creed "follow the water," NASA landed solar-powered rovers at two seemingly water-modified landscapes in Mars's northern hemisphere. The first, called *Spirit*, touched down (bounced to a stop, actually, on airbags) on January 4 in the Gusev Crater. Located at the end of the 10 × 900 km Ma'adim Vallis (valley), Gusev appeared from orbit to have once been a lake. After extensive exploration, however, *Spirit* found little but the most indirect evidence that the crater was ever filled with water. Some three weeks later, *Spirit*'s sister craft, *Opportunity*, bounced to a stop on Meridiani Planum,* a smooth, flat plain the size of the state of Colorado that had been selected because spectroscopic investiga-

*Both rovers exceeded their planned missions spectacularly. *Spirit* ran into deep sand in May 2009 and failed to get out again. In January 2010, mission control redefined *Spirit* as a stationary instrument before losing contact later that year. *Opportunity* remained active until June 2018, exceeding its projected three-month lifetime *by more than 14 years.*

Figure 9.3 In 2003, the *Opportunity* rover bounced down onto the plains of Meridiani and struck a cosmic hole-in-one by landing in a small crater dubbed Eagle. Detailed imaging and spectroscopic investigations of the bedrock suggest that it is sedimentary and was laid down from liquid water—briny, highly acidic water, but liquid water nonetheless. (Courtesy of NASA/JPL)

tions from orbit indicated it was decorated with the mineral hematite. Hematite, a type of iron oxide, can be formed by several mechanisms, but on Earth it is most commonly deposited in aqueous environments. Cushioned by its airbags, *Opportunity* bounced a dozen times on the Martian surface before finally coming to rest in a small crater. When the rover shook off its landing and turned on its cameras, it found itself staring at thick, layered beds (fig. 9.3). The *Opportunity* rover, the fifth successful Mars lander, was the first to find bedrock.

On Earth, the majority of layered rock is sedimentary, laid down over successive seasons or successive floods by liquid water. But successive volcanic eruptions (either as lava flows or as successive layers of ash) can also form layered rock, as can shifting patterns of windblown dust. So, which was it at Meridiani? The bedrock in *Opportunity*'s crater showed clear evidence of cross-bedding, or layers formed at angles, which often formed in stream beds on Earth due to turbulent flow. Some of the rock surfaces also showed clear signs of polygonal cracks, reminiscent of the hexagons that sometimes form in drying mud. The layers were also filled with the type of small, round rocks that littered the ground at Meridiani. Termed "blueberries" after the fruit they seemed to resemble in shape and size, if not color (somehow "grayberries" just sounds wrong), the small spheres littering the crater are composed of hematite—again, most likely formed in situ in the rocks by the action of water.

Perhaps even more revealing than the gross structures of the strata are their chemical compositions. The rover's α-particle and x-ray spectrometer (APXS), which can identify the elemental composition of rocks (see sidebar 9.1), indicated that the rocks at Meridiani contain large amounts of magnesium, calcium, and iron sulfates along with traces of chlorine and bromine. On Earth, such salts form preferentially by aqueous deposition, and on Mars, are a likely sign that liquid water was once present. Similarly, *Opportunity*'s Mossbauer spectrometer found evidence that some of the iron at Meridiani is tied up in the mineral jarosite, a hydrated form of potassium iron sulfate that, on Earth, is invariably formed by the aqueous leaching of iron minerals under acidic conditions. In short, it looks as if we have very firm evidence that at least parts of Mars were wet sometime in the distant past (albeit with rather acidic brine) but that Mars then took a very different environmental turn.

Following up on *Spirit* and *Opportunity*, in 2011 NASA launched a far more sophisticated rover toward Mars, the 900 kg *Curiosity*. Landing in Mars's Gale Crater in August 2012, after a journey of more

Sidebar 9.1
Sniffing out Habitats

So you're going to send a robot a hundred million miles away from home to investigate Mars for signs of water. What do you pack? In part, that depends on what type of mission you are planning. Orbiters and landers face different opportunities and demand different, if often complementary, technologies.

Because orbiters remain at a distance from the object to be investigated, they need to be equipped with *remote sensing devices*. A good camera, of course, is a mandatory accoutrement for any tourist, and the best cameras now in Mars orbit achieve submeter resolution. But in addition to pictures, you want to be able to measure the spectral properties of the atmosphere and surface. That is, you want to see which wavelengths of visible, UV, and infrared light the surface absorbs. The UV wavelengths allow for the identification of atmospheric components, whereas at infrared wavelengths, both atmospheric and surface components exhibit characteristic "fingerprints." The infrared spectrometer also provides a means of determining the temperature of the surface (calculated from the known temperature-emission relationship of a black body) and, indirectly, its structure: loose soils cool rapidly after nightfall, whereas rocks hold their heat, cooling more slowly.

The modern orbiter often also sports γ-ray and neutron spectrometers. The γ-ray spectrometer can detect the relative abundances of a few radioactive elements from the γ-rays they emit when they decay. It can detect even more elements from the cosmic ray–induced emission of γ-rays. When cosmic rays (mostly high-energy protons) strike the surface of an airless or near airless body (the Earth's atmosphere, though, would screen out most of the incoming cosmic rays), they excite atomic nuclei in the soil. The excited nuclei quickly lose their extra energy by emitting γ-rays characteristic of their element type. The neutron spectrometer runs along similar lines. Some of the cosmic rays knock neutrons out of atomic nuclei in the soil. If these high-energy neutrons strike a heavy nucleus, they bounce off without losing much speed (imagine a ping-pong ball bouncing off a bowling ball). If, instead, they strike a low-mass nucleus like hydrogen, they lose much more of their kinetic energy (imagine a ping-pong ball bouncing off a second ping-pong ball). By measuring the kinetic energy of neutrons that emerge from the surface, the neutron spectrometer indirectly detects the amount of hydrogen in the soil; if the measured neutrons are moving rapidly, the soil contains only heavy nuclei, whereas if the neutrons are moving slowly, the soil contains hydrogen. Together, the neutron and γ-ray spectrometers can map out the abundance of a few dozen elements in the first few micrometers to tens of centimeters of soil with 10 to 100 km spatial resolution.

Unlike cameras and spectrometers, which are effectively passive receivers, some remote sensing devices are active. The *Mars Global Surveyor*, for example, carried a laser altimeter that bounced an infrared laser pulse off the Martian surface in order to map the planet's topography in stunning detail. And the *Mars Express* and *Mars Reconnaissance* orbiters (both survivors from the noughties, but still active as of late 2020) carry sounding radar to prospect for subsurface water.

In contrast to orbiters, rovers and stationary landers get up close and personal with the rocks and thus require a more "hands-on" suite of instruments. Imaging, of course, is still critical (if for no other reason than the folks back home who are paying for the trip want some nice color pictures). And for more detailed mineralogical analysis, imaging spectrometers (or cameras with lots of color filters) are essential. But we also want to monitor the mineral content of rocks more directly. Four approaches have been adopted in recent Mars surface missions.

The *Spirit* and *Opportunity* rovers each brought to Mars two analytical tools: an α-particle and x-ray spectrometer (APXS) and a Mossbauer spectrometer. The APXS serves to determine the elemental composition of rocks and soils, complementing the spectroscopic analysis performed by the rovers' imaging systems. The APXS works by exposing samples to energetic α-particles and x-rays from a small amount of radioactive curium, then measuring the energy spectra of the α-particles and x-rays scattered back from the sample into the detector. The energy of the scattered α-particles is sensitive to the weight of the atomic nuclei they strike, with lighter nuclei producing larger changes in the energy of the α-particles (as in the ping-pong ball analogy above). Thus, the α-particle mode on an APXS is particularly sensitive to lighter elements such as carbon and oxygen. The x-ray mode works through the excitation of x-ray fluorescence in a sample and is particularly sensitive to important, mineral-forming elements such as magnesium, aluminum, silicon, potassium, and calcium. In contrast to the elemental analysis performed by the APXS, the Mossbauer spectrometer determines the molecular makeup of minerals containing only a single element: iron. At the heart of this device is a small chunk of radioactive cobalt-57. When this decays, the resulting iron-57 (^{57}Fe) is left in a high-energy state, which in turn decays via the emission of a γ-ray photon that can be absorbed by ^{57}Fe nuclei in the sample. The precise wavelength of the absorbed γ-radiation is sensitive to the chemical compound in which the iron resides, thus allowing Mossbauer spectroscopy to characterize the chemical nature of iron minerals. Both the APXS and Mossbauer spectrometers require physical contact with the sample and long integration times (typically overnight), and thus both are well suited to rover missions that can move the spectrometers from rock to rock.

Being much larger than its earlier cousins, the *Curiosity* rover's instrument suite is likewise much more comprehensive and includes, in addition to an APXS, a number of other analytical tools. In addition to the four cameras in the three imaging systems, *Curiosity* carries a long lens camera, the ChemCam, which can fire a laser beam at surfaces up to 7 m away, vaporizing exposed material and spectroscopically characterizing the results. In doing so, the ChemCam can rapidly analyze the elemental composition of rocks, detect water both as ice and within the crystal structure of minerals, assess the degree and depth of weathering on exposed rock surfaces, and provide help in selecting samples for other analytical tools. In addition to ChemCam's remote analysis, *Curiosity* is also equipped for "hands-on" analysis. Like any terrestrial geologist on a field trip, it carries a drill to access deeper layers of rocks and a scoop to collect samples of dust and loose soil. It can sieve the collected materials and deliver the resulting fine-powder samples to the chemistry and mineralogy instrument (CheMin), which uses x-ray diffraction and x-ray fluorescence to analyze the crystal structures and elemental composition. Finally, *Curiosity* also carries a Russian-built dynamic albedo of neutrons (DAN) tool that, by analogy to the orbital neutron spectrometers described above, samples to a depth of 1 to 2 m below the spacecraft to monitor the presence of hydrogen.

The largest part of *Curiosity*'s scientific payload, the sample analysis at Mars (SAM) suite of instruments, includes a mass spectrometer, gas chromatograph, and tunable laser spectrometer. The SAM suite is the key instrument on *Curiosity* for detecting carbon compounds and compounds of other light elements associated with life, including hydrogen, oxygen, and nitrogen, primarily via evolved gas analysis, which, as described in the main text, uses heat to drive volatiles out of on-board samples for mass spectrometric analysis.

than half a billion kilometers, it touched down within 2.4 km of the center of its intended target zone. Equipped with the most sophisticated chemical analysis equipment yet sent to another planet, *Curiosity*, whose mission continues as of late 2020, is pursuing two primary goals: determining the nature and inventory of any remaining carbon on Mars, and mapping out Mars's climatic history over the last few billion years. To achieve this, the rover was targeted to the base of Mount Sharp, an isolated peak reaching 5 km above the floor of the 150 km wide crater. Images from orbit indicated that Mount Sharp is composed of layers of exposed sediment containing, geologists hoped, a multibillion-year record of sediments washed into the crater by what appeared to be one of Mars's many ancient river channels. On its route up Mount Sharp, the rover explored clays and imaged putative, pebble-filled former streambeds (fig. 9.4). But any water in the Gale

Figure 9.4 In 2011, the *Curiosity* rover landed in Gale Crater, at the foot of the 5.5 km high Mount Sharp (top). Composed of sedimentary layers, Mount Sharp is believed by geologists to hold a multibillion-year record of the climate of Mars (bottom left). Shown is a close-up image of seemingly water-polished stones, providing putative evidence for water flowing down the mountain's slopes (bottom right). (Images courtesy of NASA/JPL)

Crater was long in the past; *Curiosity*'s Russian-built dynamic albedo of neutrons (DAN) instrument (see sidebar 9.1), which uses neutrons to probe for hydrogen within the first two-thirds of a meter of soil or rock below the rover, has identified only a single, small area along *Curiosity*'s multiyear trek to date in which the soil water content even approached 4%.

Where, then, is the water on Mars now? Some of it was lost: *Curiosity*'s mass spectrometer measured Mars's deuterium to hydrogen ratio and found it to be five times higher than the Earth's, indicating significant Jeans escape of hydrogen from the planet. Some of what remains is tied up in the polar caps, which consist of a seasonal mix of water ice and frozen carbon dioxide (Mars is so cold that a significant fraction of its atmosphere condenses out at its poles each winter). But the polar ice caps do not account for much water: the larger north polar cap is estimated to contain 1,200,000 km^3 of water ice, which is less than half the size of the Greenland ice cap and a mere 4% of the volume of the Antarctic ice cap. To search for possible reservoirs of water outside the polar caps, the *Mars Odyssey* orbiter carried a neutron spectrometer, an instrument that maps out the distribution of hydrogen within the first few tens of centimeters of the Martian surface. And what did it find? It found high concentrations of hydrogen in the soil at high latitudes and progressively less toward the equator. In fact, the soil in wide swaths of the northern and southern latitudes contains so much hydrogen that, if (as seemed likely) it is water ice, the soil must be at least 50% water by weight (fig. 9.5).

Confirmation that the hydrogen in the Martian soil is, indeed, in the form of water ice was provided by NASA's *Phoenix* lander. This relatively inexpensive craft assembled from ideas and spare components from previous failed missions (thus its name) touched down on May 25, 2008, and operated for 157 Martian days before the onset of the harsh Martian winter inevitably killed it. *Phoenix* landed in the Vastitas Borealis formation, a large, low-lying plain near the Martian arctic circle at 68° north; the Terrestrial equivalent would be landing in Barrow, Alaska, or Tromsø, Norway. Orbital images indicated a smooth landscape dotted with subtle troughs tracing out polygon shapes as is seen in the Earth's Arctic, where the alternating freezing

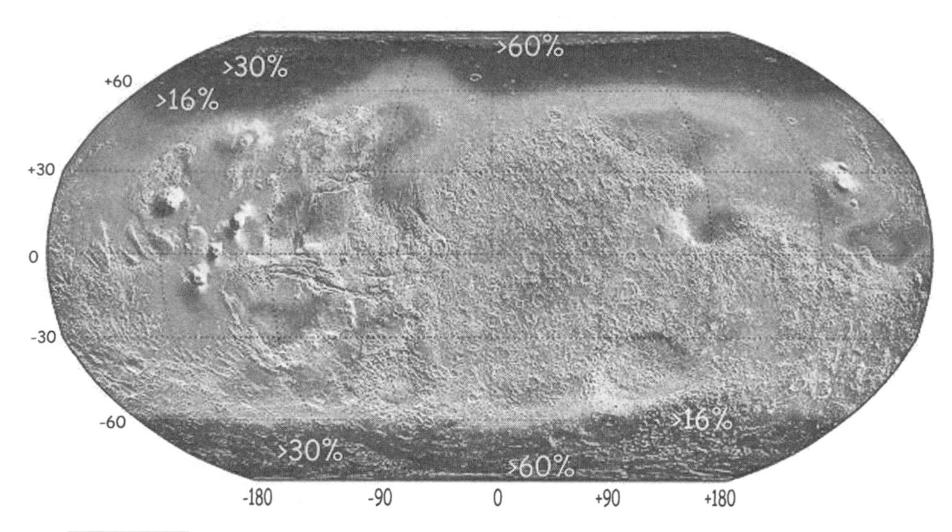

Figure 9.5 A map of the abundance of hydrogen in the first few tens of centimeters of the Martian surface as determined by neutron spectroscopy from the orbiting *Mars Odyssey* spacecraft. The soil across wide swaths of Mars's northern and southern hemispheres seems to consist of at least 50% water by weight. (Courtesy of NASA/JPL)

and thawing of the soil causes it to break into such shapes. And, of course, thermal neutron data from on orbit indicated that the first half meter of soil was dense with hydrogen. The first sign that *Phoenix* had found its prey was entirely unplanned: upon touchdown, the exhaust from *Phoenix*'s landing rockets had blown away some of the soil beneath the craft, exposing a large, white patch that resembled ice (fig. 9.6). Subsequently, *Phoenix* used its robotic arm to dig down a few tens of centimeters into the soil, where it found the long-sought ice, confirmed it to be water via its onboard mass spectrometer, and then watched it sublimate back into the atmosphere after having been exposed.

Exciting as these discoveries are, permafrost is not the stuff on which life is founded. But if there is water under the surface of Mars as ice, is there some as liquid as well? When the European Space Agency's *Mars Express*, in orbit since late 2003 and still going strong, mapped the upper reaches of several of the planet's larger volcanoes, it found them so free of craters that the terrain must be less than a few million years old. That's a long time for us, but it's less than 0.1% of

Figure 9.6 The landing rockets of the *Phoenix* lander blew away the thin Martian soil, leaving behind a white mass that that appeared to be ice. The craft's later chemical investigation of the nearby soil confirmed the presence of permafrost. (Courtesy of NASA/JPL/UA)

the age of the planet, suggesting that Mars is probably still geologically active today (after all, what are the chances that Mars was active for more than 99.9% of its existence and then its geological fires, coincidentally, died immediately before we humans started to poke around?). Consistent with this, the seismometer on NASA's *InSight* (*Interior Exploration Using Seismic Investigations, Geodesy and Heat Transport*) mission, which landed on Mars's Elysium Planitia in November 2018 and continues to operate as of this writing, has confirmed that Mars is seismically active, though far less so than the Earth. And if Mars is active, could there be liquid water, and perhaps even life, under its crust?

Putative (albeit still disputed) evidence for the existence of subsurface water on Mars today came in 2000, when NASA's orbiting *Mars Global Surveyor*, which remained active until November 2006, spotted features that point to ongoing fluvial activity. Surveying selected swaths of Mars at about 1 m resolution, the spacecraft found what appear to be rather recent (some much less than a million years old) erosional gullies emerging from cliffs throughout the higher lat-

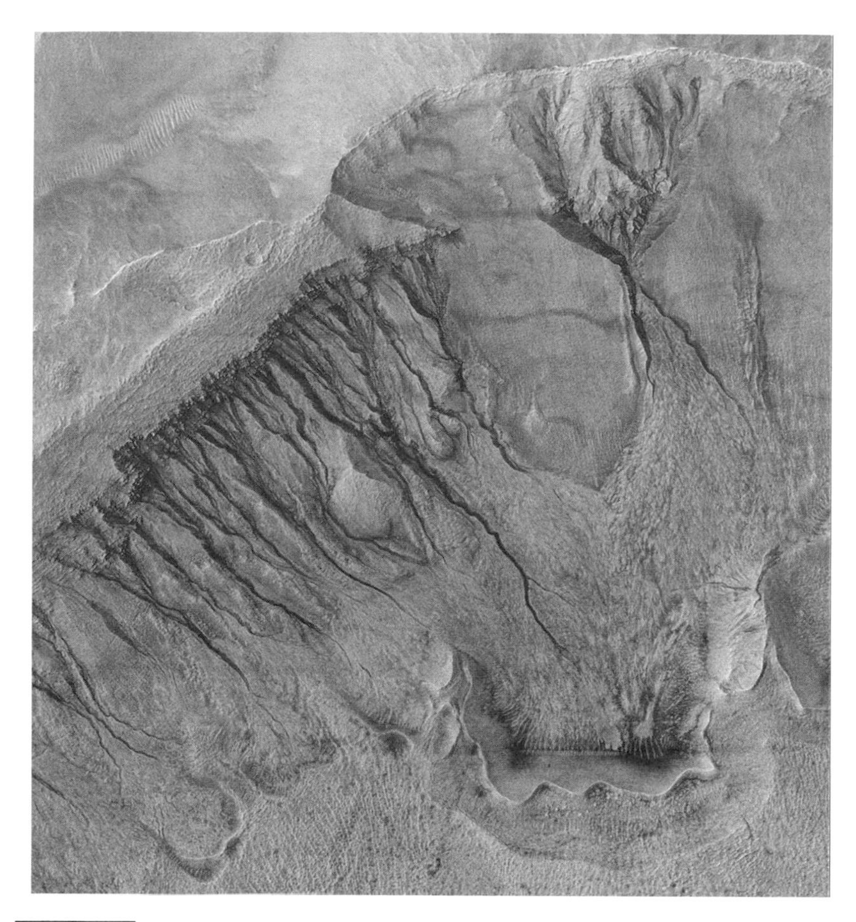

Figure 9.7 In 2000, the *Mars Global Surveyor* orbiter photographed features such as these approximately 100 m long streaks that have widely been interpreted as water-carved gullies. Whatever they are, the features are modern; a handful have been seen to change from year to year. Whether are carved by emerging liquid groundwater, however, remains contested. (Courtesy of NASA/JPL/MSSS)

itudes (fig. 9.7). And while none of these gullies, now called "recurrent slope linea," have been observed precisely in the act of forming, follow-on observations of several hundred of them have seen a half dozen that changed from year to year; therefore, whatever formed them is still active on the surface of Mars today. On Earth, such gullies are quite common and result from the spring runoff of snow that has accumulated on a cliff face. Because liquid water cannot exist on the surface of Mars for very long (seconds to minutes, depending on the volume and surface area), a different mechanism is probably occurring there. Perhaps the most popular of several current theories on the origins of these gullies is that they are formed when ice plugs on springs rupture, briefly liberating a torrent of groundwater (kept liquid by geothermal heat) before the plug freezes over again and the flow is stopped. This potential juxtaposition of groundwater and geothermal energy suggests that, although the surface of Mars is cold and dry today, the planet may still contain viable habitats deep within its crust. Others have pointed out, however, that "Mars is not Earth." Who knows whether, on its cold, exceptionally dry surface, some processes such as flowing dust or rapidly evaporating dry ice are creating these gullies? Consistent with the dust-flow hypothesis, recurrent slope linea are only seen on slopes greater than 27°, which is more or less the angle of repose for dust under the cold, dry, low-gravity conditions found on Mars.

Some of the evidence against the gullies being formed by water was provided by instruments on the *Mars Express* and *Mars Reconnaissance* orbiters, both of which carry ground-penetrating radar aimed at taking stock of underground ice layers or water tables within the first kilometer of the planet's crust. The dry Martian soil should be relatively transparent to the radio wavelengths employed by these two instruments, but discontinuities, such as hitting a water table, should cause some of the energy to reflect back to the spacecraft. Early results from these instruments confirmed that the permanent (i.e., summer) polar caps are pure water ice up to 3.7 km deep and that if they were to melt and spread evenly over the entire planet, the resulting "ocean" would be 11 m deep (for comparison, the average depth of the Earth's oceans is more than 3,600 m). The orbiters' instruments have also found evidence for remnant glaciers nearer the equator, where they are protected from sublimation by thick overlayers of dust and debris. The case for Mars as the most habitable place in the Solar System outside Earth received another boost in July 2018, when researchers using the *Mars Express* orbiter's radar instrument reported observations which they interpreted as signs of a 20 km wide lake trapped 1.5 km below the surface of the ice and dust of the polar layered terrains at 81° south. The temperature at the base of the polar deposits is estimated to be −90°C, and thus, if the interpreted lake really exists, it must be filled with brine.

So what's next on the Mars exploration docket? As we write this in mid-2020, three new rovers and two new orbiters are in the works (table 9.2). First up, with a landing date of February 18, 2021, is the *Perseverance* rover. This near twin of the currently operating *Curiosity* will explore the 45 km wide Jezero Crater, which, based on the presence of what appear to be river valleys leading into the crater, a large fan delta, and significant deposits of clays, is thought to have been a lake some 3.5 to 3.9 billion years ago (fig. 9.8). In addition to the instrument suite of its sister craft, *Perseverance* will be carrying a ground-penetrating radar to image subsurface structures and, if present, underground ice or brine down to depths of 10 m. It will also include an instrument named SHERLOC (scanning habitable environments with Raman and luminescence for organics and chemicals), which will employ spectroscopy to assess the presence of life-relevant elements (carbon, hydrogen, nitrogen, etc.), to identify potential or-

Table 9.2
Currently scheduled robotic missions to Mars

Mission	Arrival	Notes
Hope (orbiter)	February 9, 2021	A climate-focused mission lead by the United Arab Emirates
Mars Global Remote Sensing (orbiter and rover)	February 10, 2021	The stated goals of this Chinese Space Agency mission are to search for evidence of current and past life and to assess the Martian environment
Perseverance (rover)	February 18, 2021	Due to explore Jezero Crater, a landscape shaped by water, using a *Curiosity*-like instrument suite; also expected to collect and cache samples for a possible future sample-return mission
Rosalind Franklin (rover)	Mid-2023	A joint European Space Agency / Russian Space Agency lander and rover, with the latter carrying a *Curiosity*-like instrument suite to explore Oxia Planum, a valley system site with extensive deposits of clay

Figure 9.8 On February 18, 2021, NASA's *Perseverance* rover is scheduled to set down in Mars's Jezero Crater. A putative dry river channel leading into this 45 km wide crater deposited what appears to be an alluvial fan some 10 km across.

ganic biosignatures, and to provide detailed mineral analysis. In addition to exploring Jezero, *Perseverance* will also cache interesting samples onboard in the hopes that, someday, a future mission can return them to Earth for more detailed study, though the date of such a sample return mission is ill-defined and has had a habit of moving inexorably into the more distant future. And while *Perseverance* is the most sophisticated rover slated for the next batch of Mars missions, it will not be alone. In April 2021, the Chinese Space Agency is planning on landing their *Huoxing-1* rover at Utopia Planitia (near where *Viking 2* touched down almost 45 years earlier), where it will "perform chemical analyses on the soil, and look for biomolecules and biosignatures." In parallel, the United Arab Emirates has launched the *Hope* mission, a demonstration spacecraft expected to reach Mars orbit in February 2021, when it will begin to study the Martian weather. Finally, in 2022, the European Space Agency intends to launch *Rosalind Franklin*, a solar-powered rover that will also employ spectroscopy and chemical analysis to better understand organics on Mars. No landing site has yet been reported.

The Moons of Jupiter

Out beyond Mars we find the gas giants Jupiter (see fig. 3.1), Saturn, Neptune, and Uranus. The physics of these planets effectively rules out any processes that we would describe as life. Jupiter and Saturn lack solid surfaces (because of the massive bulk of these planets, even their metal-and-rock cores are thought to be liquid; they haven't cooled enough since accretion to solidify), and the solid surfaces of Uranus and Neptune, if there are such surfaces, are deep below hot, highly convective seas of supercritical water and ammonia. Life in the atmospheres of the gas giants is probably not possible (for reasons outlined in chapter 1), and life beneath the hot, turbulent seas of Uranus and Neptune is probably equally precluded by the instability of the environment. There are other environments out among the gas giants, though, that might be more conducive to life. Each of the four planets has a retinue of dozens of moons, and some of the larger ones may be among the more promising prospects for extraterrestrial life in our Solar System.

The four largest companions of Jupiter were discovered in 1610 by the Italian astronomer Galileo Galilei (1564–1642), who made himself unpopular with the church authorities by pointing out—correctly—that the existence of moons orbiting Jupiter ran counter to the Ptolemaic worldview that everything revolves around the Earth. These four "Galilean satellites," Io, Europa, Ganymede, and Callisto, remained little more than points of light for the next four centuries. Indeed, even our first robotic emissaries to the outer Solar System, *Pioneer 10* and *11*, which flew by Jupiter in 1973 and 1974, failed to return much data on the satellites. (See table 9.3 for a time line of missions to the outer Solar System.) It was not until *Voyager 1* and *2* passed Jupiter in 1979 that we began to see these moons as worlds unto themselves, and much of what we know is based on measurements made by the *Galileo* mission, which orbited Jupiter from 1996 to 2003 before being sent to a fiery death in Jupiter's atmosphere to avoid accidentally contaminating any of the Galilean satellites—potential habitats that they may be—with Earthling microbes that might have hitched a ride.

One of the more stunning discoveries of the *Voyager* missions to

Table 9.3

Past successful and future planned missions to the outer Solar System

Mission	Nature of mission	Flyby date, dates active in orbit, or launch/arrival date	Notes
Pioneer 10	Jupiter flyby	December 3, 1973	First spacecraft to venture beyond the asteroid belt; first flyby of Jupiter
Pioneer 11	Jupiter flyby Saturn flyby	December 2, 1974 September 1, 1979	Flyby of Jupiter followed by first flyby of Saturn
Voyager 1	Jupiter flyby Saturn flyby	March 5, 1979 November 12, 1980	Discovery of Io volcanoes, Titan's thick atmosphere, and Enceladus's high albedo
Voyager 2	Jupiter flyby Saturn flyby Uranus flyby Neptune flyby	July 9, 1979 August 26, 1981 January 24, 1986 August 25, 1989	First (and only) in situ exploration of Uranus and Neptune; discovery of Europa's smooth surface and Triton's geysers
Galileo	Jupiter orbiter	December 7, 1995 to September 21, 2003	First outer planet orbiter; discovered oceans under the ice of Europa, Ganymede, and Callisto
Galileo probe	Jupiter atmospheric probe	December 7, 1995	Characterized Jupiter's atmosphere for 58 minutes until the probe descended to crush depth
Cassini	Saturn orbiter	July 1, 2004 to September 15, 2017	Extensive studies of Saturn and its moons; discovery of Enceladus's geysers and Titan's lakes; intentional final descent into Saturn to avoid risk of contaminating its moons
Huygens	Titan lander	January 14, 2005	Studied Titan's atmosphere and surface during its 2.5-hour descent and for 1 hour on the surface
Dawn	Vesta and Ceres orbiter	July 16, 2011 to September 5, 2012 (Vesta) March 6, 2015 to October 31, 2018 (Ceres)	Detailed studies of the geology and minerology of the two largest asteroids; identified potential, apparently water-driven "volcanism" on Ceres
Rosetta	Comet 67P/Churyumov-Gerasimenko orbiter	August 6, 2014 to September 30, 2016	First comet orbiter; performed detailed studies of the composition, morphology, and activity of the comet as it passed the closest point in its orbit to the Sun
Philae	Comet 67P/Churyumov-Gerasimenko lander	November 12 to 14, 2014	First soft landing on a comet, albeit its anchoring harpoons failed to deploy, limiting the ability of its instruments to contact the surface; performed limited chemical analysis of the surface
New Horizons	Pluto flyby Arrokoth flyby	July 14, 2015 January 1, 2019	Investigated Kuiper belt objects Pluto and Arrokoth

(*continued*)

Table 9.3
Past successful and future planned missions to the outer Solar System (*continued*)

Mission	Nature of mission	Flyby date, dates active in orbit, or launch/arrival date	Notes
Juno	Jupiter orbiter	July 4, 2016 to present	Aiming to understand origin and evolution of Jupiter, map magnetic field, measure water and ammonia in deep atmosphere, observe auroras
Lucy	Multiple asteroid flyby	Expected launch October 2021	Will flyby one main belt asteroid and five trojans (asteroids sharing Jupiter's orbit but either leading or lagging the planet by 60°)
Psyche	Psyche (asteroid) orbiter	Expected launch/arrival 2022/2026	Orbiting the metallic asteroid Psyche, the goal of this mission is to explore the origin of planetary cores
JUICE	Ganymede orbiter	Expected launch/arrival 2022/2029	European Space Agency's *Jupiter Icy Moons Explorer* will conduct repeat flybys of Callisto, Europa, and Ganymede before finally orbiting the latter
Europa Clipper	Jupiter orbiter with multiple Europa flybys	Expected launch/arrival 2024/2030	Will conduct 44 flybys of Europa at altitudes from 25 to 2,700 km during 3.5 years in orbit around Jupiter
Dragonfly	Titan autonomous aerial vehicle	Expected launch/arrival 2026/2034	This 450 kg autonomous drone will explore Titan's surface for 2.7 years, traversing up to 8 km each 192-hour-long Titan "day"

Jupiter was confirmation of a theory regarding the innermost Galilean satellite, Io. Stan Peale (1937–2015) from the University of California, Santa Barbara, and Pat Cassen and Ray Reynolds of NASA's Ames Research Center had been pondering the ramifications of an observation made several centuries earlier: that the orbits of the inner three Galilean satellites are resonant. For every orbit of Ganymede, Europa makes two orbits and Io makes four (table 9.4 lists their orbital parameters and other physical properties). This means that the gravitational tug of Europa on Io builds up; it happens every other time at exactly the same place in Io's orbit, forcing Io into an eccentric (out-of-round) orbital path. An eccentric orbit this near Jupiter would not usually be a stable state of affairs; orbital eccentricity should lead to enormous tides in the solid material of the moon, and the friction caused as the tides flexed the moon's solid rock should dissipate orbital energy until

Table 9.4
Jupiter's Galilean satellites

Name	Orbital period (Earth-days)	Radius (km)	Mass (relative to Earth's Moon)	Density (g/cm^3)	Composition
Io	1.77	1,818	1.22	3.53	Silicates
Europa	3.55	1,561	0.65	3.01	Silicates, small amount of water ice
Ganymede	7.15	2,634	2.02	1.94	½ silicates, ½ water ice
Callisto	6.69	2,408	1.46	1.83	½ silicates, ½ water ice

the eccentricity was damped and Io returned to a circular orbit. But the resonance with Europa prevents this; while the tides of Io try to damp the eccentricity, counteracting tides raised in Jupiter tend to push the orbit of Io outward, causing the resonant interaction with Europa to kick up the eccentricity once again. In turn, a similar resonance between Europa and Ganymede forces Europa's orbit out of round, again causing tides and subsequent damping although, because tidal amplitude drops off with the third power of the distance from the source, Europa's tides are much smaller than Io's.

The net effect of all this gravitational tugging is to convert Ganymede's orbital energy into massive tidal flexing of the crust of Io and lesser tidal flexing of Europa. When the scientists worked out the precise numbers, they realized that this tidal effect dumps enormous amounts of energy into Io's crust—orders of magnitude more heat per square meter than the geothermal flux of the Earth, then thought to be the most geologically active body in the Solar System. Their conclusion was clear: "although the morphology of such a surface cannot be predicted in any detail, one might speculate that widespread and recurrent surface volcanism would occur."*

On March 8, 1979, exactly one week after the volcano prediction appeared in print, *Voyager* navigation team member Linda Morabito was looking at some images of the limb (edge) of Io. These images

*Peale, S. J., P. Cassen, and R. T. Reynolds. "Melting of Io by Tidal Dissipation." *Science* 203, no. 4383 (1979): 892–94.

Figure 9.9 Io, the innermost of Jupiter's main satellites, is intensely volcanic. The crescent rising in the upper left-hand corner of this *Voyager* image is a volcanic plume jetting some 260 km into space, and the bright spot nearer the center of the disk is a second plume rising from the dark of night into bright sunlight. (Courtesy of NASA/JPL)

were overexposed to bring out faint stars in the background for navigation purposes. Oddly, an image showed a crescent beyond the edge of the moon (fig. 9.9). At first Morabito thought it was another of Jupiter's satellites peaking over the edge, but a quick check of the locations of the other Galilean satellites nixed this idea. Instead, the crescent was a stunning confirmation of the tidal heating theory; it was the plume of an extraordinarily active volcano rising 260 km into space. In short order, seven violently active volcanoes were identified on Io (including a second obvious one in the picture Morabito was using), making it far and away the most volcanically active body in the Solar System.

While it is a fascinating place in terms of geology, our interest in Io from the astrobiological perspective is limited; the incessant volcanism—and subsequent loss of volatiles to space—has baked Io completely dry. The same, however, cannot be said for the next moon out: Europa. Measurements of the slight deflection induced in the *Voyager* spacecraft as they flew by Europa allowed scientists to estimate the mass of this moon, which, when combined with knowledge of its size, indicated that the satellite has a density of 3.02 g/cm^3. This is a little low compared with the density expected for the Solar System's typical mix of silicates and metals (see table 3.2). More perplexing still, *Voyager 1* and *2* found that, at the resolution their cameras, Europa's is probably the smoothest surface in the Solar System; the moon is almost free of craters, and its topography doesn't vary by more than a few hundred meters across its entire globe (fig. 9.10). Taken with the spectroscopic evidence on the composition of Europa's surface, these observations led astronomers to the conclusion that the surface of Europa is a thick, relatively recently reworked layer of ice.

Recently reworked, of course, once again implies "geologically active." Consistent with this, a few months after their prescient paper on Ionian volcanism, the same theory team published a second paper, with the provocative title, "Is There Liquid Water on Europa?"* Unlike their earlier prediction of volcanoes on Io, which took less than week to confirm, the accuracy of this second prediction, which also implicated tidal heating in causing subsurface melting, wouldn't be nailed down for another 21 years. In the year 2000, the chief scientist on the *Galileo* magnetometer, Margaret Kivelson of the University of California, Los Angeles, finally announced conclusive proof of an under-ice ocean on Europa. During *Galileo*'s December 1996 flyby, the magnetometer found that Europa has a magnetic field, and it is pointed perfectly opposite to the strong magnetic field of Jupiter. While the opposing field could have been coincidental (*Galileo* could have just happened by at the very moment when Europa's rotation brought its field into alignment with Jupiter's), follow-on flybys showed that the Europan field is

*Cassen P., R. T. Reynolds, and S. J. Peale. "Is There Liquid Water on Europa?" *Geophysical Research Letters* 6, no. 9 (1979): 731–34.

Figure 9.10 A crescent Europa, showing the crisscrossing lines that decorate its surface. On every other body in our Solar System, mountain tops jut into daylight and valley (or crater) floors fall into darkness, breaking up the line between night and day. On Europa, in contrast, the topography varies by only a few hundred meters, explaining the razor sharpness of the line separating daylight from the dark of night. (Courtesy of NASA/JPL)

always opposed to Jupiter's no matter where Europa is in its orbit. An opposing magnetic field can be generated by eddy currents, which are the currents induced in a conductor when it is moved through an external magnetic field. But what could the Europan conductor be? Based on the density of Europa's crust and the conductivity necessary to generate the observed magnetic field, the only compelling answer is that it's a global ocean of saltwater many tens of kilometers thick, a sub-ice ocean kept liquid by tidal heating. Even though Europa receives about one-tenth as much tidal energy as Io, that seems to be enough to prevent freezing of its ocean, insulated as it is from the cold of space by a thick layer of ice.

Although its existence is considered confirmed, details of the "Europan ocean" remain sketchy. *Galileo's* gravity measurements suggest that the combined ice and water shell can be no more than 70 to 170 km deep but cannot distinguish the liquid from the solid phase (their densities are too similar) and thus cannot estimate how much of this thickness is liquid ocean and how much is the icy shell that lays on top. Simulations of impact-crater formation in ice suggest, however, that the ice shield must be at least 3 or 4 km thick, and perhaps as much as 25 km. Still, some indirect studies suggest that the ice may, on geological time scales, sometimes become rather thin. For example, *Galileo* has photographed what appear to be "ice rafts" frozen into place, as if the surface ice had once temporarily melted and broken into icebergs (fig. 9.11). Likewise, telescopic observations from Earth have seen water vapor in the neighborhood of Europa, corre-

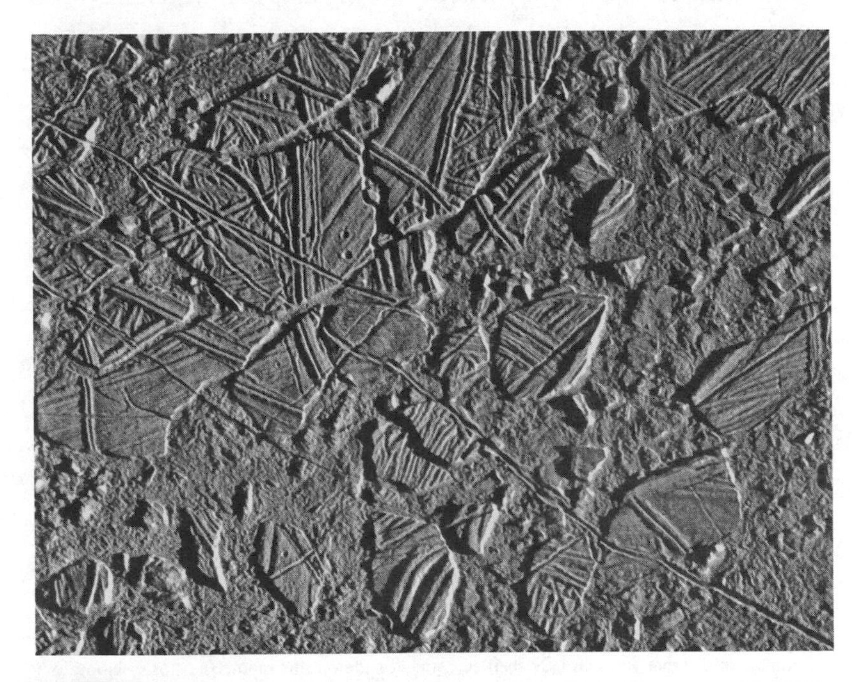

Figure 9.11 This image of raftlike elements, captured by the *Galileo* spacecraft in 1998, suggests that Europa's crust, which floats on a sub-ice ocean, is sometimes thin enough to break up like pack ice in an Arctic summer. For scale, the base of this image is 34 km across. (Courtesy of NASA/JPL)

sponding to release rates of a couple of tons per second, although whatever is causing this has proven sporadic; the vapor was detected on only a single night in 17 observations scattered over the years 2016 to 2017.

The likely presence of a liquid ocean beneath the thick, icy crust of Europa suggests that the moon may be a potential habitat, but is it truly habitable? As we have discussed in earlier chapters, water alone is not a sufficient criterion for habitability; we also need a source of energy to drive metabolism. Several possible sources have been suggested for Europan life. For example, even though the intensity of sunlight that strikes Europa is about one-thirtieth of what we receive on the Earth's surface, meltwater near Europa's surface could support photosynthetic organisms. Alternatively, and perhaps more appealingly, the geothermal (tidal) energy that keeps Europa's ocean liquid could provide a source of energy for organisms living on the ocean floor. Whether this is a sufficient source of energy, however, has been rather hotly debated in the astrobiology community. Finally, Chris Chyba of Princeton University suggests yet another potential energy source, this one rather specific to Europa. Europa, Chyba notes, orbits deep within Jupiter's intense radiation fields, and this radiation (which would kill an unprotected human in minutes) breaks down the ice chemically on the moon's surface. That is, when the high-energy protons and electrons that make up the radiation impact the ice, they tear its molecules apart and create highly reactive species. This "radiolysis" has, in fact, been observed: spectroscopic studies of Europa from the Hubble Space Telescope indicate that radiolysis provides Europa with a tenuous oxygen atmosphere (albeit only 100 billionth as dense as Earth's).* Other important products of this reaction probably include the simple organic compound formaldehyde and the oxidizing compound hydrogen peroxide. From estimates of the rate with which these species are formed on Europa, Chyba has calculated that

*Speaking of Europan spectroscopy, Corey Jamieson of the University of Hawaii claims (perhaps not too seriously) that we've already seen evidence for life on this moon: the closest laboratory match he's found to the infrared spectra of Europa is provided by Terrestrial extremophilic bacteria frozen to liquid-nitrogen temperatures! Most members of the astrobiology community are betting that a more prosaic source for the Europan spectral features will be found.

this energy source could support up to 500 tons of microorganisms. Not a huge biosphere by Terrestrial standards—the Earth's is 10 billion times larger than this—but enough to push Europa into the very short list of potentially habitable places in the Solar System.

And it appears that Europa might not be alone: there is some evidence that Ganymede, the third Galilean moon, might also hide an ocean beneath its icy crust. Slightly larger in diameter than the planet Mercury, Ganymede is the largest moon in the Solar System. Its composition, however, is a mixture of rock, metals, and ice, rendering its mass only about half that of Mercury. Magnetometer readings from the *Galileo* spacecraft have proved that, like Europa, Ganymede produces a small magnetic field counter to that of Jupiter, a clear fingerprint of a sub-ice ocean. This was unexpected as Ganymede is the outermost of the three participants in the orbital dance that pumps energy into Io and Europa and thus should be receiving far less energy than its inner two partners (remember: tides drop off with the cube of the distance). Much of the energy required to maintain Ganymede's liquid ocean instead seems to arise from within Ganymede itself. Direct evidence for this was also provided by *Galileo*'s magnetometer, which registered that, in addition to the induced magnetic field produced by the ocean, Ganymede also generates a permanent magnetic field about 2% as strong as that of the Earth. The Earth's magnetic field is thought to arise from the churning of the planet's liquid iron core, suggesting in turn that Ganymede also contains a liquid iron core. Heat escaping from this core—which is presumably created by the decay of radioactive elements working in concert with what little tidal energy the moon receives—would be sufficient to maintain a liquid water ocean, albeit only if there were significant "insulation." Consistent with this, the heavily grooved—and presumably once geologically active—surface of Ganymede (fig. 9.12) exhibits far greater topography than the smooth surface of Europa as it sports mountains up to 3 km high. Together these observations suggest a rigid, heavily insulating ice crust 100 to 200 km thick. This, in turn, suggests that the energy flux through Ganymede is small, likely far too small to support an ecosystem.

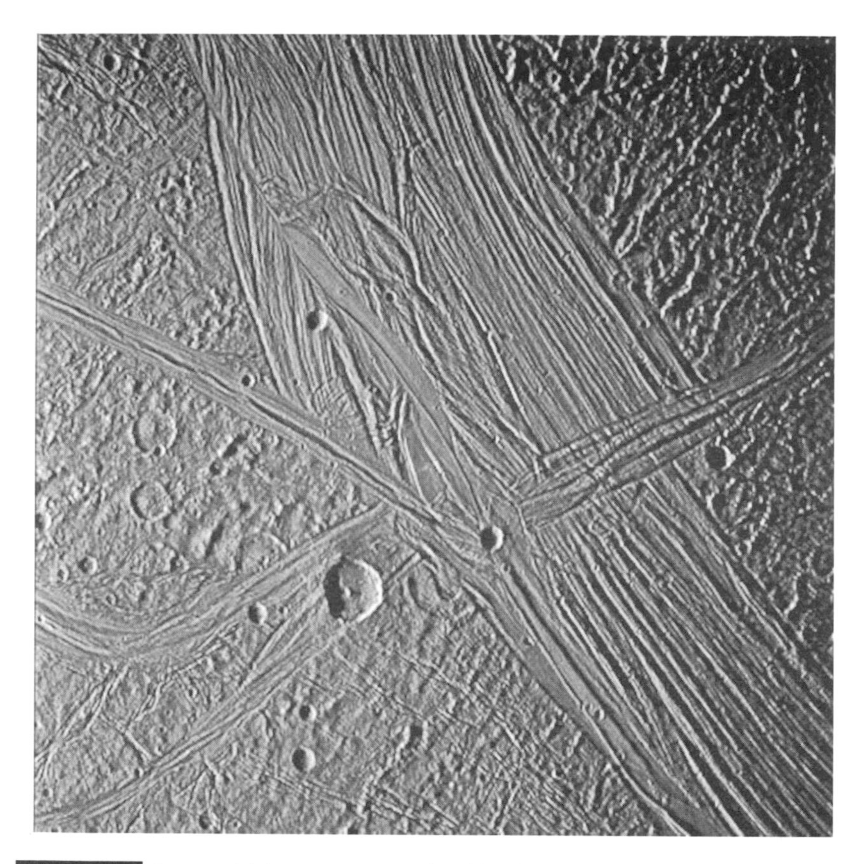

Figure 9.12 Ganymede's icy crust is crossed and hatch-marked with icy grooves that speak of a once (if long ago) active geology. This image is 390 km across. (Courtesy of NASA/JPL)

And it doesn't stop with Ganymede. Callisto, the outermost Galilean satellite and the third largest moon in the Solar System, also shows signs—in the form of an induced magnetic field—of an under-ice ocean. This is particularly odd, for a number of reasons. First, Callisto does not take part in the resonant dance of the inner Galilean satellites and thus is not subject to any appreciable tidal heating. Of course, as we just noted, Ganymede's tidal heating is also too small to account for its ocean, which instead is thought to result, at least in part, from the heat of radioactive decay. Indeed, gravitational mea-

Figure 9.13 The battered face of Callisto argues that it is the least geologically active of Jupiter's four Galilean satellites. Its icy crust is nevertheless thought to cover an ocean of liquid brine. (Courtesy of NASA/JPL)

surements by the *Galileo* spacecraft indicate that, unlike the other Galilean satellites, Callisto never differentiated into a metallic core, rocky mantle, and icy crust, suggesting that the moon's interior has always been quite cold. This observation is likewise supported by the battered appearance of Callisto's ancient surface: unlike those of the inner Galilean moons, the surface of Callisto is saturated with craters and shows no signs of having undergone any significant geological activity (fig. 9.13). The origin of Callisto's ocean, which like that of Ganymede is thought to be hundreds of kilometers below the surface, thus remains a mystery. Most probably, however, given the poor thermal conductivity of ice and the presence of salts and incorporated volatiles such as ammonia that suppress the freezing point of water, even the feeble heat released by radioactive elements in Callisto's rocky components is sufficient to maintain a liquid ocean.

The Moons of Saturn and Beyond

For all his success with the Jovian system, Galileo (the Renaissance astronomer, not the robotic spacecraft) failed to discover any of Saturn's numerous moons (the larger of which are listed in table 9.5) or, indeed, to recognize its rings for what they are. With regard to the former, his telescope lacked sufficient resolving power. With regard to the latter, Galileo had the bad luck of making his observations right before and then during the time that the Earth passed through Saturn's ring plane, rendering the rings edge-on and thus invisible. At first it appeared that Saturn had two large companions on either side—the unresolved rings—and then months later they disappeared! Clearing up this mystery, the Dutch scientist Christiaan Huygens (1629–1695) described the rings in detail in 1659. He also discovered Saturn's largest satellite, Titan, which is slightly bigger than Mercury and second

Table 9.5
Saturn's major satellites

Name	Orbital period (Earth-days)	Radius (km)	Mass (relative to Earth's Moon)	Density (g/cm^3)	Composition
Mimas	0.9	200	0.0005	1.17	Predominantly water ice
Enceladus	1.4	250	0.001	1.24	⅛ silicates, ⅞ water ice
Tethys	1.9	530	0.009	1.21	Predominantly water ice
Dione	2.7	560	0.015	1.43	⅓ silicates, ⅔ water ice
Rhea	4.5	764	0.031	1.33	¼ silicates, ¾ water ice
Titan	15.9	2,575	1.83	1.21	Predominantly water ice
Hyperion	21.3	~140	0.0002	~1.2	Water ice with voids
Iapetus	79.3	718	0.022	1.21	Predominantly water ice

only to Ganymede in size among the Solar System's moons. Titan, though, was long considered the more interesting of the two from the astrobiological perspective. The reason for this interest dates back to 1925, when James Jeans (of the "Jeans mass" limit to star formation described in chapter 2) used his "dynamical theory of gases" to predict that, even given Titan's modest gravity, the low temperatures found at Saturn's distance from the Sun would allow the moon to retain gaseous molecules as heavy as or heavier than methane. Measuring the infrared spectrum of Titan, the Dutch American astronomer Gerard Kuiper (1905–1973) confirmed Jeans's prediction in 1944 by identifying methane at a pressure of a few percent that of total atmospheric pressure on Earth. Among the many dozens of moons in our Solar System, Titan alone retains a thick atmosphere.

Our detailed understanding of Titan and its atmosphere began during the *Voyager* flybys in 1980 and 1981. In images taken by the *Voyager* craft, the moon presented only a featureless yellow-orange orb—Titan's atmosphere is filled with a thick haze that prevents observation of its surface at visible wavelengths (fig. 9.14). Refraction of *Voyager 1*'s radio signal as it passed behind Titan (from the perspective of Earth) provided a means of characterizing the moon's atmosphere. Surprisingly, Titan's surface pressure is 1.5 times that of Earth's; given Titan's weaker gravity, this corresponds to an atmospheric density some three times ours.* Kuiper's results, however, indicated that the pressure of methane in Titan's atmosphere is far lower than this, suggesting that the bulk of the atmosphere is made up of a species invisible to his infrared observations. We now know this to be nitrogen, which makes up 97% of Titan's atmosphere, with the remainder largely being the methane that Kuiper had observed. The discovery of methane and nitrogen in Titan's atmosphere solved the mystery of its atmospheric haze: Carl Sagan, then at Cornell University in upstate New York, studied the effects of UV light and ionizing radiation on such a mixture and found that it reacts to form a high molecular weight mix of what he named "tholins," from the Greek *thol*, for "mud."

*The atmosphere on Titan is so dense, and the gravity so low, that were you visiting it, you could strap wings to your arms and fly like a bird. You would need to dress warmly, though.

Figure 9.14 The haze-filled atmosphere of Titan, Saturn's largest moon. (Courtesy of NASA/JPL)

Tholins, which are largely composed of long-chain hydrocarbon molecules terminated by nitrile groups (−C≡N), are generally reddish-yellow and probably constitute the orange fog that renders the surface of Titan impossible to image from space using visible light.

The radio data from *Voyager 1* also provided a means of estimating the temperature of Titan's atmosphere, which was found to be a chilly −179°C at the surface. This is curious because it is near the temperature at which the atmosphere's methane should condense out as a liquid. Indeed, at a concentration of a few percent, the methane "relative humidity" in Titan's atmosphere is approximately 50%, suggesting that the gas could condense and fall as rain as water does on Earth. Together with the observation that the haze-producing, photolytic creation of tholins is so rapid that the resulting loss of hydrogen to space should deplete Titan's atmospheric methane in a few tens of millions of years, this suggests that the methane is being replenished from some reservoir on the surface. And that reservoir? By analogy to water on Earth, the *Voyager* data were taken to imply that, beneath its thick covering of haze, Titan's surface is dotted with lakes or even oceans of liquid methane. Indeed, the *Voyager* data suggested that, since Titan has just the right range of temperatures for methane to exist as a liquid, solid, or gas, methane might drive weather on Titan closely analogous to Earth's water-driven weather.

Methane clouds billowing up above a frozen landscape; methane storms raining down to form methane rivers that in turn cascade into methane seas—it's a very pretty image, but one that would have to wait more than two decades for confirmation. Not until the arrival of *Cassini*, which fell into orbit around Saturn in late 2004, and its companion, the *Huygens* lander, which parachuted to the surface of Titan in January 2005, would our knowledge of Titan take its next significant leap forward.

The first hints about the nature of Titan's surface came when, during its 2.5-hour descent, *Huygens* snapped pictures of what look very much like dry river channels and dry lakebeds (fig. 9.15) before finally gently settling down on a flat, cobble-covered landscape reminiscent of a flood plain (fig. 9.16). Indeed, during the hour or so that it continued to relay data from Titan's surface, *Huygens*'s warmth drove methane out of the surface in such large volumes that the ground must be saturated with the stuff, just as sand on a beach is saturated with water. Further hints about the nature of methane on Titan came from *Cassini* as it periodically flew by on its orbit around Saturn. Using *Cassini*'s radar to peer through the thick haze, the first thing scientists noted was that the moon bears very few impact scars, perhaps because craters erode rapidly under the onslaught of the hypothesized methane rains and rivers. Likewise, radar images of the moon's north

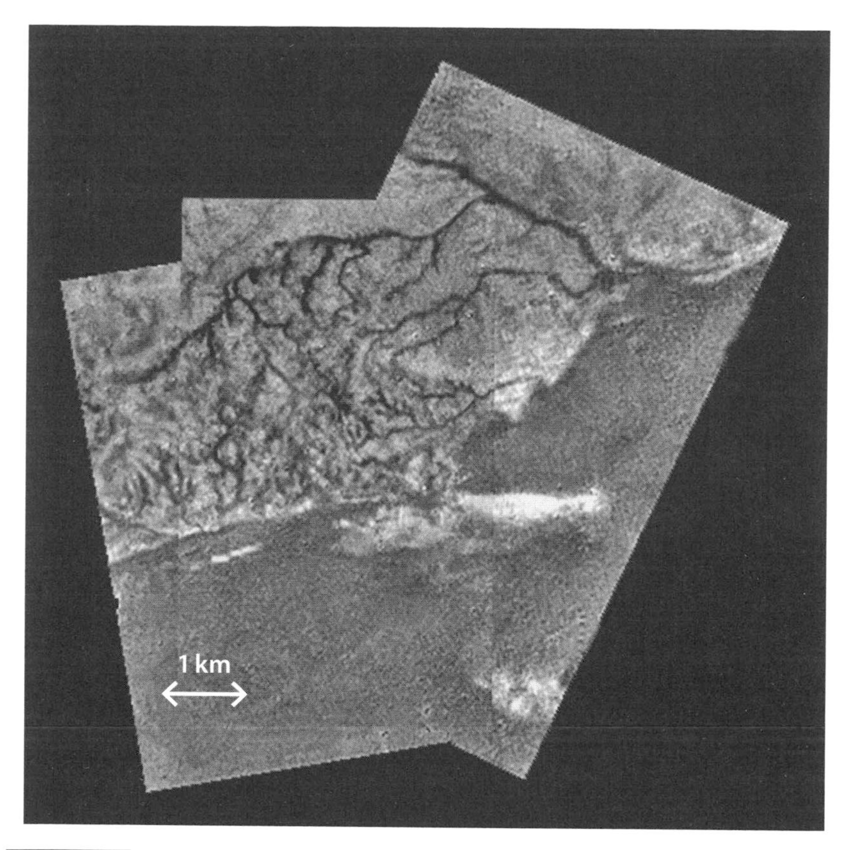

Figure 9.15 Titan seems to have a complex, liquid methane–based fluid cycle reminiscent of the Earth's hydrological cycle. During its 2.5-hour descent to Titan's surface, for example, the *Huygens* probe spied what are probably river channels carved by methane rains. (Courtesy of NASA/ESA)

pole, which was cloaked in winter darkness at the time, showed that wide swaths of its surface were dark to radar and perfectly smooth (to within the measurement precision of just 3 mm) over hundreds of kilometers as would be expected for a windless lake (fig. 9.17). Finally, in 2009, during the Saturnian equinox, Titan's north pole moved from night to day and, as it did so, *Cassini* captured a spectacular photograph of low-angle, early morning sunlight glinting off a mirror-like surface (fig. 9.18). It is now clear that, at least at its poles, Titan's surface is spattered with lakes, and the physical properties of its atmo-

Figure 9.16 After a seven-year, 5-billion-kilometer trip, *Huygens* landed on Titan's frozen surface to become, for an hour and a half, humanity's most distant outpost on solid ground. In the distance, beyond scattered cobbles of water frozen to rock-hardness by the deep cold, lies the Titanian horizon. (Courtesy of NASA/ESA)

sphere and surface are dominated by a "methane cycle" analogous to the hydrological (water) cycle in the Earth's atmosphere.

As fascinating as the river-carved and lake-dotted surface of Titan may be, it is Titan's atmospheric chemistry that is of interest to the astrobiology community. The reason is that Titan's atmosphere, although extremely cold, is in some ways a rough analog of the early Earth's as postulated by Harold Urey. That is, the atmosphere is reducing and nitrogen rich. Under the influence of UV light from the Sun (and perhaps lightning in those methane clouds—although, despite significant effort, none has yet been detected), these dominant atmospheric components, so it is thought, must be reacting to pro-

Figure 9.17 Ligeia Mare, one of the largest of the many methane lakes on Titan, covers about twice the surface area of Earth's Lake Superior. (Radar image courtesy of NASA/JPL/USGS)

duce life's precursors, which form the thick haze and no doubt rain down to the surface. Over billions of years, perhaps hundreds of meters of complex organics have accumulated on the surface of this frozen world. The *Huygens* probe, alas, was not equipped to search for such molecules on the surface. Still, some of the *Cassini* flybys have been close enough for its mass spectrometer to detect, as the craft passed through the moon's outermost atmosphere, the high molecular weight hydrocarbons and nitriles predicted by Sagan.

Figure 9.18 Early morning sunlight glinting off a methane lake near Titan's north pole. (Courtesy of NASA/JPL/UA/DLR)

But is Titan a potential habitat? In that light, the moon's excessive cold is a real problem. Temperatures on the surface of Titan are so low that all oxygen-containing compounds, which of course play an absolutely critical role in life on Earth, are locked into their solid form and would not be available to participate in Miller-Urey-type chemistries, much less in life. But perhaps not all is lost. Sagan argued, for example, that meteorite impacts must have imparted enough energy to the moon to ensure that every part of it has seen liquid water during at least some part of its history.

And what of the rest of Saturn's retinue of 82 (and counting) moons? Saturn's second largest moon, Rhea, is quite small, coming in at only one-third the diameter and one-sixtieth the mass of Titan. Moreover, judging by its heavily cratered surface, it appears to be geologically dead. Moving down in size from Rhea, Saturn has a bevy of other, still smaller icy moons, all of which, with two exceptions, lack any signs of activity. The first of these exceptions is fairly minor: the surface of Saturn's fourth largest moon, Dione, sports a few "fresh"-looking cracks hinting at some sort of tectonic activity. The other exception, in contrast, appears far more active. That exception? Enceladus.

Enceladus is small; at only 500 km in diameter, it is about one-tenth the diameter and 1/2,000 the mass of Titan, suggesting that it, too, should be a dead world. Oddly, though, when *Voyager 1* finally observed the moon up close in 1980, Enceladus was found to be as white as freshly fallen snow, reflecting almost 100% of the light that strikes it. In fact, Enceladus has the most highly reflective surface in the Solar System. Given that meteorite impacts and "space weathering" (the cumulative effects of radiation damage and meteorite impacts) tend to darken anything exposed to space, the snow-white appearance of Enceladus suggests that its surface is constantly being replenished. A much closer flyby by *Voyager 2* the next year only added to the mystery: while parts of Enceladus's northern hemisphere are at least modestly cratered, much of the rest of the moon is nearly crater free, and its south pole is crisscrossed with tectonic cracks. The two *Voyager* missions, however, failed to provide any insights as to how the surface of Enceladus's southern hemisphere is replenished. The answer would have to wait for the next mission to Saturn.

Unlike the *Voyager* spacecraft, the next mission to Saturn, *Cassini*, was an orbiter, so we had the pleasure of viewing Saturn's moons from a variety of perspectives as it slowly orbited the giant planet. In 2005, while behind Enceladus (from the perspective of the Sun), *Cassini* spied light scattered from geyser-like plumes of ice particles streaming from cracks at the moon's south pole (fig. 9.19). Later flybys took the orbiter directly through these plumes, where *Cassini*'s mass spectrometer confirmed that, apart from water as their dominant component, they also contain a bit of ammonia and traces of several simple organic

Figure 9.19 The geysers of Enceladus as they slip into a long winter's night. (Courtesy of NASA/JPL/MSSI)

compounds, including methane, propane (C_3H_8), acetylene ($HC{\equiv}CH$), and formaldehyde ($H_2C{=}O$). *Cassini*'s mass spectrometer also detected salts (NaCl, KCl, $MgSO_4$) and silica (SiO_2) in the plumes, implying that their source must be pools of liquid water interacting with rock, and not just solid ice sublimating directly into water vapor.

Infrared measurements from *Cassini* have shown that that temperatures at the south pole of Enceladus are far milder than the −210°C temperature found elsewhere on the moon's icy surface. Indeed, some areas reach −90°C, but as these measurements were averaged over the smallest length scales resolvable by the instrument, it is likely that there are still smaller spots at much higher temperatures. All told, the geysers on Enceladus are pumping out 5 GW of heat (a gigawatt is equal to 1 billion watts), which is about 1,000 times the output of Old Faithful, the best-known geyser in America's Yellowstone National Park. The source of this energy remains subject to debate. Enceladus is in a weak orbital resonance with Dione, but the resonance should be able to produce only about 1 GW of tidal heating over the long term, and Enceladus is so small that radiogenic heating isn't thought to produce more than a fraction of a gigawatt. A possibility is that Enceladus's orbit was, until recently, more eccentric than it is currently and that the geysers are driven by leftover tidal heating from that era.

Later flybys of Enceladus during *Cassini*'s second and final mission extension, which lasted from 2010 to its intentional destruction in September 2017, showed that the plumes are extended curtains of relatively constant emission, presumably from a crack in the surface, rather than geyser-like jets. The closer investigation of the plumes also brought the discovery of larger organic molecules, such as benzene (C_6H_6), which added to the appeal of Enceladus as a target for future astrobiology-focused missions.

Any more habitats out there beyond Saturn? Perhaps. Though demoted to "dwarf planet" status, Pluto is still fairly large, as is Neptune's moon Triton, which is only slightly smaller than our Moon and ranks as the seventh largest satellite in the Solar System. Moreover, at the approximate 38 K (−235°C) temperatures found at these distances from the Sun, even Triton's and Pluto's weak gravities are sufficient to maintain nitrogen-rich atmospheres, albeit at densities of only 10 and 15 *millionths* those of the Earth's, respectively.

When *Voyager 2* flew past Triton in 1989, it observed more than a dozen geyser-like plumes spraying 8 km above the moon's icy surface before they drifted 150 km downwind (fig. 9.20). Given that all the plumes were observed in a narrow latitude belt around 50° south, which during the *Voyager* flyby was the region of Triton pointing directly toward the Sun, it is thought that the likely energy source behind these geysers is solar heating; that is, visible light penetrating the ice, which absorbs infrared and thus serves as a greenhouse "lid," warming the subsurface.

Pluto too, seems to be more active than one might have imagined; when *New Horizons* flew past in 2015, it found a surprisingly fresh surface marked by mountains of rock-solid ice (fig. 9.21). And while Pluto lacks any tidal heating (as it is tidally locked to its moon Chiron), theoretical studies indicate that, like Jupiter's Callisto, which also lacks a tidal heat source, Pluto may harbor an ocean of briny or ammonia-containing water under its icy crust. For example, simulations run by

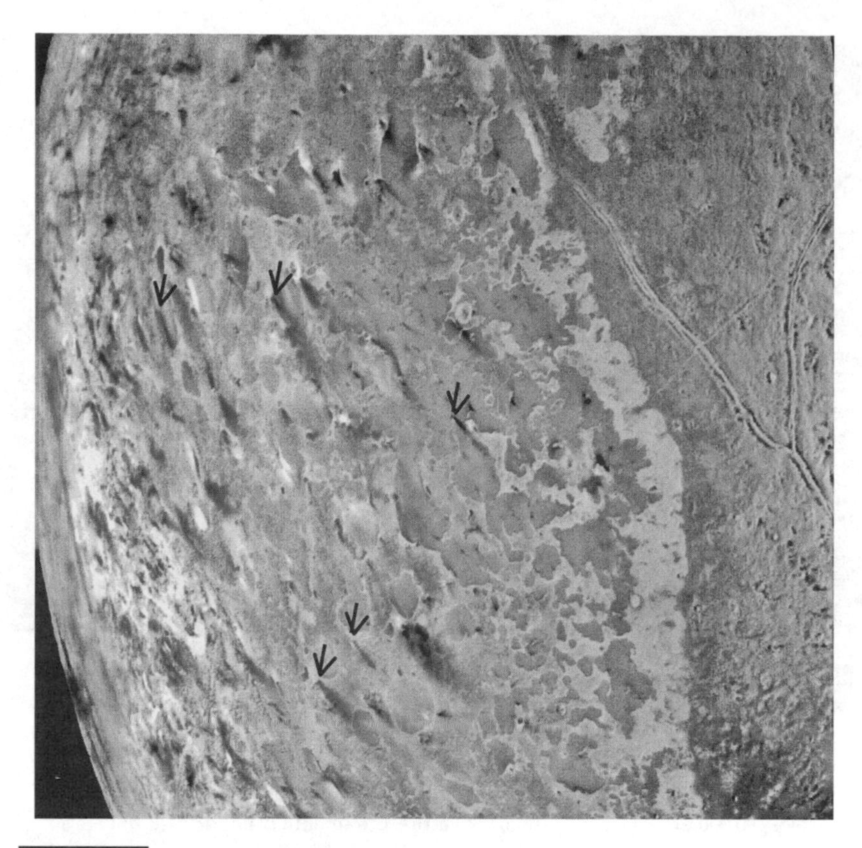

Figure 9.20 When *Voyager 2* flew past Neptune's largest moon, Triton, in 1989, it observed more than a dozen geyser-like plumes (arrows) erupting 8 km above the icy surface and drifting up to 150 km downwind, leaving dark streaks of dust in their wakes. (Courtesy of NASA/JPL)

Shunichi Kamata of Hokkaido University in Japan suggest that, despite Pluto's being just one-eighth of Callisto's mass (increasing the rate with which it cools and reducing its sources of radioactive heat), a thick layer of gas hydrates—crystalline solids formed by gas molecules trapped within molecular water cages—could insulate Pluto's interior enough such that some liquid remains there since accretion. But could distant Pluto harbor life? The paucity of available energy sources suggests that this is unlikely. And given the good many decades that will pass before a spacecraft lands on Pluto, we can make this claim without fear of being proven wrong anytime soon.

Figure 9.21 On July 14, 2015, exactly 50 years *to the day* after *Mariner 4* sent back the first close-up images of another planet, the *New Horizons* spacecraft flew past Pluto, completing the initial, robotic exploration of the Solar System. This near-sunset image taken a few minutes after closest approach is about 380 km across at the base. Across the top are faint bands of haze in Pluto's thin nitrogen atmosphere, which is approximately 100,000 times less dense than the Earth's. The smooth plains on the right are a glacier of frozen nitrogen; Pluto's temperature range is such that its nitrogen cycles between the atmosphere and the crust. The mountains on the left and on the skyline, named Tenzeng Montes and Hillary Montes, respectively, after the first team to successfully summit Mount Everest, are comprised of water ice and rise up 6 km above their bases. They are thought to be less than 100 million years old; otherwise their weight should have caused them to "relax" back into Pluto's icy crust. The energy source driving their relatively recent formation remains a mystery. (Courtesy of NASA/JHUAPL/SwRI)

Planets beyond Our Solar System

Although long popular on television and movie screens, even the very existence of planets around stars other than our own was mere conjecture until the 1990s. The conjecture usually revolved around arguments we've seen before: given the many billions of stars that are very much like our own, it seems highly unlikely that the Sun is the only one that has planets, blah, blah, blah. The indirect arguments were based on the observation that young stars are often observed to be surrounded by gas disks resembling the one from which our planetary system is thought to have formed (as discussed in chapter 3). Thus, based on the assumption that the gas disks around young stars would tend to consolidate into planetary systems, most scientists believed that planets are common around other stars, even in the absence of any direct evidence.

The dearth of direct evidence came to an end in 1995, when Michel Mayor and Didier Queloz at the Geneva Observatory announced that they had discovered a planet half as big as Jupiter whizzing around the yellow dwarf 51 Pegasi, orbiting some eight times closer to its star than Mercury does to the Sun.* They made this discovery by analyzing the spectra of 51 Pegasi for Doppler shifts induced when the massive planet traveled around it, alternately pitching the star a little bit toward and a little bit away from the Earth as the two bodies orbited around their common center of mass once every 4.2 days (fig. 9.22).

Mayor and Queloz's discovery of an extra-solar planet, or "exoplanet," unleashed a rush. Within weeks, other astronomers had used their "radial velocity method" to not only confirm the existence of a planet at 51 Pegasi, but to also identify two further candidate solar systems. Within a few years, the list of exoplanets grew to more than a hundred, and by 2019, when Mayor and Queloz received the Nobel Prize in Physics for their discovery,** it had surpassed 4,000.

The radial velocity method is not the only way exoplanets can be detected. A second is to observe their *transits*. This is when the planet, its star, and the Earth line up such that the planet is in front of its star from our perspective on Earth, producing a slight, but measurable, dimming of the star's light (fig. 9.23). The transit approach has proven able to detect smaller planets than the radial velocity method. Observ-

*More precisely, this was the first confirmed exoplanet around a "normal" star. In 1992, after a similar, false report the year before by another group, Aleksander Wolszczan and Dale Frail announced the discovery of two planets in orbit around a neutron star, the pulsar PSR B1257+12. These and the half dozen "pulsar planets" discovered since likely formed from the debris blasted out in the supernova explosion that created the pulsar.

**They shared the prize with Jim Peebles, the cosmologist whose work we discussed in chapter 2.

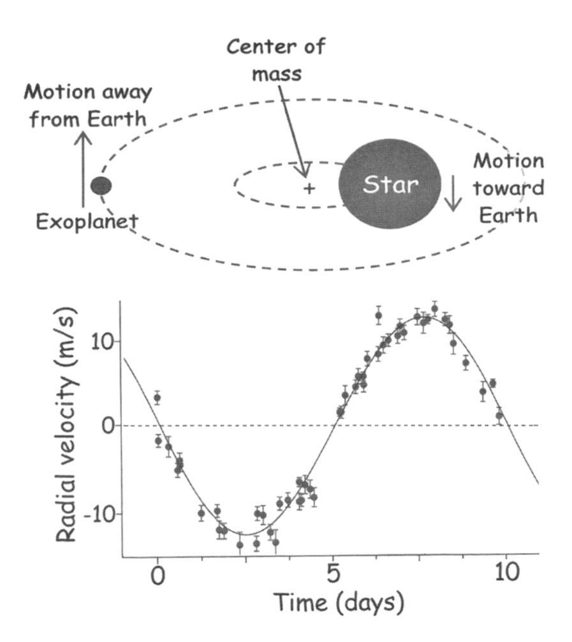

Figure 9.22 A planet and its star both orbit around their common center of mass. Thus, if a planet is moving away from us along its orbital path, the star will move toward us. This causes the velocity of the star along our line of site, its *radial velocity*, to oscillate sinusoidally over the course of the planet's orbit. Monitoring this via the Doppler effect provides one of the means by which we can detect exoplanets. Shown are radial velocity data for the star Gliese 581, a red dwarf about 20 light-years from Earth. (Lower panel courtesy of European Southern Observatory)

ing a transit, however, necessitates a bit of luck. Specifically, it requires that the Earth lie in the planet's orbital plane, which, using a bit of trigonometry, we can calculate would happen only about one in 210 times for a planet in an Earth-like orbit around a Sun-like star. Likewise, transits are brief and occur only once during a planet's orbit, and thus the approach requires large amounts of observation time and automated data analysis. And while this can be done from the Earth (David Charbonneau of Harvard reported the first example in 1999), it is better done from space, where the lack of a day/night cycle renders uninterrupted observations straightforward and the lack of atmosphere makes it easier to detect small fluctuations in a star's light (an Earth-size planet passing in front of a Sun-like star dims its light by just 84 parts per mil-

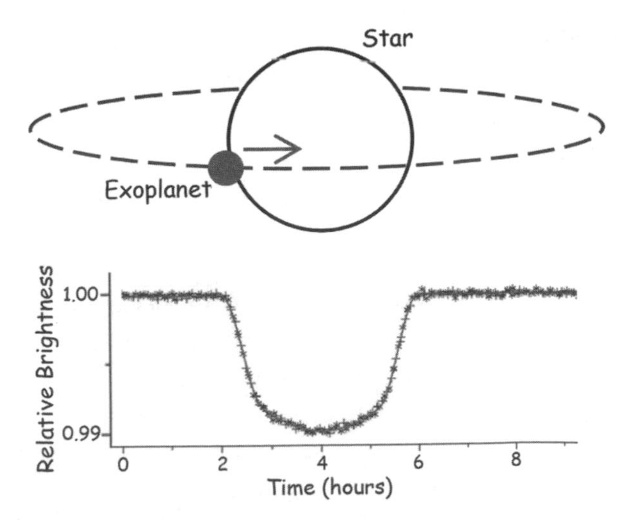

Figure 9.23 *Transits*, in which a planet passes in front of its star relative to our line of site, dim the star's light, which provides a second means of detecting exoplanets. The magnitude of the dimming tells us about the radius of the planet relative to that of its star, and the frequency of the transit tells us the orbital period. Shown are data for the transit of Kepler 6b, a gas giant two-thirds the mass of Jupiter in a 3.2-day orbit around the star Kepler 6. (Lower panel courtesy of NASA/Kepler)

lion). To this end, NASA's *Kepler* mission, a space telescope launched into solar orbit in March 2009 and named after the discoverer of the laws of orbital mechanics, was commanded to stare continuously at patch of sky about 50 times larger than the full Moon. During its nine-year mission, *Kepler* monitored more than a half million stars and identified at least 2,662 planets in orbit around them.

Radial velocity and transit measurements do more than just detect planets; taken together, they can be used to tell us quite a bit about the planet's nature. When coupled with the knowledge of the host star's physics (we know enough about stars to infer their diameter from their color and gravity), the length of the transit (how long a planet takes to pass) tells us the planet's orbital velocity, and the extent to which the star's light dims tells us the planet's radius. Given knowledge of the orbital velocity, the mass of the star (again derived from our well-established understanding of stellar mass and evolution), and Kepler's laws of orbital motion, we can derive the planet's orbit. Using the mass of the star and its Doppler-derived radial velocity, we can determine the planet's mass. Finally, from mass and diameter we can calculate density, which we can use to constrain models of their bulk composition.

Clearly, we humans have gotten the hang of planet hunting and have even garnered some clues regarding the bulk compositions of some of the exoplanets in our neighborhood. Nevertheless, as valuable as they've proved to be, radial velocity and transits are *indirect* detection methods. That is, while they provide proof of a planet's existence, they don't allow us to actually *see* the planet. The problem is that planets normally shine by reflected light and thus are easily lost in the glare of their parent star. Note, though, we said "normally." Newly formed planets emit copious infrared radiation as they cool down from accretion (even after 4.5 billion years, Jupiter still emits 1.6 times more energy as heat than it receives from the Sun). At infrared wavelengths, a hot, freshly accreted, Jupiter-sized planet is "only" a million times dimmer than a Sun-like star. And while blocking out a million photons from a star for each photon ejected from its planet is not easy, it can be done—provided the planet orbits far enough away from its star.

To date, a dozen or so gas giant planets in a half dozen solar systems have been imaged directly, and, in a few cases, had their spectra recorded. Among the best characterized of these are the four known planets orbiting HR 8799, a young (approximately 30 million years) star in the constellation Pegasus that, under ideal conditions, is just barely visible to the naked eye. Using a coronagraph coupled with adaptive optics, which corrects for atmospheric flicker, Christian Marois of Canada's Herzberg Institute of Astrophysics and his team have collected multiple images of four planets circling HR 8799 at orbital radii ranging from 16 to 68 AU (by comparison, Pluto orbits at 39 AU) and with masses ranging from five to 10 times that of Jupiter (fig. 9.24).

Tallying up the results of radial velocity measurements, transit observations, direct imaging, and a few other, more esoteric methods, as of March 2020, there were 3,105 confirmed solar systems beyond our own, containing a total of 4,187 well-established exoplanets.* To-

*For the latest numbers, go to: http://www.exoplanet.eu/.

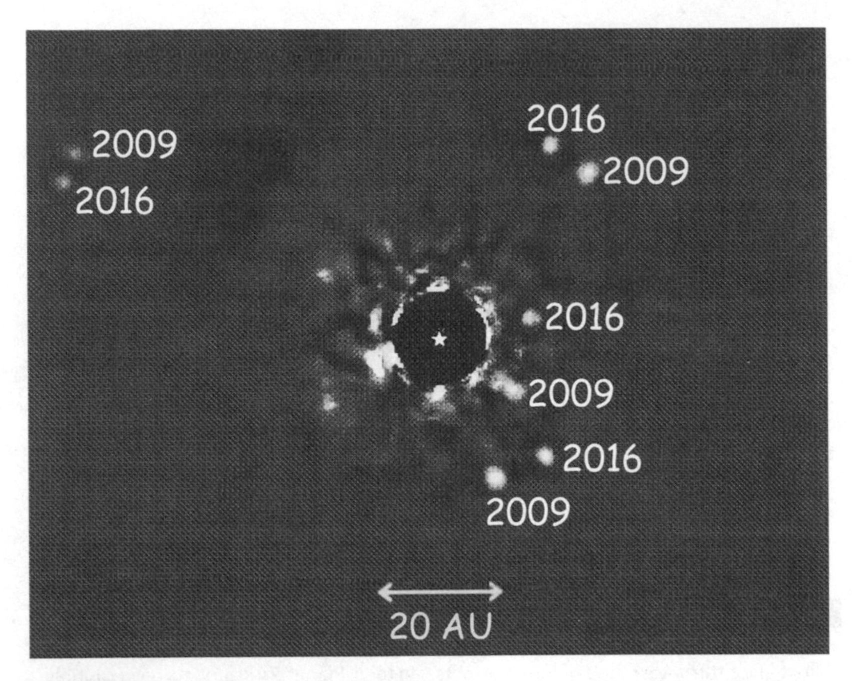

Figure 9.24 Shown are merged infrared images taken in November 2009 and May 2016 of four young (and thus still glowing) planets orbiting the star HR 8977, illustrating their orbital motion. To prevent the light from the star overwhelming that of the planets, it was suppressed using a coronagraph; the position of the star is indicated. (Courtesy of Christian Marois, NRC Canada)

gether, these planets form a veritable zoo, ranging from small, rocky worlds through "super-Earths" and "mini-Neptunes" to the ice and gas giants (fig. 9.25). Terrestrial planets, ice giants, and gas giants we've seen since these are well represented in our Solar System. Super-Earths and mini-Neptunes are not. The former, which range from about 1.1 to about 1.7 times the diameter of Earth and about 1.4 to five times the mass of Earth, exhibit densities—and thus a bulk composition—similar to that of our planet. Planets weighing in at more than about five times the Earth, however, are generally of much lower density, suggesting that this is the approximate cutoff above which a protoplanet can start to collect hydrogen and helium from the protoplanetary disk, ultimately to become an ice or gas giant.

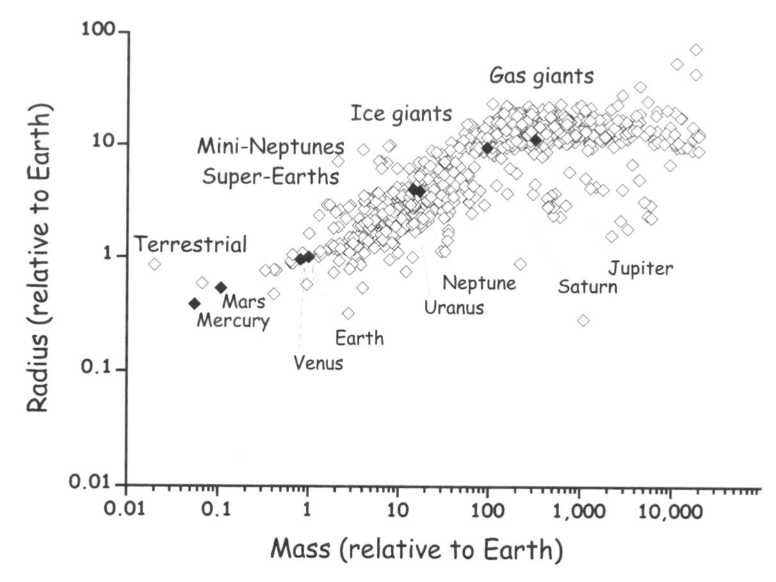

Figure 9.25 Plotted are the diameters and masses of the first few hundred planets (both exoplanets and planets in our Solar system) for which both values are known, illustrating the wide range of forms planets can take. Planetary scientists have roughly ordered this planetary zoo into terrestrial planets, super-Earths and mini-Neptunes, ice giants, and gas giants. Super-Earths and mini-Neptunes are without representation in our Solar System. The largest of the gas giants weigh in at 1,000 times the size of the Earth. Objects still greater in mass support nuclear fusion and thus are stars (the smallest being "brown dwarfs") and not planets. (Data courtesy of http://www.exoplanet.eu/)

A noteworthy example of the many solar systems that have been characterized is that of the star TRAPPIST-1. An amazing seven rocky planets have been identified in orbit around this extremely small, dim red dwarf, which weighs in at less than a tenth of the mass of the Sun. But although the luminosity of TRAPPIST-1 is 2,000 times less than that of the Sun, this planetary system is so compact (even the outermost known planet orbits six times closer to its star than Mercury does to our Sun) that two or three of the TRAPPIST-1 planets are thought to be in the habitable zone (fig. 9.26). Moreover, all seven are perhaps surprisingly Earth-like, with diameters ranging from 20% less to 15% greater than that of the Earth and masses ranging from 33% less to just 10% greater than the Earth. Consistent with this, the bulk

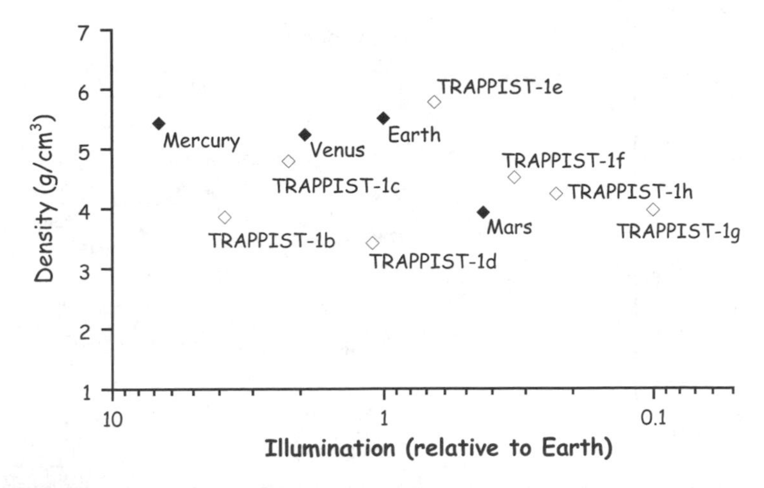

Figure 9.26 The TRAPPIST-1 solar system is comprised of seven, fairly Earth-like planets; their densities, for example, suggest that they are of similar bulk composition to the terrestrial planets in our Solar System (shown for comparison). And despite the dimness of their star, their close orbits mean that at least three of the planets are thought to reside within the habitable zone, where liquid water, if it is present, would be stable on their surfaces. This said, their close orbits also render it likely that all seven of these planets are tidally locked, greatly reducing their potential habitability. (Data courtesy of http://www.exoplanet.eu/)

densities of the TRAPPIST-1 planets range from slightly higher than that of Earth to slightly lower than that of Mars, implying typical, terrestrial bulk compositions. Finally, maintaining stability in a solar system containing seven planets all crammed closer to their star than Mercury is to the Sun requires orbital resonances akin to those that keep Jupiter's Galilean satellites in line. Thus, as is the case for the Galilean satellites, the tidal energy flux on these planets is greater than the geothermal energy flux of the Earth, providing another potentially life-supporting source of energy. This said, given their extremely close orbits, it is almost certain that all seven planets are tidally locked, greatly reducing the habitability of their surfaces.

Unfortunately, few if any of the 4,000 plus currently confirmed exoplanets are likely inhabitable. The tidal locking of the TRAPPIST-1 system illustrates one reason: a significant majority of the exoplanets discovered to date orbit within tidal locking distance of their parent

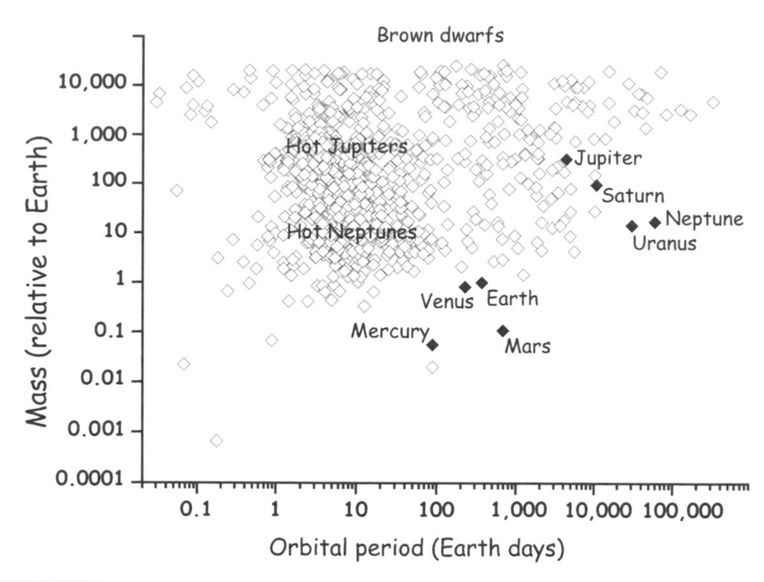

Figure 9.27 A plot of the masses and orbital periods of the first few hundred planets for which both values are known illustrates the strong "observer bias" in the data. That is, radial velocity and transit methods are most sensitive to large planets orbiting close to their stars, and thus "hot Jupiters" and "hot Neptunes" are overrepresented in the data. The extent to which the paucity of Earth-like exoplanets in Earth-like orbits reflects the rarity of such planets versus the difficulty of detecting them remains unclear. (Data courtesy of http://www.exoplanet.eu/)

stars (fig. 9.27), including four of the five exoplanets known to date to be reasonably Earth-like in size and temperature (counting two in the TRAPPIST-1 system). Likewise, the large majority of known exoplanets are ice or gas giants. This said, the seeming preponderance of planets that are (too) close to their stars and/or (too) massive to support habitable surfaces is due to observational biases and does not necessarily reflect the true diversity of plants and orbits. Specifically, both the radial velocity and transit methods of detecting exoplanets are more likely to identify large planets that are in close orbits. Larger, closer planets cause their companion stars to move more rapidly, improving the chances of observing a statically significant Doppler shift. Similarly, the probability of a planet being positioned such that it is seen to transit its star from the perspective of Earth is inversely proportional to its orbital distance (planets in closer orbits are more likely to pass in front of their stars), and larger planets block more of their star's light. For these reasons, the transit method is, like the radial velocity method, biased toward large planets in close orbits.

Things may be looking up, though. Specifically, the database of confirmed exoplanets is large enough that we can now start to use statistics to correct for the effects of the known planet-hunting biases in order to derive a truer picture of planetary populations. Doing so, we find that, for example, although so-called "hot Jupiters" (a gas giant planet with an orbital period of less than a week) are quite common among the *known* exoplanets (see fig. 9.27), they are actually rather rare, with only about 1% of Sun-like stars sporting them. In contrast, mini-Neptunes and super-Earths in intermediate orbits (10 to 100 days) are much more common; although our Sun does not host planets in either class, about half of all stars are thought to.

But what about potentially habitable planets? That is, on average, how many Earth-sized planets orbit in the habitable zone of a Sun-like star? Estimates of this vary widely. In part, this is because the "big-and-close" biases in exoplanet observation methods render the statistics poor: look again at fig. 9.27 and you'll see how few planets of Earth-like mass and orbit we've identified to date. But it also varies depending on how you define *Earth-sized* and *habitable zone*. For example, one survey of data collected by *Kepler* suggested that between 15% and 30% of Sun-like stars harbor in their habitable zone at least one Earth-sized planet, where *habitable zone* was defined as the region receiving between one-quarter and four times as much light flux as the Earth receives from the Sun, and *Earth-sized* was defined as having a planetary diameter falling between one and two times that of the Earth's. This definition of *habitable zone*, however, seems far too broad as it would place the inner edge of our Solar System's habitable zone inward of Venus's orbit and its outer edge beyond Mars. Likewise, this definition of *Earth-sized* is too broad since a planet larger than 1.7 Earth diameters is more likely to be a mini-Neptune than a super-Earth. Given these caveats, it's not surprising that other surveys use other, often narrower definitions of habitable zone and Earth-sized and, in doing so, come up with other probabilities. Indeed, even a quick review of the literature turns up values ranging from 124% (i.e., on average slightly more than one Earth-sized planet per habitable zone) to just 1.3%. Still, the argument that even 1% of all Sun-like stars harbor least one Earth-sized planet in their habitable zone is one of the more optimistic things we've written in this book relative to the possibility of life elsewhere, no?

And, of course, scientists are not done looking. NASA's *Transiting Exoplanet Survey Satellite* (*TESS*), which launched in 2018, is monitoring 200,000 bright stars, covering 85% of the entire sky, for indications of planetary transits with the aim to discover a range of exoplanets, including Earth-sized ones as well as gas giants. And while *TESS* is less sensitive than *Kepler* was, rendering it even more biased toward large, close-in planets, the greater brightness of the stars it is investigating renders ground-based follow-up studies far more accurate. By early 2020, *TESS* observations had already yielded more than 1,700 "objects of interest," including at least one Earth-sized planet, TOI700d, orbiting within the habitable zone of its star. A year after the start of the *TESS* mission, the European Space Agency launched its *Characterizing Exoplanets Satellite* (*CHEOPS*), whose goal is not to discover new exoplanets but instead to determine the size of known exoplanets, which will improve our knowledge of their density and composition and, with that, our understanding of planet formation and habitability. Looking farther ahead, in 2026, the European Space Agency anticipates the launch of the *PLATO* (*Planetary Transits and Oscillations of Stars*) mission, which will use 34 separate small telescopes to scan a million stars, searching for new exoplanets with sensitivity approaching that of *Kepler*.

Conclusions

What, then, is the bottom line on habitable places in our Universe? It is clear that, in the Solar System, neither Venus nor Mars lies in the continuously habitable zone. And while Mars (and perhaps even Venus) may have hosted liquid water once upon a time, the surfaces of both planets are far too dry (and cold and hot, respectively) to support life now. Still, big questions remain. Did life arise on Mars when it was more clement and then, perhaps, flee to still-populated habitats beneath the surface? And when nonsolar energy sources (e.g., tidal heating) are available to keep things warm and moist, can life arise in places, like Europa, that fall far outside the classical habitable zone? And what of life around other stars? In the next chapter, we detail humanity's efforts to answer these exciting questions by searching for evidence of life in these far-flung places.

And what about the *Apollo* astronauts? Obviously, they—and the rest of the biosphere—survived. Six months after the *Apollo 11* mission, the astronauts of *Apollo 12* were similarly quarantined. When they emerged from their three-week isolation unscathed, the decision was made that the Moon posed no threat, and none of the four remaining *Apollo* missions subjected their astronauts to the same isolation. Not that these "planetary quarantine" issues are behind us. While the highly tentative launch date of NASA's Mars sample-return mission is "sometime before 2030," its proposed date keeps moving forward at more than one year per year (i.e., it is receding rapidly into the future), but someday, we presume and hope, we'll face a decision about the quarantine of samples from our neighboring planet. As far back as 1997, a National Research Council report argued that, although the probability that such samples will contain pathological or environmentally dangerous organisms is low, we should not simply assume the risk is zero. The consensus then and now is that such samples should be delivered to a combined quarantine and research facility unlike any other in existence—one that is capable of protecting the scientific integrity of the extraterrestrial samples (preventing their contamination with Terrestrial substances) while protecting Earth's environment from exposure to potentially dangerous organisms from Mars.

Further Reading

Venus

Grinspoon, David H. *Venus Revealed.* New York: Helix Books, 1996.

A Lake on Mars

Orosei, R., S. E. Lauro, E. Pettinelli, A. Cicchetti, M. Coradini, B. Cosciotti, F. Di Paolo, et al. "Radar Evidence of Subglacial Liquid Water on Mars." *Science* 361, no. 6401 (2018): 490–93.

The Galilean Satellites

Showman, A. P., and R. Malhotra. "The Galilean Satellites." *Science* 286, no. 5437 (1999): 77–84.

Europa as a Potential Abode for Life

Pappalardo, Robert T., James W. Head, and Ronald Greeley. "The Hidden Ocean of Europa." *Scientific American* 281, no. 4 (1999): 54–63.

Titan

Lorenz, Ralph, and Jacqueline Mitton. *Titan Unveiled*. Princeton, NJ: Princeton University Press, 2008.

Exoplanets

Petigura, Erik A., Andrew W. Howard, and Geoffrey W. Marcy. "Prevalence of Earth-Size Planets Orbiting Sun-Like Stars." *Proceedings of the National Academy of Sciences USA* 110, no. 48 (2013): 19273–78.

The Extrasolar Planets Encyclopaedia. Exoplanet.eu. http://exoplanet.eu/. Accessed August 24, 2020.

Chapter 10

The Search for ET

The spacecraft was, at the time, the most sophisticated robot that its creators had ever built. Traveling through the planetary system of a yellow star a third of the way out from the center of the Milky Way, it skimmed just 1,000 km above the surface of a medium-sized, rocky planet. Visually, the planet was unusual. For example, fully three-quarters of its surface was blue-green and appeared perfectly flat to the onboard cameras, which could resolve details as small as a hundred meters over the patch of ground it approached most closely. The remaining quarter of the planet was varied in color and quite rugged. Oddly, however, and unlike nearly every other solid body in this planetary system, the surface of the planet was devoid of any obvious impact craters. Clearly, erosion was erasing craters faster than the rate with which they were formed. Much of the planet, particularly the regions near its equator, was covered with a large quantity of some material that strongly absorbed both red and blue light, giving these regions a greenish tinge. The onboard infrared spectrometer detected water vapor in the atmosphere at concentrations that would be saturating given the planet's temperature, which the probe's spectrometers had determined to be approximately 20°C. Maintaining saturation meant that the planet must have significant reservoirs of liquid water in equilibrium with its atmosphere. Could the flat, blue-green areas be this reservoir? The same onboard spectrometer detected clear spectral fingerprints of a fraction of a percent of carbon dioxide and parts-per-million quantities of methane. Meanwhile, the craft's UV-visible spectrometer detected large amounts of molecular oxygen (O_2) and some ozone (O_3).

The question on everyone's mind was, not surprisingly, did the planet harbor life? The high oxygen content of its atmosphere hinted that it might. Oxygen is extremely reactive; it is the second most strongly oxidizing element in the periodic table and thus is unlikely to remain in a planetary atmosphere unless it is constantly replenished. This is all the more true when, as was the case here, geology is constantly churning the surface of the planet and exposing fresh rocks for the oxygen to react with. Still, oxygen alone is not proof of life. Nonbiological processes such as the photolysis of water also produce this reactive gas, although they would be stretched to produce so much of it. Moreover, photolysis also produces reactive hydroxyl radicals (•OH) that would have destroyed any ozone had the oxygen been created solely by this mechanism. Still more telling was the simultaneous presence of oxygen and methane. Given the avidity with which the two react, even the trace amounts of methane detected were a hundred orders of magnitude higher than would be expected at chemical equilibrium; at equilibrium, not a single methane molecule should exist in the planet's atmosphere, and yet here it was in parts-per-million concentrations. And at these temperatures, the half-life of methane in an oxygen-rich atmosphere is only about a decade, suggesting that something was replenishing the planet's supply on a timescale quite rapid relative to geology. This massive, actively maintained disequilibrium seemed the surest signature of life on the planet.

And if there was life, was there intelligent life? The imagers aboard the spacecraft saw no signs of roads or cities or other artificial creations, but perhaps this was to be expected given the modest resolution of the probe's cameras and the fact that its closest approach was over what appeared to be a sheet of ice at the south pole. Likewise, because the craft passed over the daylight side of the planet, it had no opportunity to look for artificial lights. But its radio-science instruments, built to monitor the motions of charged particles in planetary magnetic fields, did record powerful, perfectly regular radio pulses—a sign of intelligence? The case for this was less clear than the case for life itself, but even the skeptics had to wonder.

Looking for Life

Considering our natural environment with its meadows, trees, and large animals, a naive observer might be forgiven for thinking that life should be easy to detect. But these highly visible, macroscopic forms of life appeared relatively late in the evolution of life on Earth. During the 70% or 80% of its multibillion-year residency on Earth, life was represented exclusively by microbes. And in many extreme environments, microbes still are the *only* forms of life. In many cases, finding and identifying these bacteria and archaea remains a serious technical challenge to microbiology. For example, metagenomic analysis (see, again, sidebar 7.2) of large collections of DNA randomly collected from samples of ocean water indicates that more than 90% of the bacteria in any given sample are species that have never been cultivated in the laboratory and formally identified. It is probably a good bet that an even smaller fraction of the myriad of microbes living in cracks in rocks deep beneath the Earth's surface, or perhaps even suspended in clouds, have been identified. And if it is so hard to identify these organisms here, on our very own planet (see chapter 8), how feasible will it be to identify life elsewhere?

There are both obvious and not-so-obvious reasons that the identification of extraterrestrial life will be difficult.* At the obvious end of the spectrum, we have the problem that opportunities to bring extraterrestrial samples into the laboratory for intensive study are rather rare (although they do exist, as we'll show later). A second reason is perhaps less obvious but much more profound: how would we recognize extraterrestrial life if we saw it? Given the far-reaching biochemical homologies among all Terrestrial species, it is clear that all life on Earth shares a common origin. And because all known cellular life on Earth employs, for example, ribosomes, the search for life in new Terrestrial environments (once the possibility of contamination has been excluded) can be reduced to the detection of novel ribosomal RNA

*Or even the identification of a "shadow biosphere" on Earth composed of microorganisms that arose from an independent origins-of-life event or split off before the last common ancestor of all known life and thus are based on biochemistry that differs radically from that of all known life.

genes. The assumption that extraterrestrial life shares this trait, however, is extremely weak at best; after all, the biochemistry of a truly alien life form may not be similar to ours. So how do we detect life while making the fewest possible assumptions about what the chemistry of that life will be? This is not an easy question to answer. To delve into it, let's look at how it has been done (on the few occasions when it has been done!) in the past and what has been proposed for the future, starting with our neighboring planet that has so often hosted imaginary life forms.

Life in the Clouds of Venus

In September 2020, astrobiology rose to the top of the news agenda for a few short days when radio astronomer Jane Greaves of Cardiff University announced the detection of 20-parts-per-billion phosphine (PH_3) high in the Venusian atmosphere. Joining in the announcement, planetary scientist William Bains of MIT argued that (1) no known chemical process that could occur on the surface or in the atmosphere of Venus could account for this much phosphine, and (2) phosphine is a known biosignature on Earth. But, seriously, *life on Venus*?

As we discussed in chapter 3, it is thought that early Venus was likely far cooler and wetter than the oven-hot, bone-dry surface conditions we see today but that photolysis and the resulting Jeans escape of hydrogen to space caused Venus to dry out, oxidize, and, ultimately become a greenhouse gas–driven hell. Conversely, though, high in the Venusian clouds, many tens of kilometers above the surface, conditions are more clement, approaching the temperatures and pressures seen at the surface of the Earth. On this basis, there have been suggestions that if life arose on Venus during some earlier, more habitable period, it may have evolved to survive in the cloud droplets as some bacteria do on our planet.

Given the theoretical possibility of microorganisms in the cloud droplets of Venus, is the detection of atmospheric phosphine a biosignature? The argument has two potential weaknesses. First, the detection of phosphine via microwave spectroscopy from Earth is not easy and requires significant data interpretation that is controversial and may prove wrong. Consistent with this, none of the many atmospheric entry probes ever detected the gas in Venus's atmosphere (see, also, the controversial story of methane in the Martian atmosphere later in this chapter). The second weakness is phosphine as a biomarker. On Earth, phosphine is produced by anaerobic microorganisms living in environments lacking oxygen. These organisms harvest electrons from local reductants, such as decaying plant material, and transfer them to phosphates (typically PO_3^-), producing phosphine and a small amount of energy that the cells can use to drive growth and reproduction. But although the atmosphere of Venus does contain some phosphates (phosphorus trioxide, P_4O_6, was detected in the atmosphere by the Soviet *Vega* landers at a level of around 2 parts per million), the Venusian atmosphere is quite oxidized. Enough so that, under the conditions found, the reduction of phosphates to produce phosphine would not provide energy, suggesting that, even if its presence is confirmed, any phosphine found there is unlikely to be the product of Venusian cloud dwellers.

Life on Mars?

Orbiters such as *Mariner 9*, the *Viking* orbiters, and the more recent *Mars Global Surveyor* (see table 9.1) have been invaluable in charting Mars's topography and, crudely, its mineralogy. The detailed search for past or present life on Mars, in contrast, has focused almost entirely on landing missions. The Soviets were the first to attempt to land craft on the Martian surface, but they failed to get scientific data back from any of them. In November 1971, for example, the landing probe of the *Mars 2* mission became the first human-made object to reach the surface of the Red Planet, but it crashed and returned no data. A few days later its sister craft, *Mars 3*, touched down softly but fell silent 20 seconds later. Equipped with a landing program that could not be altered during its approach to the planet, the craft landed in the middle of a major sand storm and lost contact. Two years later, the *Mars 6* lander stopped transmitting some 12 km above the Martian surface, and *Mars 7* accidentally released its lander some four hours too early, put-

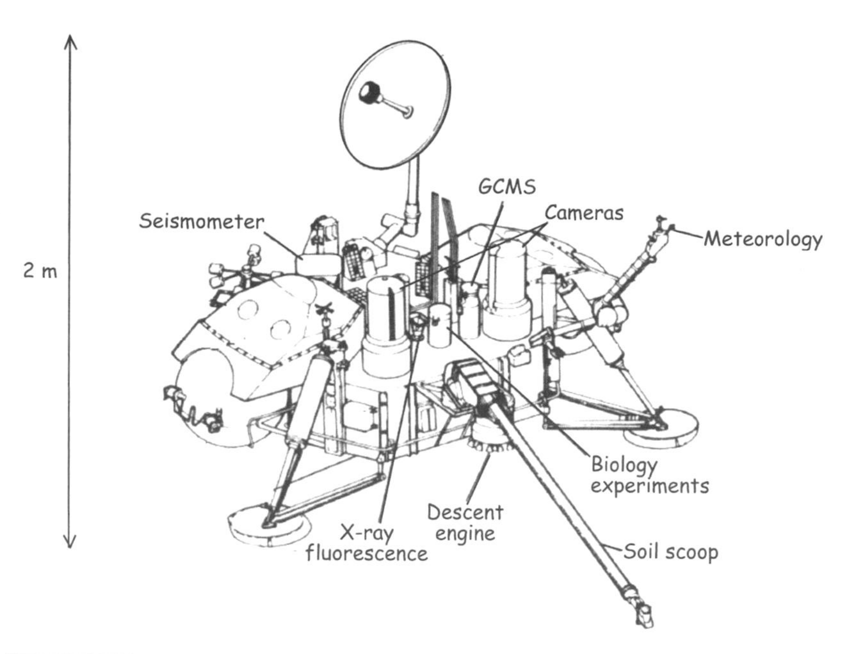

Figure 10.1 The *Viking* landers, which crammed a complete biology laboratory into a volume equivalent to a carry-on suitcase, carried out what remain the most sophisticated search-for-life studies yet conducted by humans on another planet. GCMS stands for gas chromatograph / mass spectrometer. (Courtesy of NASA)

ting it in a solar orbit that missed Mars by 1,300 km. Finally, in 1976, humanity obtained its first close-up look at the surface of Mars, when twin American landers, *Viking 1* and *2*, successfully touched down. The first images radioed back from Mars showed nothing more than sweeps of rocky red desert, but as the first images ever taken on the surface of that planet,* they were greeted with enthusiasm and graced the front pages of newspapers around the world.

Exciting as the first images from the surface of Mars were, perhaps more intriguing were the *biology* experiments conducted by the twin

*They were not, however, the first photos ever taken from the surface of another planet. The Soviet Venus landers *Venera 9* and *10* had snapped two photos each of the surface of Venus (before being fried by Venus's lead-melting surface conditions) just eight months earlier. Cold War competition between the United States and the Soviet Union did wonders for the exploration of the Solar System.

landers, starting just a few days after arrival. The life-search systems of the *Viking* missions, developed by a large, interdisciplinary team led by NASA, included three different biochemical experiments in what was, more or less, an entire state-of-the art analytical laboratory crammed into 27 L of space on a mass- and power-limited spacecraft (fig. 10.1).

The first *Viking* biology experiment, the labeled release (LR) experiment, was headed by Gilbert Levin, a one-time sanitary engineer in California who developed techniques to detect bacterial contaminants in drinking water and went on to become a prominent astrobiologist. The LR experiment was aimed at detecting catabolic metabolism (fig. 10.2)—that is, metabolism in which organic molecules supplied as food are broken down into still simpler, lower-energy carbon compounds that are then "exhaled" into the environment. In the LR experiment, small amounts of a dilute aqueous "nutrient broth" were added

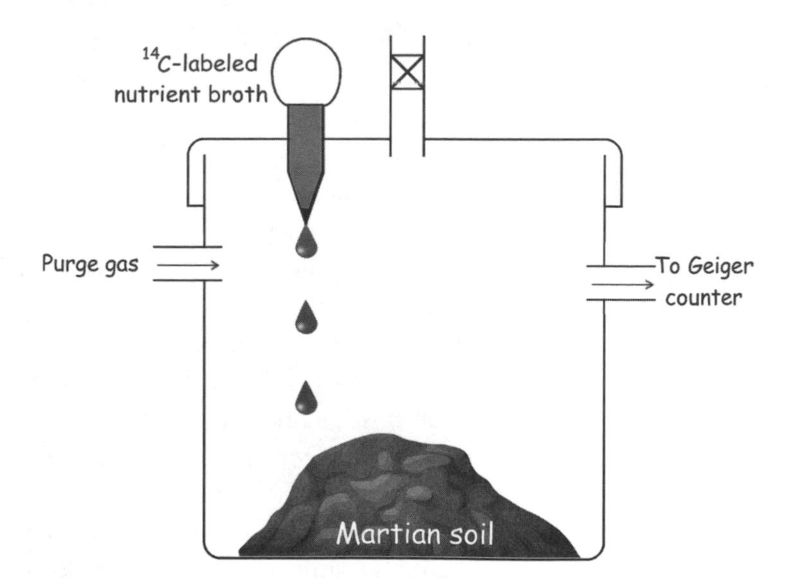

Figure 10.2 The *Viking* labeled release experiment searched for catabolic metabolism—that is, the oxidation of simple, reduced carbon compounds. To do so, it wet a sample of Martian soil with a dilute "nutrient broth" comprised of seven Miller-Urey products labeled with radioactive carbon-14 (^{14}C) and then monitored for the production of ^{14}C-labeled carbon dioxide.

to a sample of Martian soil. The broth contained seven simple organic molecules—formate, glycolate, glycine, as well as both enantiomers of alanine and lactate—each of which had been labeled with the radioisotope ^{14}C. The thought behind the experiment was that Martian microbes in the soil would eat the labeled nutrients and exhale radioactive carbon dioxide, carbon monoxide, or perhaps methane. To monitor for such catabolic reactions, the atmosphere over the soil sample was tested using a Geiger counter to detect any radioactive gases released. A potential failing of the experiment is that it assumed Martian organisms would eat the same materials as Terrestrial organisms. To reduce this concern, all the molecules in the broth were common Miller-Urey products; given the "universal" nature of that chemistry, the assumption that these molecules would be amenable to Martian biology seems reasonable.

The initial LR results radioed back to Earth seemed to support the idea that the Martian soil contained life (fig. 10.3). Specifically, after the first addition of nutrient broth to the soil sample, the level of radioactive gas slowly rose before leveling off after several days. Heat-treating a soil sample at 165°C abolished the effect, and no labeled gas was observed. Heating to 50°C did not entirely abolish the effect but led to very significant decreases in gas output. Similar effects had been seen with Terrestrial soils, with gas production leveling off after a few days when the bacteria in the sample had consumed all the added nutrients. Terrestrial samples similarly failed to emit gas if sterilized at 165°C and produced much less gas if first heated to 50°C. With Terrestrial samples, though, the addition of more nutrients after the initial incubation would produce still more radioactive gas as the dormant bacteria sprang into action to consume the new dose of food. This was not true of the Martian soil; on Mars, the second and third nutrient injections did not produce any further release of labeled gas.

The pyrolytic release (PR) experiment, headed by Norman Horowitz (1915–2005) from Caltech, also aimed at carbon metabolism, but this time researchers were looking for metabolism running in the opposite direction: metabolism in the sense of organisms taking up carbon dioxide (or carbon monoxide, a small amount of which is present in the Martian atmosphere) and producing higher molecular

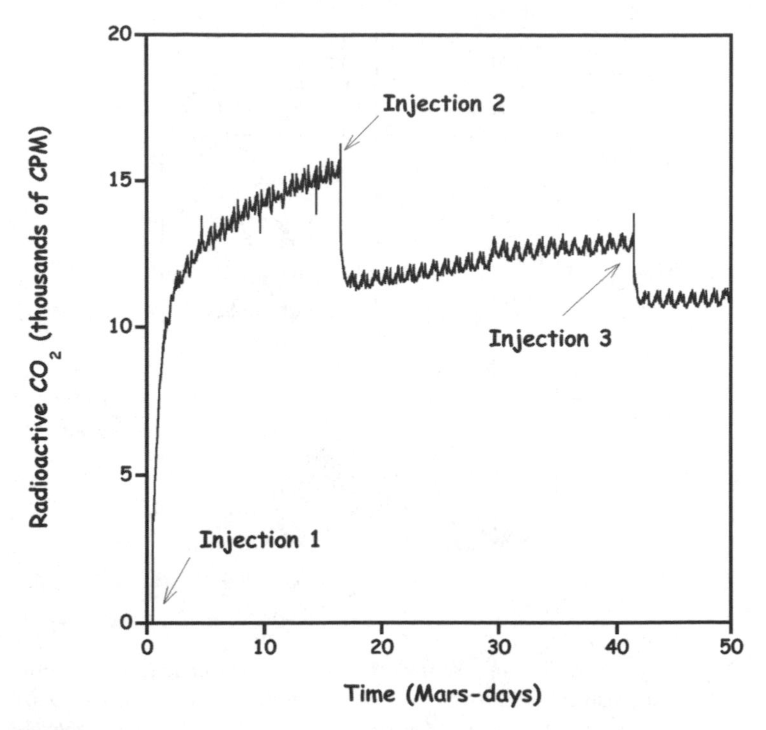

Figure 10.3 The first *Viking* labeled release results seemed to suggest that the Martian soil contains life: after the first addition of radiolabeled nutrient broth, the level of radioactive carbon dioxide in the chamber (in counts per minute, CPM) rose slowly for several Mars-days before leveling off. With Terrestrial samples, a second nutrient injection would have led to the production of a yet larger burst of radioactive gas as the now larger population of bacteria consumed the newly offered food. However, this was not true of the Martian soil: no further release of labeled carbon dioxide was seen upon either the second or third nutrient injections.

weight carbon compounds of their own (fig. 10.4). Soil samples were placed in the test chamber and, in various experiments, were incubated from five to 139 days in the presence of ^{14}C-labeled carbon dioxide and carbon monoxide in a ratio close to that observed in the Martian atmosphere. Given that the most obvious carbon-fixing metabolism on Earth is photosynthesis, the experiments were run both with and without light mimicking the Sun. After the incubation period, the chamber was purged with helium to evacuate the labeled

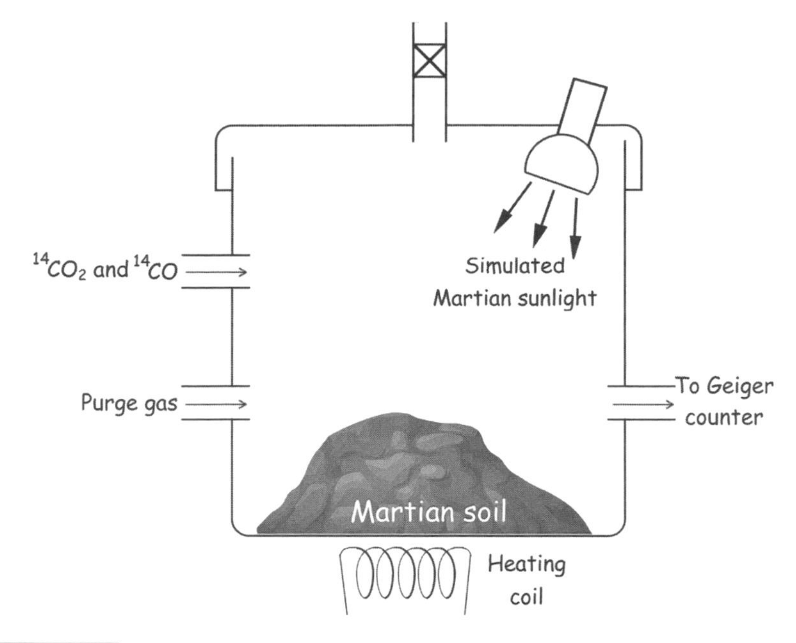

Figure 10.4 The *Viking* pyrolytic release experiment searched for anabolic metabolism—that is, the fixing of carbon dioxide or carbon monoxide into reduced carbon compounds. After incubation with ^{14}C-labeled carbon dioxide and carbon monoxide (in the presence or absence of light), the soil was heated to 635°C to pyrolyze (char) any fixed carbon into volatile compounds that could be detected by a Geiger counter.

carbon monoxide and carbon dioxide and then heated to 635°C to pyrolize any fixed carbon compounds into volatile gases. Any radiolabeled gases released were detected using a Geiger counter.

In total, the two *Viking* landers investigated nine soil samples with the PR experiment. After incubation, seven of the nine experiments produced detectable peaks in the quantity of radioactive carbon atoms released from the soil on heating (fig. 10.5). The peaks in question were small (roughly equivalent to the metabolic intake of 1,000 bacterial cells), but they were much larger than the peaks seen for sterile Terrestrial soil samples, suggesting that the experiment had, indeed, detected life. In contrast, soil samples that were sterilized by heating to as little as 50°C before addition of the radiolabeled gas did not produce any detectable peak, again consistent with a biological interpretation of the PR results. But how firm was the conclusion that the PR experiment had detected life? A weakness in the conclusion was that the conditions under which the PR experiment detected life were different from those of the LR experiment. The putative organisms in the LR experiment were not killed even when heated to 50°C, whereas the putative organisms responsible for the PR result were killed at far lower temperatures.

The third *Viking* biology experiment, the gas exchange (GEx) experiment, headed by Vance Oyama (1922–1998) from NASA's Ames Research Center in California, was also aimed at the gaseous products of metabolism. However, this experiment used a gas chromatograph to detect not only carbon dioxide (as in the LR experiment)

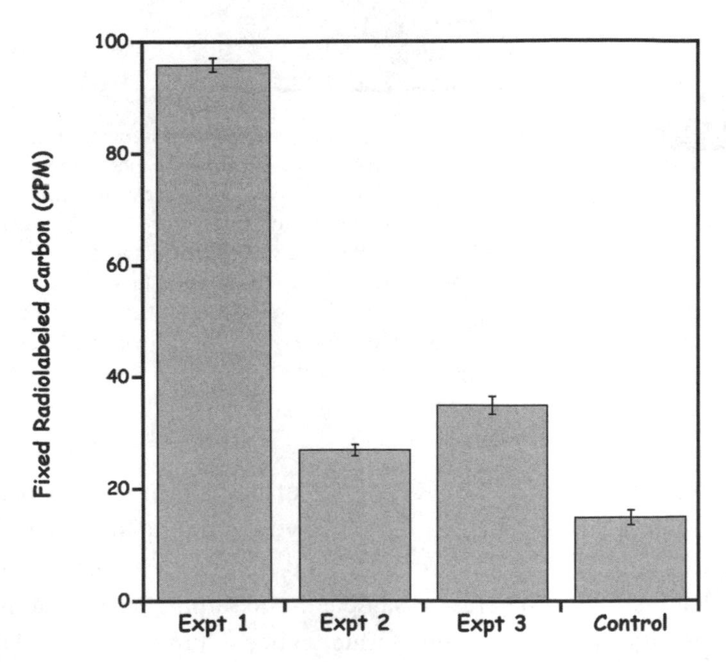

Figure 10.5 The results of the *Viking* pyrolytic release experiments conducted with four different soil samples from the *Viking 1* site show that small and varying but statistically significant (error limits indicated by the "antennas" at the tops of the bars) amounts of fixed radiolabeled carbon (in counts per minute, CPM) were detected in three experiments. The peak produced by a fourth sample, which served as a control and had been "sterilized" before incubation by heating to 175°C for three hours, was significantly smaller.

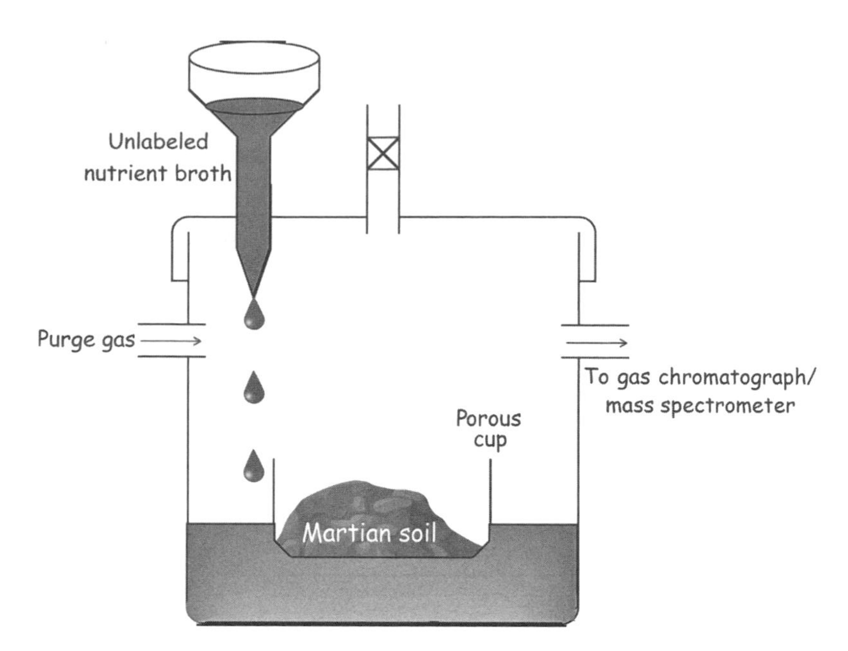

Figure 10.6 The *Viking* gas exchange experiment measured the gaseous products of metabolism. It used a gas chromatograph / mass spectrometer to monitor any exchange of gases, such as carbon dioxide for oxygen or vice versa, that might indicate metabolic activity. The soil sample was introduced into the test chamber and the chamber's atmosphere was sampled. The soil was then moistened with water vapor ("humid mode") and incubated, and the atmosphere was reexamined; any differences between the two atmospheric samples were interpreted as a sign of activity. Later, aqueous nutrient broth was added to the chamber to physically wet the soil ("wet mode," as illustrated here), and the gas analysis was continued.

and fixed carbon (as in the PR experiment) but also metabolic products lacking carbon, such as oxygen, hydrogen, and nitrogen (fig. 10.6). First, a soil sample was introduced into the test chamber and the chamber's atmosphere was sampled. The sample in the chamber was then moistened with a bit of water vapor ("humid mode"). After a period of incubation, the atmosphere in the chamber was reexamined, and any differences between the two atmosphere samples would be viewed as a sign of activity. Later, a significant amount of aqueous nutrient broth was added to the chamber such that the soil became physically wet ("wet mode"), and the gas analysis was continued. As a control experiment, some samples were "sterilized" at 145°C before analysis.

The results of the GEx experiment were surprising (fig. 10.7). In the humid mode, a large quantity of oxygen emerged from the soil sample almost immediately, but then the rate of oxygen production quickly subsided. In the wet mode, about one-third of the carbon dioxide in the chamber disappeared, presumably absorbed by something in the soil. In contrast to the humid-mode experiment, however, no oxygen was observed in the wet mode. In fact, the trace amount of oxygen initially present in the chamber dropped, suggesting that oxygen, too, was be-

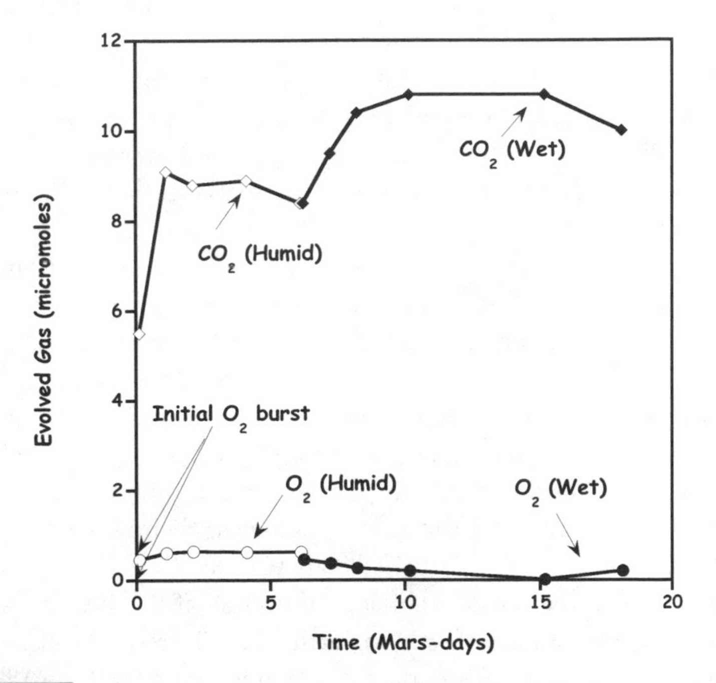

Figure 10.7 The *Viking* gas exchange experiment detected a modest increase in carbon dioxide (upper curve) and a huge surge of oxygen (from initially undetectable levels; lower curve) as soon as water vapor was added to the soil-containing chamber. But the rate of oxygen production promptly slowed. In the wet mode, initiated on the sixth Mars-day, about one-third of the carbon dioxide in the chamber disappeared, presumably absorbed by something in the soil, but no oxygen was produced. In fact, the trace amount of oxygen initially present in the chamber dropped, suggesting that oxygen, too, was being absorbed or consumed under these conditions.

ing absorbed. Was this consistent with life? The humid-mode production of oxygen was clearly reminiscent of photosynthesis. But it happened in the dark! Moreover, the oxygen was released in a big, sudden burst; bacteria-laden Terrestrial samples, in contrast, start releasing gases slowly and then more rapidly as the microorganisms multiply. Perhaps most tellingly, the heat-treated Martian soil sample produced just as much oxygen as the untreated samples.

What does this jumble of conflicting results mean? For most of the *Viking* scientists, the final conclusion was that the missions failed to detect life in the Martian soil. In large part, these doubts were driven by the results of yet another *Viking* experiment: a detailed investigation of the soil composition using a combination of gas chromatography and mass spectrometry (GCMS). The GCMS experiment utterly failed to uncover any organic molecules, even at a parts-per-billion level. Indeed, in the end, the instrument's sensitivity was turned up so high that it finally detected traces of the solvents used to clean the soil chamber a year earlier, when the spacecraft was still on the ground in Florida.

On our "life-infested" planet, one finds high concentrations of organic molecules in most soil samples (save, as noted in chapter 8, the most hyperarid regions of the Atacama Desert). For every living organism in Terrestrial soil, there are at least traces of thousands of dead ones, and thus life is always associated with large amounts of carbon compounds. But, if "no carbon" is the equivalent of saying "no life," what accounts for the seemingly positive LR and GEx results? Could they just be the result of some odd abiological chemistry occurring on the cold, dry surface of the Red Planet?

Hints as to the possible abiological origins of the *Viking* results have come from studies of soils on Earth. Many Terrestrial soils (even sterile ones) take up carbon dioxide when warmed (and the *Viking* biology experiments were conducted at temperatures higher than the ambient temperature on Mars), and this adsorbed carbon dioxide can often be liberated by heating. Thus, simple physical processes readily account for the results of the PR experiment. But what about the LR and GEx results? They might well be due to chemistry. For example, Albert Yen of the Jet Propulsion Laboratory has shown that, under extremely cold and dry conditions in a carbon dioxide atmosphere, UV light (remember: Mars lacks an ozone layer, so the surface is bathed in UV) can cause carbon dioxide to react with soils to produce various oxidizers, including highly reactive superoxides (salts containing O_2^-). When mixed with small organic molecules, superoxides readily oxidize them to carbon dioxide, which may account for the LR result.

Superoxide chemistry could also account for the puzzling results seen when more nutrients were added to the soil in the LR experiment; because life multiplies, the amount of gas should have increased when a second or third batch of nutrients was added, but if the effect was due to a chemical being consumed in the first reaction, no new gas would be expected. Lastly, many superoxides are relatively unstable and are destroyed at elevated temperatures, also accounting for the "sterilization" seen in the LR experiment. Superoxides might also explain the GEx humid-mode result: many superoxides react with water to produce oxygen, which would account for the initial burst of oxygen production, but the superoxides are quickly consumed, and this would account for the lack of any additional oxygen production after the initial burst.

Given the above observations, most of the experts involved at the time begrudgingly concurred that the *Viking* results were negative—no life detected—even though several of the experiments had produced results that, before the mission's launch, would have been considered telltale signs of life. Most scientists, that is, but not all. More than 40 years later, Gilbert Levin, the designer of the LR experiment, continued to argue (in, for example, the October 2019 issue of *Scientific American*) that the positive results of that experiment are strongly compelling evidence for life on Mars. For example, despite some significant efforts, no one has ever precisely duplicated the Mars LR results with any abiological oxidizer. Moreover, in 2008, the *Phoenix* lander identified high concentrations of the oxidizer perchlorate (ClO_4^-) in the soil of its cold, northerly landing site. And while perchlorate is too poor an oxidizer to reproduce the LR results (under the conditions of that experiment perchlorate does not oxidize organics), it does oxidize, and thus destroy, organics at the higher temperatures used in the *Viking* GCMS experiment. NASA astrobiologist Chris McKay has estimated, in fact, that if *Phoenix*-like levels of perchlorates were present in the *Viking* samples, the organic content of the Martian soil could have been as high as 0.1% and still would have produced the (false) negative result that the GCMS experiment returned. Thus, while the conventional wisdom regarding the *Viking* biology experiments still points to "no evidence for life," recent years have seen at least a small shift toward "inconclusive evidence."

Even if we assume the *Viking* experimental results truly are negative, what does this really mean in terms of life on Mars? The *Viking* experiments, like almost everything in the search for extraterrestrial life, were prejudiced by what we know about life here on Earth. The landers, for example, were designed to look for carbon-based life (as we noted in chapter 1, this is not an unreasonable assumption) exposed on the surface of the planet (perhaps a poorer assumption, given the intense solar UV and cold, dry conditions of the surface). And not just any carbon-based life; the LR experiment assumed that we know something about the Martian microorganisms—that they eat the same simple carbon compounds loved by Terrestrial bacteria. Likewise, the GEx experiment assumed that photosynthesis occurs on Mars as well. While these may (or may not) be reasonable assumptions, we have to be aware that any such assumption constrains our chances of finding extraterrestrial life.

Post-Viking Exploration of Mars

Although the *Viking* probes continued to send data back to Earth for more than five years and thus stand out as a major success story among the many failed Mars missions of the twentieth century, the disappointment of their biology experiments, along with the measurements suggesting that general conditions are too hostile to support life, discouraged further Mars exploration for almost two decades. Only after the discovery of chemosynthetic food webs in Earth's deep-sea and underground ecosystems (see chapter 8) did scientists realize that Martian life might in fact be hidden underground or within its ice caps. By the mid-1990s, the astonishing discoveries of life in extreme conditions on Earth had helped scientists regain some optimism about the possibility of past or present life on Mars.

Inspired in part by this new perspective on the history of the planet, NASA's Mars program resumed in the early 1990s, but it suffered an immediate setback in August 1993 when the probe *Mars Observer* fell silent as it attempted to brake into Mars orbit. In response to the loss of this billion-dollar, multiyear mission—not to mention further troubles with a crippling antenna problem on the otherwise successful, decade-long Jupiter project *Galileo*, as well as the initially near-sighted mirror on the multibillion-dollar Hubble Space Telescope—NASA launched a new class of missions, the Discovery missions, that were intended to be "faster, better, cheaper." The first of these, the *Mars Global Surveyor* orbiter and the *Mars Pathfinder* lander, arrived at Mars in 1997 after only a few years of design, fabrication, and testing (and for a relatively paltry $154 million and $256 million, respectively). In doing so, they opened a new and still ongoing era of intensive Mars exploration. And while neither of these missions was specifically designed to look for traces of life, they both laid important foundations for the later explorations of the planet and search for life.

Follow-on missions run by NASA as part of the quest to understand whether Mars has, or had, what it takes to harbor life have focused on the mantra "follow the water." As described in chapter 9, the *Mars Odyssey* orbiter carried a neutron spectrometer that detected abundant low-energy neutrons signifying hydrogen, no doubt as water, on the Martian surface. And the Mars Exploration Rovers, *Spirit* and *Opportunity*, which, as we noted in the last chapter, roamed the Martian surface for more than six and 14 Earth-years, respectively, used spectroscopy and imaging to probe for geological signs of past water on the Martian surface. But none of these missions were equipped to detect the signatures of life itself. Instead, they were aimed primarily at improving our knowledge of Martian geology and climate history, thus providing a clearer picture of where on the planet the remnants of Martian life might be hiding.

In contrast, the European Space Agency's probe *Mars Express*, which dropped into Mars orbit in 2003 (and was still going strong as of late 2020), has been using the most advanced infrared spectrometer sent to the planet to date to monitor for the potential methane biosignature in the Martian atmosphere. A year into the mission, its researchers reported the detection of methane at a level of about 10 parts per billion. This was quickly confirmed by studies using ground-based telescopes, which reported a plume of methane developed over 60 days in 2003, reaching a peak value of 45 ± 10 parts per billion, with methane levels falling to undetectable levels (≤8 parts per billion) both before and after this event. And in 2005, a second ground-based detection was reported at 10 parts per billion.

As is true in the Earth's atmosphere, methane is unstable in the Martian atmosphere. Specifically, due to the photolysis of water and carbon dioxide by UV light, the Martian atmosphere contains 0.17% oxygen, which should oxidize any Martian methane to carbon dioxide with a half-life of 200 to 300 years. Given this, the 10-parts-per-billion methane levels found by *Mars Express* would correspond to a production rate of approximately 150 tons per year, a far cry from the 15 million tons per year produced by life on Earth. Indeed, 150 tons per year is approximately the amount of methane produced by a mere 1,500 cows. Is this evidence, then, of at least a meager biosphere on the Red Planet? We don't know. And, unfortunately, the situation has only grown less clear with time.

First to follow up on the results of *Mars Express* was the *Curiosity* rover, which included a tunable laser spectrometer in its "sample analysis at Mars" (SAM) instrument suite that was suitable for monitoring atmospheric methane down to a few tens of parts per trillion. Most of its measurements fell in the range of 0.2 to 0.7 parts per billion, with the data showing a weak seasonal peak at the end of the northern Martian summer. But over two months, from the end of 2013 through early 2014, the level spiked to 7 parts per billion before again plunging tenfold back to baseline. And, in June 2019, the instrument detected its largest peak yet at 21 parts per billion, but by the time of the next measurement, performed a week later, things had again dropped to baseline. Such results suggest that something on Mars is producing methane episodically and that something else is probably destroying it a lot faster than the 200- or 300-year half-life estimated above. Looking from above at the same time, however, *Mars Express* failed to see the putative methane spike, as did the European Space Agency's newest Mars spacecraft, the *Trace Gas Orbiter*, which arrived at Mars in 2016 with the specific aim of characterizing, as the name implies, the trace gas components of the Martian atmosphere. More perplexing still, as of 2019, the latter orbiter had failed to *ever* detect *any* methane in the Martian atmosphere down to the 50-parts-per-trillion limit of its exquisitely sensitive spectrometer. In short, the jury is still out regarding the amount of methane in the Martian atmosphere and the nature (and even existence) of its strange fluctuations.

Placing aside for a moment our concerns regarding Martian methane, might it reflect biology? Perhaps. In 2020, for example, Deborah Maus, working in Dirk Schulze-Makuch's group at Technische Universität Berlin, showed that methanogenic archaea could produce methane when fed hydrogen and carbon dioxide under conditions reasonably approximating a summer day on Mars (provided the dreaded perchlorates found in the Martian soil by the *Phoenix* mission were omitted). Still, biology is not the only hypothesis that might explain the presence of parts-per-billion concentrations of methane: geological processes also produce this gas via the reaction of carbon dioxide and hot water with olivine, an igneous, reducing mineral that is also known to be reasonably plentiful on Mars. In fact, although not common, the gas exhaled by at least some Terrestrial volcanoes is up to 0.1% methane. A key question is thus whether the geology of Mars is active enough to produce the observed amounts of methane. As we described in chapter 9, images taken by the *Mars Express* orbiter indicate that some of the Martian volcanoes have been active in the past few million years, suggesting that the planet remains at least a little bit geologically active today, a claim supported by *InSight*'s confirmation that Mars remains seismically active. It appears that we cannot rule out geology as a source of the methane.

Methane, of course, is not the only potential biomarker one might hope to find on Mars. In this light, one of the more exciting instruments on the *Curiosity* rover, the SAM, is able to perform "evolved gas analysis," in which a soil or ground rock sample is heated to extreme temperatures to drive off any volatile components and to convert higher molecular weight organics into their more volatile breakdown products, all of which can then be identified using mass spectrometry. Using this approach, *Curiosity*'s SAM identified a number of low molecular weight, organic compounds when it investigated some approximately 3-billion-year-old sediments in the Pahrump Hills on the lower flank of Mount Sharp. These included a range of short alkanes (straight and branched molecules of the generic formula C_nH_{2n+2}), benzene (a cyclic compound of formula C_6H_6), toluene (benzene with an attached methyl group), chlorobenzene (benzene with a single chlorine attached), and the sulfur-containing organics methanethiol ($CH_3{-}SH$), dimethylsulfide ($CH_3{-}S{-}CH_3$), and thiophene (a cyclic compound with the formula C_4H_4S).

Are these molecules signs of past life on Mars? Perhaps. But, once again, perhaps not: when the same experiment was run back on Earth using samples of a Martian meteorite (more on these below) and of the Murchison meteorite (which does not come from Mars), both produced organics similar to those seen by *Curiosity*. This said, the relative amounts of the various compounds varied from sample to sample. Given this, perhaps all options are still on the table: Martian biology, Martian geology, or delivery to Mars from space.

Have Martians Landed on Earth?

The in situ exploration of Mars has seen spectacular successes in the first two decades of this century, courtesy of the three indefatigable rovers and the *Phoenix* lander, but plans for a mission that collects samples on Mars and returns them to Earth for detailed analysis have been postponed again and again with, as we noted in chapter 9, the current—if rather tentative—proposal being a joint US/European mission that launches in 2026 and returns to Earth 5 years later. Fortunately, however, while they are waiting for this perpetually delayed project to materialize, researchers interested in Martian samples already have at least some material to work with.

A class of meteorites called the SNCs have long been of particular interest to the scientists who study these things. The SNC meteorites were named after Shergotty (India), Nakhla (Egypt), and Chassigny (France), where the first examples of each of the three types were observed falling from the heavens.* The SNCs attracted attention for two reasons. The first was that, while they differed significantly from one another in terms of their mineralogy, their oxygen-isotope ratios clearly indicated that they were all chips off the same celestial body. The second reason was their surprising youth. Isotopic dating placed the age of these rocks between 1.2 billion and as little as 160 million years—far, far younger than any other meteorite that's been dated. Most meteorites, remember, come from asteroids, and asteroids are so small that they cooled off and stopped forming new rock shortly after the birth of the Solar System, so the isotopic ages of most meteorites are quite close to 4.56 billion years. The SNC meteorites must instead have come from a body that did not cool off and thus did not become geologically dead so rapidly. But which body? The observed isotopic abundances pointed to a single source for the SNCs, but they also ruled out the Earth and the Moon as that source. By elimination, Mars seemed a logical choice. But how to prove this? By comparing the SNCs with the results from the *Viking* missions. Small bubbles of entrapped atmosphere found in one of the SNC meteorites were analyzed and proved to have a composition in such extraordinarily good agreement with that of the current Martian atmosphere (fig. 10.8) that there is now no doubt: the SNCs are samples of Mars that we can hold in our hands and examine. Albeit not cheaply; samples of the Tissint Mars meteorite, which was found after it was seen falling on the Moroccan desert in July 2011, are currently going for about $1000 a gram on eBay.

How were the SNC meteorites delivered to our backyards? Occasionally, meteorite impacts are violent enough to catapult material into space (as described in chapter 3, our Moon is an extreme example!). Mars being relatively small (its surface gravity is one-third the Earth's) and having a thin atmosphere, it is not too difficult for an impact to launch Martian rocks into space. And given that Mars orbits between Earth and Jupiter, rocks that escape Mars's gravity and enter

*The Nakhla fall in 1911 killed a dog, which is the only record of death by meteorite on the books.

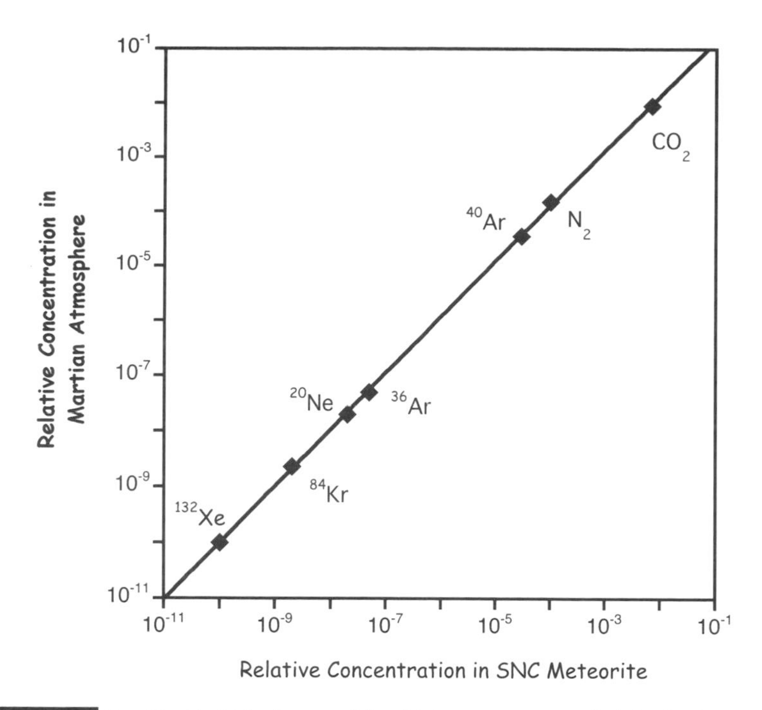

Figure 10.8 Gas-filled bubbles in one of the SNC meteorites provided the final confirmation of their origin on Mars. The composition of the bubbles is a near perfect match with the composition of the Martian atmosphere as defined by the *Viking* landers.

solar orbit can be fairly quickly (geologically speaking) perturbed by Jupiter's gravity into orbits that intersect with the Earth's and thus fall at our feet.*

And what does this have to do with our story of life in the cosmos? The link is a Martian meteorite that caused a major media frenzy in 1996 when, in August of that year, David McKay (1936–2013) claimed in the journal *Science* that his team at NASA's Johnson Space Center had identified evidence of life in the rock named ALH84001. ALH84001, so named because it was the first meteorite found in the Allen Hills

*The transport goes both ways. In 2006, Brett Gladman and his team at the University of British Columbia calculated that the impact that wiped out the dinosaurs would have launched 600 million rocks into solar orbit. Simulating the trajectories of these, they found that, within a million years, about 20 would make it all the way to Titan. If we find life elsewhere in the Solar System, it may have come from Earth long ago. Conversely, it's likewise possible that Earthly life started on Mars before making its way here.

ice field of Antarctica during the 1984 collecting season, is much older than all the other Martian meteorites: it solidified almost 4.5 billion years ago and thus represents a piece of Mars's oldest crust. Studies of the radioactive elements produced while the rock was exposed to cosmic rays in space indicated that it spent about 16 million years on its trip from Mars to Earth (see sidebar 10.1), and based on the decay of those isotopes (once it was on the Earth's surface and spared exposure to any further cosmic rays), ALH84001 is thought to have landed in Antarctica around 13,000 years ago. Like many other meteorites

Sidebar 10.1

From Mars to the Earth

The advent of conclusive proof that Mars rocks can travel through space to Earth revitalized, in a small way, the panspermia (transspermia?) theory that life travels between the stars to seed new planets. If rocks can travel between the planets, could they not bring life with them? Perhaps they can.

Computer simulations of the transit of rocks between Mars and Earth suggest that, on average, a rock blasted from Mars will orbit the Sun several million times before its orbit is perturbed enough that it crosses the Earth's. Studies of the transit time of the SNC meteorites confirm this result. The transit time, or lifetime, of meteorites in space can be determined because cosmic rays alter the composition of the rock minerals in measurable ways. When on a planet's surface, a rock is protected from exposure to cosmic rays. When in space, however, these energetic elementary particles produce characteristic noble gas isotopes in the rock, such as helium-3, neon-21, and argon-38. By monitoring the buildup of these three isotopes in the meteorite ALH84001, researchers have estimated the length of time it spent in space while in transit to the Earth at 16 to 17 million years.

If rocks can make it from Mars to Earth, could Martian life survive the journey? Living in a crack within a rock would help in one way: it would protect the microbe from the deathly effects of solar UV and, to a lesser but still significant extent, from cosmic rays. Still better, the same computer simulations that predicted a several-million-year mean transit time indicate that rare transits could take as little as a few years. Coupling this with estimates that, on average, several tons of Martian rocks make it to Earth each year, it is possible that Martian life—if there is any—could have hitched a ride and survived the trip. And although lifting material out against Earth's much stronger gravitational pull is more difficult, it is probable that Mars and Earth have been engaging in a two-way exchange of material since the origin of the Solar System.

Could Earth life have seeded Mars? Could Mars have seeded life on Earth? Obviously, short of identifying Earth-like organisms on Mars, we cannot answer this question. But should we ever find living organisms on Mars, we must keep this point in mind. They may be our long-lost relatives.

Figure 10.9 The rock shown here, ALH8004, was blasted off the surface of Mars some 16 million years ago, landed in Antarctica about 13,000 years ago, and was collected from the ice in 1984. In 1996, David McKay and colleagues stunned the world by announcing that the 3.6-billion-year-old carbonate globules in the meteorite (upper inset) contained evidence of past life on Mars. This evidence included "nanofossils" (lower inset), submicrometer-sized objects that somewhat resemble Terrestrial bacteria, though far smaller. (Images courtesy of David McKay/NASA)

found in or on the ice of the frozen continent, it is well preserved and has not suffered significant weathering (fig. 10.9).

The evidence described in the original *Science* paper was essentially as follows:

- Organic molecules: Polycyclic aromatic hydrocarbons (PAHs) were detected at fracture surfaces within the meteorite. And while PAHs are relatively common in Terrestrial environments (in diesel fumes, for example), the concentration of PAHs in

ALH84001 increased deeper into the rock, suggesting that they came with the meteorite and were not contaminants picked up while the meteorite sat in the Antarctic ice. PAHs can form from the breakdown of biological organics, and in Terrestrial rocks they are considered a compelling signature of biological activity.

- Carbonate globules (see fig. 10.9): In contrast to the other SNC meteorites, ALH84001 contains spherical inclusions of carbonate (a mineral often laid down from aqueous solution) with diameters ranging from 1 to 250 μm. At an estimated age of 3.5 billion years, these carbonate globules are significantly younger than the surrounding rock and, as judged by several lines of evidence, were formed while the rock was still on Mars. Using oxygen isotopic signatures, the authors estimated that the carbonates were deposited from liquid water at temperatures well below 100°C.
- "Nanofossils" (see fig. 10.9): Structures resembling Terrestrial microfossils of cells were found close to the carbonate globules. These, however, are less than 100 nm long, one-tenth or so the length of the smallest, well-established Terrestrial microbes.
- Unusual iron chemistry: The meteorite also contains small particles of iron sulfide and magnetite (an iron oxide) in close conjunction—minerals of differing oxidation states that do not normally exist together in equilibrium.

The authors of the report admitted from the start that each of these features might have arisen from nonbiological causes. They concluded, however, that the simultaneous presence of these features in the Martian rock constituted very strong circumstantial evidence for the existence of life on early Mars. Falling in with this conclusion, follow-on studies suggested that the structure of the magnetite crystals is similar to that of the small, intracellular, magnetic-sensing "organs" of Terrestrial magnetotactic bacteria. No abiological process was known that could produce these oddly shaped crystals. And others noted that the sulfur isotope fractionation on the meteorite hints at biological processes as well. Short of a test tube full of living, breathing (well, respiring at least) Martian microorganisms, what more could one ask for?

Sadly, though (especially for those of us who would really like to get our hands on some Martian life to "lift the hood" and see what makes it work), the scientific consensus seems to have tilted fairly strongly away from the conclusion that ALH84001 contains signs of life. This tilt was caused by the slow but seemingly inexorable raising of questions about each of the original arguments. First, it has been pointed out that carbonaceous chondrite meteorites contain PAHs, as do even interstellar clouds of gas. Since it is unlikely that these PAHs are the products of living processes, it is clear that the ALH84001 PAHs might also be abiological in origin. This seems even more likely when the structures of the PAHs are investigated in detail; PAHs come in a wide variety of molecular structures, and biologically produced PAHs are generally quite diverse. The pattern of PAHs in ALH84001, in contrast, is rather bland, which argues more clearly for an abiological origin than an origin linked to life.

Significant controversy has erupted, too, over the original claim that the carbonates were deposited from relatively low-temperature water. On the one side, studies by Joseph Kirschvink and by Edward Stolper, both of Caltech, on the magnetic properties of the carbonates and their oxygen-isotope ratios, respectively, are said to support the low-temperature claim. On the other side, scientists Edward Scott of the University of Hawaii and Harry McSween of the University of Tennessee have argued equally forcefully for deposition at temperatures far too high to support even extremophilic life. Similar controversy has focused on the nanofossils. There are some (widely disputed) claims of nanometer-sized bacteria here on Earth, but the possibility of such organisms is generally dismissed.* The reason is that the volume of these putative cells would be less than one-thousandth the volume of a typical bacterial cell, which is certainly too small to contain the metabolic machinery of even the simplest free-living organisms on Earth. Steve Benner has pointed out, however, that ribosomes take

*A small number of research groups claimed to have identified so-called nanobacteria in Terrestrial habitats, and some even believed they had medical relevance. However, John D. Young and Jan Martel could show that these morphologies believed to be bacteria arise from interactions between inorganic crystals and proteins. They tell the story of their myth-debunking mission in the January 2010 issue of *Scientific American*.

up much of the space inside a bacterium, and thus if the Martian organisms date back to the RNA world (before the existence of ribosomes and, with them, proteins), it is just conceivable that they could have been as small as the putative fossils. Still, while each of these points may be debatable—and indeed have been vigorously debated in the years since the original announcement—the first three lines of evidence initially put forth as evidence of life in ALH84001 have generated very little in the way of scientific consensus.

In contrast, the observation of odd bits of the mineral magnetite in ALH84001 was, for some time, rather better received. For example, the truncated hexa-octahedral shape of these crystals was considered probably the best evidence for fossil life. In terms of size and composition, the magnetite looks exactly like the magnetite beads from magnetotactic bacteria (which use the beads to sense the Earth's magnetic field and, in turn, to sense the directions "up" and "down"; when you are as small as a bacterium, gravity doesn't provide much of a cue!). So much so that if the specimens had been found on Earth, they would have been considered uncontroversial "magnetofossils." In fact, Kirschvink, the prime proponent of this argument, has claimed that the magnetite crystals provide such strong evidence that the case for life on Mars some 3.7 billion years ago is much stronger than the case for life on Earth at the same time! In contrast, however, a team led by D. C. Golden of Houston, Texas, has shown that similar magnetite particles can be synthesized by the thermal decomposition of iron carbonates, a decidedly abiological process. In short, for every claim of "evidence of life" there is an equally—if not more—compelling counterhypothesis. What is one to make of this? In a nutshell, it's probably safe to say that the broad scientific consensus is that geology is a better explanation than life for the features found in ALH84001.

The concerns about ALH84001 aside, chemical signatures are not the only signs that microorganisms sometimes leave behind. A case in point was the putative nanofossils found in ALH84001, which were largely dismissed when it was realized that they are probably far too small to have contained a complete metabolism. In 2008 and 2009, though, David McKay and his collaborators were in the news again, this time with new, nonmolecular signs of life. This story starts with

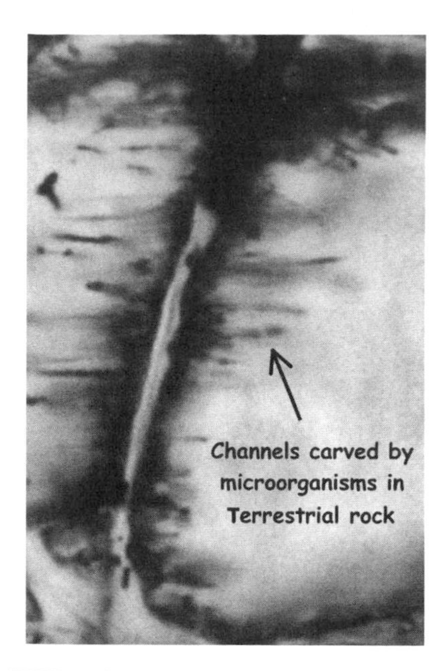

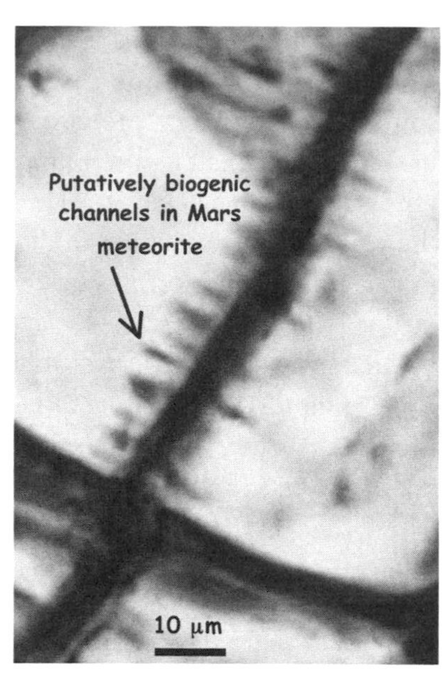

Figure 10.10 As evidenced by the fact that they contain DNA, the micron-scale channels emanating from larger cracks in this fresh, basaltic Earth rock (left) are thought to have been etched by lithotrophic microbes harvesting the rock's chemical energy. Although (possibly due to their great age) similar channels identified in Nakhla meteorites (right) from Mars lack DNA, they may be our best evidence to date of past Martian life. (Images courtesy of Lauren Spencer/David McKay/NASA)

Martin Fisk, a professor of marine geology at Oregon State University, who in 2008 reported that another Martian meteorite, one that landed in Nakhla, Egypt, in 1911, contained microscopic "tunnels" similar in size, shape, and distribution to the tracks left behind by lithotrophic (rock-eating) bacteria (fig. 10.10). On Earth, these tracks are easily identified because they contain both DNA from dead bacteria and clay minerals left behind after the rock is metabolized. And while the Martian meteorites lacked any DNA—even if Martian microbes did use DNA, this lack would be expected given the rock's great age—McKay and colleagues identified both carbon (perhaps as organics) and clays in the putative fossil tunnels, both of which are similar to the materials found in bacterially produced tunnels on Earth. Finally, and perhaps tellingly, effectively indistinguishable tunnels have been found in Yamato 000593, a Nakhla-type Martian meteorite found in Antarctica, suggesting that, whatever caused these tunnels, they are not of Earthly origins. The argument is that a rock that fell to Earth in early twentieth-century Egypt is very unlikely to undergo the same types of Earthly contamination as a rock that fell on Antarctica several tens of thousands of years earlier before being collected and stored in a sterile laboratory. Tiny, micron-diameter, clay-filled etched tubes in two meteorites is not a lot to go on when trying to rule whether or not Mars has (or had) life, but these results have opened a lot of eyes. Once again, stay tuned.

Astrobiology in the Outer Solar System

Although much of the public attention on Solar System exploration has focused on Mars, the renewed optimism in astrobiology has also raised the profile of a few other solid bodies in our neighborhood, especially the icy moons of the gas giants. Jupiter's satellite Europa, with its ice shield covering large amounts of a conducting liquid presumed to be saltwater, is now a prime candidate for extraterrestrial life, as we discussed in chapter 9. Current estimates, however, are that Europa's ocean lies beneath a 10 to 100 km crust of ice; unfortunately, it will be quite some time before anyone lands a craft on that icy moon that can melt its way down to have a look. The *Cassini* spacecraft was spectacularly successful in exploring the Saturnian system, but the mission was terminated in September 2017 when, on its last dregs of fuel, it was intentionally crashed into Saturn to avoid its ever possibly contaminating Enceladus or Titan. The next visit to Titan is likely to be NASA's *Dragonfly* mission, slated to launch in 2026, which will have a drone hopping around that moon for two Earth-years after its arrival in 2034. Thus, if there is any exotic kind of life on Titan, we definitely won't find it before then. Similar considerations hold for Enceladus, which has emerged as one of the possible habitats in the Saturnian system (see chapter 9) but will remain unvisited for decades to come.

The Search for Life beyond the Solar System

As we described in the previous chapter, astronomers have already confirmed more than 4,000 extrasolar planets. The first 100 or so were all gas giants orbiting in exceedingly close proximity to their host stars—not the kind of places where life is likely to have gained a foothold. But after 25 years of successful extrasolar planet hunting, we have identified a number of at least somewhat Earth-like, and thus potentially inhabitable, planets. But are they inhabited?

Considering how difficult it is for us to prove or disprove the existence of life on Mars, it might seem an impossible challenge to do the same for exoplanets. For example, given the insurmountably long distances involved in actually visiting a planet beyond our Solar System, the only viable approach is to characterize their atmospheres spectroscopically in a search for molecular signatures that might be indicative of life. Here the transit method can help us. Specifically, some of the starlight that reaches the Earth during a transit has first passed through the distant planet's atmosphere, allowing us to perform spectroscopic studies of its composition. Using this approach, researchers would love to characterize exoplanet atmospheres well enough to spot the kinds of chemical disequilibrium indicative of life. But this is not easy; for an Earth-sized planet in transit from the habitable zone around a Sun-sized star, only one in a few billion of the photons that reach our telescopes will have passed through the planetary atmosphere.

The first studied exoplanet atmospheres were those of "hot Jupiters" orbiting small stars as this choice maximizes the fraction of photons that has passed through the atmosphere. And even this work required careful subtraction of the light from the star itself, which is collected when the planet is not in transit. To date, however, the technology has advanced to the point where the atmospheres of super-Earths and mini-Neptunes can also be characterized, provided that they're in close orbits around small enough stars. The exoplanet atmospheric components that astronomers have so far identified include, not surprisingly, the second and third most common molecules in the Universe: water and methane. But they've also identified a few surprises, including atmospheres of sodium and what appear to be high clouds of condensed silicates. Some of these planets are hot! Studies of more temperate exoplanets have been less successful. Using the *Hubble Space Telescope*, for example, astronomers have been able to rule out primordial, hydrogen- and helium-rich atmospheres for the six innermost of the TRAPPIST-1 planets, but they have not been able to identify what, if any, atmospheric components are present.

To truly understand exoplanet atmospheres, we're going to need better instruments. Fortunately, some of the most sophisticated instruments currently in action or under development are going to boost this line of research. The *James Webb Space Telescope* (*JWST*), a space-based infrared telescope with a primary mirror 6.5 m in diameter (compared to *Hubble*'s 2.4 m), is a collaborative project of NASA and the Canadian and European space agencies that is currently scheduled to launch in October 2021 (some 14 years past its originally expected launch date). Among a wide range of other astronomical investigations, it will use infrared spectroscopy over a wide range of wavelengths to study the atmospheres of known exoplanets. The telescope's superb sensitivity in the infrared, which is the spectral range over which low-mass stars are brightest, will allow observations of smaller, cooler stars, rendering the signals from planetary atmospheres proportionally larger. Crunching the numbers, the expectation is that the *JWST* could discover water vapor, methane, and other molecules associated with life in the atmospheres of planets as small as super-Earths and perhaps even terrestrial planets provided that they orbit closely enough to small, dwarf stars. *JWST* will also carry a coronagraph, enabling the direct imaging of at least a few exoplanets. The planet, of course, would just be a single-pixel spot on the image, but that will likely suffice to ascertain seasonal and variations in color, measure the rotation rate, and monitor changes in cloud cover.

Looking further ahead, the European Space Agency's *ARIEL* (*Atmospheric Remote-Sensing Infrared Exoplanet Large*) survey mission, featuring a space telescope with an infrared spectrometer, is due to be launched in mid-2028. Based at the L2 Lagrangian point, which is a gravitationally stable spot 0.01 AU further away from the Sun than we are, it will study what exoplanets are made of, how they formed, and how they evolve by surveying a diverse sample of about 1,000 large exoplanets simultaneously in visible and infrared wavelengths. It is the first mission dedicated to measuring the chemical composition and thermal structures of the atmospheres of hundreds of transiting exoplanets and thus might even pick up the give-away chemical signatures of life on some of those distant worlds.

SETI: The Search for Extraterrestrial Intelligence

The search for extraterrestrial life as we've described it so far consists of two major approaches: finding habitable celestial bodies, and checking for any chemical traces of life, past or present, primitive or evolved. We left out one significant aspect of the search: the search for intelligent life—that is, for any kind of life form advanced enough to be able to communicate with us *directly* across space. Myths and a lot of bad television aside, no such life has yet been detected, so the contents of this chapter are bound to be based on probabilities, speculation, and philosophy. But the question still has to be asked: is anybody out there, and if so, why haven't we heard from them?

A fundamental problem here is that, while some of our technology may seem pretty amazing (especially to those of us who won't see 50 again), human technology is extremely limited in comparison with the dimensions of the Universe. And we're talking about serious limitations, not obstacles that we can overcome next year or even in a decade or two. For example, some have speculated that the first person to set foot on Mars has already been born. But the first person to travel to α Centauri (at 4.3 light-years away, our nearest stellar neighbor) and come back alive has certainly not been born yet, and the people who will boldly go to visit other galaxies will not be born for a long time, if ever. Given the enormous times and distances that would be involved in interstellar travel, any exploration that requires instruments to be in situ will be limited to our Solar System for the foreseeable future.

The limitations imposed on us by the vast distances of space are somewhat relieved when we consider the search for civilizations advanced enough to communicate by light or radio waves. As the tireless astronomer and science popularizer Carl Sagan illustrated in his novel (and later film) *Contact*, the first major television transmission, the opening of the Olympic Games in Berlin in 1936, is still winging out into space at the speed of light (implying that by the time this edition appears in print it has been broadcast to everybody living within a radius of 85 light-years). Of course, that broadcast was pretty weak and would be well-nigh impossible to detect from truly galactic distances. In contrast, it is said that the 305 m dish antenna at Arecibo, Puerto Rico, for well on 50 years the Earth's most powerful radio telescope and radio transmitter (for radar studies of planets and asteroids; see fig. 3.9), could have heard a similar antenna from clear across our galaxy, albeit at a paltry one-bit-per-hour bandwidth. One can only speculate what other civilizations with much more advanced technologies could do. Quite probably, they could communicate even from beyond our galaxy. Therefore, in the search for advanced life, widely known as SETI (search for extraterrestrial intelligence), the Universe is our haystack.

The first attempts to search for radio signals from intelligent extraterrestrials date back to 1899, when, in early experiments with radio, the inventor Nikola Tesla (1856–1943) detected signals he believed had arrived from extraterrestrials on Venus or Mars (the "signals," we now know, arise due to lightning in the Earth's atmosphere). Modern SETI, however, only goes back to 1960, when the astronomer Frank Drake, then at Cornell University, started Project Ozma, which he named after a princess in Frank Baum's imaginary land of Oz. Ozma used a 24 m radio telescope at the National Astronomy Observatory in Green Bank, West Virginia, to listen for messages from the nearby stars τ Ceti and ε Eridani at a frequency of about 1,420 MHz (megahertz). This frequency is in the so-called "watering hole" between the frequencies at which hydrogen atoms and hydroxyl radicals emit radio waves and thus would seem to be a logical place for at least water-based life to broadcast messages if it intended them to be found amid the myriad of frequencies on the radio dial. Drake recorded for 150 hours over the course of several months, and at one point found some nonrandom signals that appeared intelligent in nature. This surprised even him; despite his being a perhaps extreme optimist with regard to the number of communicating civilizations in our galaxy (see sidebar 10.2), even Drake estimated that the chance of any single, seemingly suitable star harboring intelligent life is much less than one in 10,000. His surprise, though, didn't last long as the signals turned out to be emanating from a high-flying aircraft. Project Ozma ended before the year was out, without detecting any apparently intelligent extraterrestrial signals.

In the 1970s, SETI proponents grew more active. In 1971, for example, John Billingham (1930–2013) of NASA's Ames Research Center in California authored a detailed study of the feasibility of building an "antenna farm," called Cyclops, consisting of fifteen hundred 100 m dishes. Cyclops, Billingham argued, could detect routine television and radio signals, not to mention intentional attempts to communicate with us, from any of a large number of neighboring stars. The Cyclops proposal didn't get very far, though, perhaps because of its $10 billion price tag. Instead, in 1972, Patrick Palmer and Benjamin Zuckerman used the National Astronomy Observatory's new 91 m telescope in a much more modest effort to monitor 670 nearby stars from time to time over the course of four years, again to no avail. And, in 1974, NASA initiated formal SETI research programs at Ames and at California's Jet Propulsion Laboratory. But the good times were not to last: serious congressional unease at the costs of "a silly search for aliens that was unlikely to yield results" hampered efforts to fund SETI, quickly closing the project down. Undaunted, in 1979, Jill Tarter and Stuart Bower of the Berkeley SETI Research Center initiated SERENDIP (Search for Extraterrestrial Radio Emissions from Nearby Developed Intelligent Populations),* a still-active SETI program that employs the 90 m telescope at Green Bank, West Virginia, and the 305 m dish at Arecibo, Puerto Rico. SERENDIP uses these antennas in "piggyback" mode, passively sitting on the telescope and collecting data from wherever it is pointed, riding on the back of ongoing astronomical observations.

**Serendip*, the old Persian name for Sri Lanka, features in the ancient tale of the three princes of Serendip, who succeeded in their endeavors by fortuitous coincidences and accidental discoveries. Based on his memory of this story, Horace Walpole (1717–1797) coined the word *serendipity*, which still has an important place in science.

Sidebar 10.2

Weighing the Probabilities

Will the SETI attempts ever find intelligent life out there? Assessments that focus solely on the number of stars in our galaxy generally lead to the conclusion that there must be somebody somewhere. On the other hand, Enrico Fermi (1901–1954), the Italian-American physicist and 1938 Nobel Prize winner (for his studies of nuclear fission), argued that we seem to be alone.

Fermi calculated that a civilization with even a modest amount of rocket technology could colonize the entire Milky Way within, say, a few tens of millions of years even if its top speed were limited to but a small fraction of the speed of light. His argument was that, even if it took each new colony planet half a million years to set up two colonies of its own, this exponential growth would lead to more colonies than there are stars in just 20 million years. And while that may seem like a long time, it is extremely short compared with the age of our galaxy. Clearly, Fermi realized, aliens have had plenty of time to colonize a galaxy. So, if there are a lot of aliens out there (as almost everyone assumes), why aren't they here?

Fermi argued that the fact that aliens don't seem to be hanging out with us here on Earth (tabloid stories aside) strongly contradicts the assumed existence of intelligent life elsewhere in our galaxy, a problem that came to be known as the Fermi Paradox. And it's a hard paradox to break; you can argue that the aliens can move at 10% of the speed of light or 1%, and you still come up with more or less the same answer. Namely, that the entire Milky Way should be colonized over a period vastly shorter than its age. That this has not happened implies that we are probably alone.

On the other side of the argument are many astronomers, a community that is generally optimistic about these things. A prominent member of this camp is the radio astronomer and pioneering SETI researcher Frank Drake, who called together the world's first scientific conference on the topic in 1961. Among the dozen participants were a 28-year-old postdoc named Carl Sagan and the rather more senior scientist Melvin Calvin, whose work on prebiotic chemistry we mentioned in chapter 4 and whose Nobel Prize–winning elucidation of the dark reactions in photosynthesis we discussed in chapter 7. (Amusingly, Calvin received the call notifying him of the award while he was attending Drake's meeting.)

To Drake, the agenda of his meeting was "what do we need to know about [in order] to discover life in space?" To capture this, on the first day of the meeting Drake stood at a chalkboard and wrote an equation that he felt quantitatively framed the topics this question spanned:

$$N = R^* f_p n_e f_l f_i f_c L$$

where:

- N = number of extraterrestrial civilizations with whom we can communicate
- R^* = rate at which potentially life-sustaining stars are formed in our galaxy (per year)
- f_p = fraction of these stars that host planets
- n_e = average number of planets per planet-hosting star that could support life
- f_l = fraction of those "qualifying planets" that actually develop life
- f_i = fraction of planets with life that develop intelligent life
- f_c = fraction of intelligent life-forms that are willing and able to communicate
- L = average lifetime of a communicating civilization (in years)

(*continued*)

Sidebar 10.2 (*continued*)

The values for most of these variables, though, were unknown then and remain unknown today. Specifically, we can probably estimate the first parameter, R^*, to within a factor of two or so, and, in recent decades, have started to constrain f_p, but after that we begin to lose almost all contact with known reality. Nevertheless, many people, starting with Drake, have estimated N and used the Drake equation as a sounding board with which to describe their optimism or pessimism about life elsewhere in our Universe.

To this day, Drake remains an optimist, as one would have to be to persevere in SETI research for more than half a century. His estimates of the various parameters place the value of N at about 10,000 communicating civilizations in the Milky Way alone, which makes SETI seem like a reasonable effort but also essentially reinforces the Fermi Paradox. But is Drake's optimism well founded? As we've discussed in detail in chapters 4 and 5, we do not yet understand how life arose on Earth in anywhere near enough detail to support such an optimistic scenario. Admittedly, since we do not know how life arose here, we cannot rule out the possibility that f_l is near one, but the very best theories we currently have concerning the origins of life suggest that f_l might be *tens or hundreds or thousands of orders of magnitude less than one*. Similarly, since the anthropic principle requires that intelligent life must form before we can discuss the probability of intelligent life forming, all we know about f_i is that it is not zero. Its value too, however, may be infinitesimally close to zero (it took on the order of *3 billion years* before it happened here) and, at the very least, seems likely to be far lower than the near unity assigned to it by Drake.

Lastly, as Carl Sagan pointed out, advanced civilizations may tend to eradicate themselves more rapidly than they can send out colonies via interstellar travel. Given the ratio of the number of rockets built for killing fellow Earthlings to the number built for sending them off to other planets, it may well be that advanced civilizations are more likely to self-destruct than to travel to the stars. Perhaps less pessimistically, even here on Earth we have started to use up radio-frequency bandwidth (for cell phones and the like) so rapidly that within a few years the radio signals emanating from our planet may look much more like white noise than a sign of intelligent life. If so, L could also be infinitesimally close to zero (the period over which human radio sources have been beaming intelligible "we are here" signals across the heavens is negligibly small compared with cosmological ages). The range of values that each of Drake's parameters could adopt is so great that, despite the huge number of stars in the observable Universe, current scientific knowledge is *entirely consistent with* $N = 1$. That is, Fermi was right and we are alone.

Drake, Frank, and Dava Sobel. *Is Anyone Out There? The Scientific Search for Extraterrestrial Intelligence*. New York: Delacorte Press, 1992, page 51.

The ups and downs of SETI continued in the 1980s. After significant lobbying by Carl Sagan, by then a well-known and well-respected astronomer and popularizer of science, Congress reinstated NASA's SETI funding in 1983. The first hardware for the program, which, in an attempt to get past the "giggle factor," was named the High Resolution Microwave Survey, was built in 1988, and the system began searching for signals in 1992. This, too, however, was not to last: the congressional rhetoric over "little green men" eventually heated up once again, and the program was canceled after less than a year of observations. "This hopefully will be the end of Martian hunting season at the taxpayer's expense,"* remarked Senator Richard Bryan, a major SETI critic.

But all was not lost; although the NASA searches were short-lived, their professionalism transformed SETI, which had previously been a rather ad hoc affair. When NASA's SETI effort was unexpectedly cancelled, the SETI Institute, a privately funded institution founded in 1984 and based in Mountain View, California, stepped into the void to become the endeavor's primary sponsor. Acquiring much of the NASA SETI equipment, in February 1995, the institute launched Project Phoenix, which was the continuation of the defunded NASA program. The project focused on 1,000 stars deemed most likely to harbor alien civilizations: Sun-like stars older than 3 billion years and within 200 light-years of Earth.

To make the most of its limited resources, Project Phoenix built only receivers and data analysis equipment, not the enormous and expensive antennas that are needed to collect the radio waves in the first place. Instead, the project placed its custom hardware into a truck trailer that can be parked next to any of the world's major radio observatories. Its first stops, in 1996, were the 64 m dish at the Parkes Observatory in Australia and the 43 m dish at Green Bank, West Virginia, just a few hundred meters from where Drake had performed the first search in Project Ozma. Finally, from 1998 to its end in 2004,

*As quoted in S. J. Garber, "Searching for Good Science: The Cancellation of NASA's SETI Program," *Journal of the British Interplanetary Society* 52, no. 1 (1999): 3–12.

the project used the 305 m dish at Arecibo, Puerto Rico. After eight years of searching, they came up with naught: as the project's leader, Peter Backus, dryly noted, "we live in a quiet neighborhood."*

The failure of Phoenix, and its predecessors, aside, the SETI Institute remains undaunted and continues to play a major role in SETI research and education. For example, in collaboration with the University of California, Berkeley and backed by $30 million in funding from Microsoft cofounder Paul Allen (1953–2018), the institute built a 42-element array of 6.1 m dishes in the mountains of Northern California. To this day, the SETI Institute runs searches using the Allen Array from 6 p.m. to 6 a.m. every night, turning it over to radio astronomy studies during the day.

Antenna time and funding are not the only hurdles SETI faces. Sifting through the huge amounts of radio data looking to extract intelligent signals from the noise is also a major challenge. A pioneering solution to that problem was provided by the SETI@home project, which farmed out an automatic data-sifting process to thousands of personal computers as screen savers, using these to comb first through the SERNDIP data set, through their own, independent recordings from Arecibo, and eventually data from the Breakthrough Listen project described below. The idea is that computers that are always on and always online (as, for example, most of the desktop computers in universities) provide an enormous computational resource that can be exploited, during their idle time, for SETI data processing. In the first decade after its 1999 launch, computers connected to the SETI@home project accumulated 2.3 million years of CPU time. So far, though, despite this massive firepower, all of the SETI searches have recorded only noise. The occasional "unexplained signal" has always turned out to have a Terrestrial or "non-ET" astronomical explanation. At the end of March, 2020, SETI@home stopped queuing new data for processing in a planned "hibernation" phase. This is to allow the project to catch up with the analysis of the existing data and focus on writing a research paper about the efforts so far.

*David Whitehouse, "Radio Search for ET Draws a Blank," BBC News, March 25, 2004.

Moving forward, in July 2015, a stellar cast of scientists including astrophysicist Stephen Hawking (1942–2018) announced an ambitious new SETI project. Funded with $100 million for the next 10 years by the Russian internet entrepreneur Yuri Milner, Breakthrough Listen began operating in January 2016. The funding is split into three roughly equal parts for instrument time, for equipment and method development, and for academic staff and graduate students. The project, which is scanning a much wider part of the radio spectrum than any previous SETI effort, is also an enormous number of stars, including the million closest to Earth, the center and central plane of the Milky Way, and even the hundred closest galaxies to ours. And all this, the initiative promises, will be done 100 times faster than in previous efforts.

In support of its vision, the project is renting vast amounts of instrument time on two of the world's largest radio telescopes, the 100 m telescope at Green Bank, West Virginia, and the 65 m telescope at Parkes in New South Wales, Australia. In parallel, the program is conducting searches for optical signals using the automated planet finder at the Lick Observatory near San Jose, California, which is sensitive enough to detect an ordinary 100 W laser across interstellar distances. On April 12, 2016, the project made its first batch of observation data accessible via its website. This encompassed data collected from most of the stars within 16 light-years of Earth, including stars such as 51 Pegasi that are known to host extra-solar planets, a sampling of stars between 16 and 160 light-years away, and 40 of the nearer spiral galaxies.

So much effort and we still haven't heard from ET, at least not by the time this edition went to press. Committed SETIologists keep their spirits up, however. Addressing an astrobiology meeting of the Royal Society in London in 2010, just after the fiftieth anniversary of his first search for intelligent aliens, Drake remained optimistic that other civilizations are "out there" and that modern technology, with listening devices now 10^{14} times more powerful than the radio receiver with which he first started the search, will eventually pick up a signal. He also admitted, however, that the progress in communications technology seen on Earth in the past few decades has rendered our radio emissions into space quieter rather than louder, so other civilizations may be more difficult to spot than he once thought. Put simply, the task ahead is not an easy one.

Conclusions

What about the planet that we visited at the beginning of this chapter—the one that showed the intriguing indications of chemical disequilibrium that seemed to point to life? The scientists in control of this spacecraft concluded that the planet was indeed inhabited. Even, probably, with intelligent life (the regular radiofrequency pulses seemed too regular to be of natural origin). The scientists, led by Carl Sagan, excitedly wrote up this work and published it in the prestigious British journal *Nature*, which ran the story on its cover on October 21, 1993, under the caption "Is there life on Earth?"

So, ahem, yes, this question may look silly at first glance, as we wouldn't be here to discuss it if the answer was no. The spacecraft in question, NASA's *Galileo*, had to swing by its home planet for a gravity assist on its way to Jupiter with all its instruments blazing away (for calibration purposes). Sagan and his coworkers realized that the flyby provided an ideal opportunity to find out how difficult it is to detect life on a planet that has been teeming with it for billions of years. The resulting article is interesting in that it reveals the difficulties in obtaining unequivocal evidence for life on a planet during a flyby. In the end, however, the evidence of serious chemical disequilibrium in combination with the radiation and absorbance characteristics of the planet was counted as sufficient to conclude that there is life on Earth. Thus, even though the result obtained by the spacecraft swinging by our planet only confirmed the "ground truth" that we knew all along, it was a good excuse to think about how life shapes our planet, and whether we might see life similarly shaping other planets and moons out in the vast gulf of space beyond our home.

Further Reading

The Search for Life on Earth

Sagan, Carl, W. Reid Thompson, Robert Carlson, Donald Gurnett, and Charles Hord. "A Search for Life on Earth from the Galileo Spacecraft." *Nature* 365 (1993): 715–21.

Life on Mars (General)

Johnson, Sarah Stewart. *The Sirens of Mars: Searching for Life on Another World.* New York: Crown Publishing Group, 2020.

Evidence for Life in ALH84001

McKay, David S., Everett K. Gibson Jr., Kathie L. Thomas-Keprta, Hojatollah Vali, Christopher S. Romanek, Simon J. Clemett, Xavier D. F. Chillier, Claude R. Maechling, and Richard N. Zare. "Search for Past Life on Mars: Possible Relic Biogenic Activity in Martian Meteorite ALH84001." *Science* 273, no. 5277 (1996): 924–30.

SETI

Garber, Stephen J. "Searching for Good Science: The Cancellation of NASA's SETI Program." *Journal of the British Interplanetary Society* 52 (1999): 3–12.

SETI Institute. https://www.seti.org/.

Shklovskii, I. S., and Carl Sagan. *Intelligent Life in the Universe.* New York: Dell, 1966.

Epilogue

When *Voyager 1* flew by Jupiter in 1979 and by Saturn in 1980, the gravitational slingshots provided by these gas giants flung the spacecraft out of the Solar System at a blistering 17 km/s (a half *billion* kilometers per year). A decade later, after crossing the orbit of Pluto, the spacecraft looked back over its shoulder and snapped a family portrait of the Sun and her retinue of planets. From this distant vantage, some 6 billion kilometers away, the Earth hangs in the inky blackness of space, a pale blue dot not even a single pixel wide (fig. E.1).

Astrobiology provides us with an armchair equivalent of that perspective, the chance to step back and look at our home planet in the context of the rest of the Universe. Is that tiny speck, floating like a dust mote in a vastly larger cosmos, the only inhabited planet in the Solar System? In the Milky Way? In the observable Universe? On the one hand, the Universe is a mighty big place; it contains an estimated 50 sextillion (5×10^{22}) stars (go back and stare at fig. P.1) and probably at least that many planets and moons. On the other hand, the best-founded theory of the origins of life, the RNA-world hypothesis, requires the spontaneous formation of RNA polymers that are so rare that the probability of their forming spontaneously dwarfs the number of stars in the Universe (see sidebar 5.1). And then there is the anthropic principle (see sidebar 2.3) to confuse things: if there were only one inhabited planet in our Universe, we questioning beings would necessarily find ourselves on that planet, and thus the existence of Earth as a habitable planet does not by any means prove that habitats—much less life—are common in the cosmos.

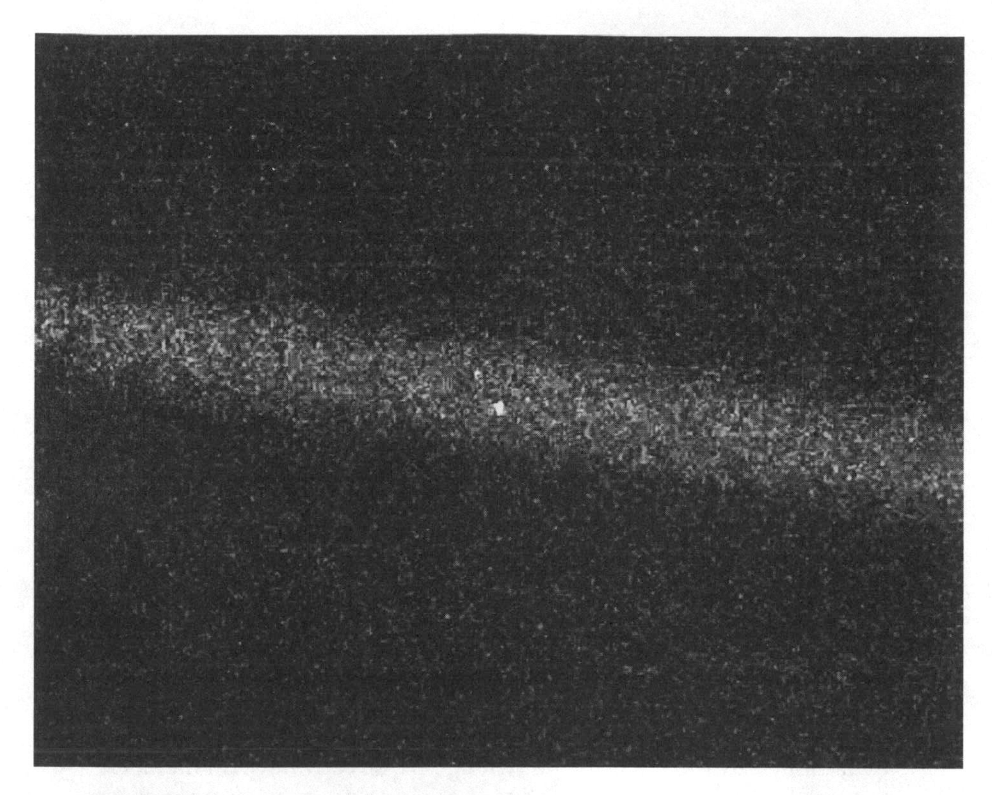

Figure E.1 You are here. That dot, floating in a beam of stray sunlight scattered into the camera, is our home planet as viewed on Valentine's Day 1990 by *Voyager 1* from a vantage point farther from the Sun than Pluto. (Image courtesy of NASA/JPL-Caltech)

Where does this leave us? Of course, it is always possible that life arose here on Earth through some more facile, more highly probable mechanism than those described in this book; there is certainly much that remains to be discovered. Indeed, we are reminded of a comment by J. B. S. Haldane, the man who coined the phrase *primordial soup*: "the Universe is not only queerer than we suppose, but queerer than we can suppose."* Conversely, though, if life did arise through chemistry that we, more or less, currently understand, then the probabilities suggest that we most likely are alone in the Universe.

*J. B. S. Haldane, *Possible Worlds and Other Papers* (London: Harper and Brothers, 1927), 286.

Obviously, we do not yet know the answer to the central question of whether life on Earth is an anomaly, and we must face the fact that it may be centuries before we find out—if we ever do. Still, it is perhaps helpful to look back: some 400 years ago, it took an enormous leap of the imagination for pioneering scientist Johannes Kepler to consider what the movement of the celestial bodies would look like to an observer based on the Moon. He addressed this question in a work of fiction, published posthumously as *Somnium* ("The Dream"). Thinking outside the box of his own limited planetary world enabled Kepler to develop the laws of planetary motion, which we still use today to predict the movements of the planets and to send our spacecraft to visit them. Kepler could not possibly have guessed how many centuries it would take for humans to make a trip to the Moon and explore the Solar System beyond.

Similarly, while we can think outside the box of our own limited planetary world to discover and understand planets even far beyond our Solar System, visiting them is as remote to us as visiting the Moon was to Kepler, if not more so. But, as we have shown in this book, even speculative considerations, provided that they are based on and constrained by the scientific knowledge we already have on hand, enlighten us about our place in the Universe. And, who knows? Perhaps someday we will have the ability to go and look for ourselves. Until then, as the only intelligent life we know of, all we humans can do is keep asking questions that further our understanding of our place in the Universe.

Glossary

accretion. The growth of an object by the gravitational attraction of more matter; typically used to describe the origin and growth of planets.

amino acid. A molecule containing an amino group ($-NH_2$) and a carboxylic acid group ($-CO_2H$) in the general formula $H_2N-CHR-CO_2H$, where the R represents a grouping of atoms, called the *side chain*, that varies from one amino acid to another. Proteins are polymers of amino acids formed by linking the individual amino acids together using peptide bonds.

anabolism. The set of metabolic reactions that build small molecules into larger molecules.

Archaea (archaea; singular: archaeon). A domain of single-celled organisms that lack a nucleus. One of the three domains of life on Earth, along with the Eukarya and the Bacteria. Although their outward appearance may be very similar to bacteria, genome studies have clearly proved that the two domains are as distant from each other as they are from the Eukarya. The term is capitalized when referring to the domain but lowercased when referring to individual species or organisms.

ATP (adenosine triphosphate). A high-energy ribonucleotide that serves as the energy equivalent of "currency" in the cell; that is, energy-liberating reactions such as the oxidation of carbohydrates synthesize ATP, and energy-consuming reactions such as the polymerization of amino acids into proteins consume it.

AU (astronomical unit). One AU is equal to the mean distance between the Earth and the Sun, which is about 150 million kilometers.

Bacteria (bacteria; singular: bacterium). A domain of single-celled organisms that lack a nucleus. One of the three domains of life on Earth, along with the Eukarya and the Archaea. Although their outward appearance may be very similar to archaea, genome studies have clearly proven that the two domains are as distant from each other as they are from the Eukarya. The term is capitalized when referring to the domain but lowercased when referring to individual species or organisms.

big bang. The term introduced by astronomer Fred Hoyle to ridicule the idea that the Universe had a discrete beginning, in which it expanded from an originally superdense, superhot state. This name was then adopted as the quasi-official name for what soon became the dominant paradigm describing the origin of the Universe.

black smoker. The unofficial name for a class of deep-ocean hydrothermal vents that spew super-heated, mineral-rich water. The precipitation of compounds insoluble in the cold seawater produces both the chimney and the characteristic "smoke."

Calvin cycle. The chemical reactions that reduce carbon dioxide to glucose in photosynthesis. Also known as the dark reactions as they do not directly use photons.

catabolism. The set of metabolic reactions that break down large molecules into smaller molecules.

catalysis. The acceleration of a chemical reaction by lowering the energy barrier that it has to overcome. In biology, catalysis plays an important role and is typically carried out by proteins but in some cases is carried out by RNA molecules (called ribozymes).

chirality. Handedness. Most biologically relevant molecules are chiral, meaning that they cannot be superimposed on their mirror image. This is often due to the asymmetric arrangement of four different binding partners around a single carbon atom. However, any helical structure is also chiral.

cosmic microwave background. Radiation consisting of relic photons from the big bang that have now red-shifted into the microwave region of the spectrum. The radiation intensity is nearly equal in all directions, but subtle differences in its intensity at the parts-per-100,000 level provide valuable insights into the early history of the Universe.

DNA (deoxyribonucleic acid). A biological polymer (a chain-like molecule) built from four types of building blocks, the nucleotides, which differ in the chemical nature of the nucleobase protruding from the chain. The nucleobases are adenine (A), cytosine (C), guanine (G), and thymine (T). These bases are pairwise complementary (A pairing with T, and C with G) and can form specific bonds that enable the DNA to form its characteristic double-helical structure. DNA is the carrier of genetic information in all cellular life forms on Earth.

enzyme. Any biomolecule (protein or RNA) that performs catalysis.

Eukarya (eukaryotes). One of the three domains of life on Earth, along with the Bacteria and the Archaea. This domain, formerly referred to as Eukarya, includes all multicellular species and some single-celled organisms, such as yeast. Its defining feature is that eukaryotic cells store their genetic material in a nucleus, a membrane-bound organelle.

exoplanet. A planet orbiting a star other than our Sun.

extremophile. A species or organism adapted to particularly challenging environmental conditions, including high temperatures (thermophiles), high pressures (barophiles), high salt concentrations (halophiles), and extremes of pH (acidophiles, alkaliphiles). When contrasted with an extremophile, an organism living under more "normal" conditions is referred to as a *mesophile*.

gas giant. A large planet consisting predominantly of hydrogen and helium and typically of a mass >70 times that of Earth. Examples in our Solar System are Jupiter and Saturn.

genetic code. The (nearly) universal set of rules by which combinations of three nucleobases in a DNA or RNA sequence (codons) are assigned to specific amino acids (see table 6.1).

genome. The genetically encoded information of a particular organism or species.

glycolysis. The nonoxidative biochemical pathway by which sugars are broken down into smaller organic molecules, liberating energy.

helix. A structure wound along an axis like the thread on a screw. In Terrestrial biochemistry, helices are common in proteins (α helix) and nucleic acids (double helix). In both cases, the helices are right-handed (like the screws you'd find in a hardware store)—that is, if you look along the helix axis and trace the helix in a clockwise direction, the movement will lead away from you.

heteropolymer. A chain-like molecule (polymer) composed of more than one type of subunit (monomer).

homologous. Describing biological features that are similar because they have arisen from a common ancestor.

homopolymer. A chain-like molecule (polymer) composed of either multiple copies of a single type of subunit (monomer) or a regular, repeating pattern of a small set of monomers.

hydrolysis. The breaking of a bond via its reaction with one or more water molecules.

hydrophilic. Water-loving; easily soluble in water. The term can be used to describe the properties of molecules or parts of molecules. Most proteins contain regions that are hydrophilic and regions that are the opposite (hydrophobic).

hydrophobic. Water-avoiding; the opposite of hydrophilic. The term typically refers to "grease-like" molecules such those that make up fats and oils.

hydrothermal vent. A hot spring occurring in volcanic regions of the ocean floor (see, also, *black smoker*). It emits a constant stream of overheated fluid rich in heavy-metal ions and hydrogen sulfide; metal sulfides precipitate as the fluid mixes with cold seawater, which produces the black "smoke" plume. The vent chimney builds up from precipitated materials, including gypsum and sulfides.

ice giant. A large planet, typically 10 to 70 times larger than the Earth, consisting mainly of molecular compounds including water and methane. Examples are Uranus and Neptune.

Krebs cycle. A circular arrangement of metabolic reactions that enables aerobic organisms to "burn" food substances such as carbohydrates, producing carbon dioxide and chemical energy. Conceptually, the Krebs cycle is complementary to the *Calvin cycle*, which enables plants to fix carbon dioxide into carbohydrates through the input of solar energy. Also called the *citric acid cycle* or *TCA (tricarboxylic acid) cycle.*

Kuiper belt. A large population of icy bodies around and beyond the orbit of Pluto, which is the most massive of the known Kuiper belt objects.

LUCA. The last universal common ancestor; the last ancestor shared in common among all living things on Earth today. From the features that all of today's species have in common, we can conclude that LUCA stored genetic information as DNA and produced a couple of hundred proteins (enzymes, receptors, transporters) using the same RNA-based machinery, the same genetic code, and the same 20 standard amino acids used in biology today.

metabolism. The set of chemical reactions that maintain life, allowing organisms to respond to their environments, grow, and reproduce. There are two broad classes of metabolism. In *catabolism*, larger molecules are broken down into smaller molecules (e.g., glycolysis), typically liberating the energy that drives the rest of metabolism. In *anabolism*, energy is used to construct larger molecules from smaller precursors (e.g., protein synthesis).

metal. In astronomy, a term often used to describe all chemical elements heavier than helium, in blatant violation of the chemical definition, which focuses on the ability of electrons to move through the solid phase of an element (conductivity).

messenger RNA (mRNA). A short-lived RNA molecule that carries the genetic information stored in DNA to the ribosome, where it is used to direct the synthesis of proteins.

metagenomics. The identification of microbial species in a sample (e.g., a soil sample) via the wholesale sequencing of DNA; as opposed to the identification of bacteria via culturing in the laboratory.

mini-Neptune. An ice giant planet (i.e., of density <3 g/cm^3) with a mass less than about 10 times that of the Earth.

nucleic acid. A broad term including both individual DNA or RNA nucleotides and polymers of them.

nucleoside, nucleotide. The components of RNA and DNA. Nucleosides consist of a ribose or deoxyribose sugar covalently linked to one of four nucleobases (adenine, guanine, cytosine, or uracil in RNA, with thymine instead of uracil in DNA). Nucleotides are nucleosides with one or more phosphate groups covalently attached.

orbital. In atoms and molecules, the specific region of space occupied by electron pairs. Molecular orbitals are responsible for the bonding together of atoms to form molecules.

oxic atmosphere. An atmosphere containing free molecular oxygen (O_2).

oxidation. A reaction involving the transfer (or partial transfer) of electrons from the reaction partner in question to another; that is, a reaction in which an atom, molecule, or part of a molecule either gives electrons away outright or ends up with lower electron density because it is bound to a partner that attracts the shared electrons more strongly. The reaction partner that accepts electrons or increases its electron density undergoes *reduction* in a process often referred to as a *redox reaction*. The term derives from the first known examples of oxidation reactions, such as the combustion of carbon compounds to form carbon dioxide and the rusting of iron, which involve oxygen as the oxidizing reagent.

peptide bond. The type of covalent bond that links amino acids together to form polypeptides or proteins.

photolysis. The light-induced breaking of a molecular bond.

polymer. A chain-like molecule built from a large number of identical or similar building blocks (monomers). Polymers consisting of either a single type of monomer or a regular, repeating pattern of a small set of monomers are called *homopolymers* (e.g., nylon). The biological polymers DNA, RNA, and proteins, in contrast, are *heteropolymers*; they consist of irregular sequences of monomers that are identical in the functions needed for chain propagation but have separate chemical identities in the molecular parts (the nucleobases or amino acid R groups) protruding from the polymer chain.

polypeptide. A chain of amino acids covalently linked together via peptide bonds. Rigorously speaking, proteins are polypeptides. Still, the word is generally used to describe shorter polymers of amino acids, especially those that lack a well-defined, folded structure. It is also used if the one-dimensional chain aspect of a polymer of amino acids is being discussed. Polymers containing more than about 50 amino acids and exhibiting a well-defined three-dimensional structure are generally called *proteins*.

primordial soup. A mixture of simple organic compounds dissolved in water, presumed to have been present on the early Earth and to have somehow enabled the origin of life.

protein. A biopolymer consisting of one or more chains each containing 50 or more amino acids and, typically, folded into a well-defined, three-dimensional shape. Proteins carry out a wide variety of tasks in the cell, including the catalysis of metabolic reactions, the transport of small molecules and ions, the generation of mechanical work, and the regulation of gene expression and metabolism.

protein folding. The process in which the linear chain of amino acids folds into its functional, three-dimensional structure via the stabilizing effects of a large number of weak, intrachain interactions.

proteinogenic. Describing the 20 amino acids from which proteins are made.

radical. A molecule or atom with an odd number of electrons, thus leaving one electron unpaired. Because the main driving force of chemical reactions is pairing up electrons to fill molecular orbitals, most radicals are highly reactive.

recombination. A phase in the early history of the Universe. Until recombination, matter was present as a plasma, a random mixture of free protons, nuclei, and electrons. Recombination took place when matter had dispersed and cooled down sufficiently to allow electrons, protons, and nuclei to form stable atoms, mainly hydrogen and helium. It also led to an uncoupling of matter and the photons that now make up the cosmic microwave background.

reduction. A reaction involving the transfer (or partial transfer) of electrons from the reaction partner in question to another; that is, a reaction in which an atom, molecule, or part of a molecule either receives electrons outright or ends up with higher electron density because it is bound to a partner that attracts the shared electrons more strongly. The reaction partner that accepts electrons or increases its electron density undergoes *oxidation*. The term derives from the use of carbon (as charcoal or coal) to "reduce" ores (usually metal oxides) to the free metal.

ribosomal RNA (rRNA). The RNA component of the protein-synthesizing ribosomes. Ribosomal RNAs are now known to serve as the catalyst in protein synthesis.

ribosome. An intracellular complex of more than 50 proteins and several RNA molecules that carries out the synthesis of proteins according to the genetic instructions read from messenger RNA, with the help of transfer RNA and various protein factors. The core function of the ribosome, attaching a new amino acid to the growing polypeptide chain, is catalyzed by its RNA, so the ribosome is a key piece of evidence for the importance of RNA in the early evolution of life.

ribozyme. An RNA molecule that performs catalysis.

RNA (ribonucleic acid). A biological polymer composed of nucleotides consisting of the nucleobases adenine (A), guanine (G), cytosine (C), and uracil (U), each linked to a ribose sugar, with the sugars linked together by phosphate groups. In general, RNA acts as a mediator between genetic information (carried by DNA) and function (performed by proteins). The most important types of RNA are messenger RNA (mRNA), transfer RNA (tRNA), and ribosomal RNA (rRNA).

RNA world. A hypothetical stage in the early evolution of life in which RNA molecules were the only biological polymers, playing the roles of both information-carrying molecules (genes) and functional molecules (catalysts).

scattered disk. A broad disk of icy objects scattered to orbits beyond that of Pluto by the gravitational effects of the gas and ice giant planets. Eris is the largest known scattered disk object. The inner edge of the scattered disk overlaps with the Kuiper belt, and its outer fringes extend to more than 100 AU.

side chain. A group of atoms that is attached to the backbone of a polymer.

solar wind. A stream of high-energy protons and electrons emitted by the Sun and other stars.

super-Earth. A terrestrial planet (i.e., composed of rock and metal and thus of mean density >3 g/cm^3) that is more massive than the Earth.

transfer RNA (tRNA). An RNA molecule that mediates between the codon (the sequence of three nucleotides in a messenger RNA that encodes for a specific amino acid) and the amino acid that the codon encodes, thus supporting the translation of genetic information into a specific sequence of amino acids (a protein).

transit. The passage of an astronomical object (e.g., a planet) through the sightline between an observer (e.g., an astronomer on Earth) and a more distant object (e.g., a star).

tRNA synthetase. An enzyme that links the appropriate amino acid to its specific tRNA.

Index